Springer

*Tokyo*
*Berlin*
*Heidelberg*
*New York*
*Barcelona*
*Budapest*
*Hong Kong*
*London*
*Milan*
*Paris*

Editors: T. Nakano
D.J. Hartshorne

# Regulation of the Contractile Cycle in Smooth Muscle

With 92 Illustrations, 1 in Color

Springer

Takeshi Nakano, M.D., Ph.D.
Professor
Mie University School of Medicine
2-175 Edobashi, Tsu, Mie, 514 Japan

David J. Hartshorne, Ph.D.
Professor
University of Arizona
Tucson, AZ 85721, USA

ISBN 978-4-431-70149-1 Springer-Verlag Tokyo Berlin Heidelberg New York
ISBN 978-3-540-70149-1 Springer-Verlag Berlin Heidelberg New York Tokyo
ISBN 978-0-387-70149-3 Springer-Verlag New York Berlin Heidelberg Tokyo

Library of Congress Cataloging-in-Publication Data
Regulation of the contractile cycle in smooth muscle/editors, T. Nakano, D.J. Hartshorne.
p. cm.
**ISBN** 978-4-431-70149-1(Tokyo: alk. paper).—**ISBN** 978-3-540-70149-1 (Berlin: alk. paper).—**ISBN** 978-0-387-70149-3 (New York: alk. paper)
1. Smooth muscle—Congresses. 2. Muscle contraction—Congresses. I. Nakano, T. (Takeshi), 1942- . II. Hartshorne, David J.
[DNLM: 1. Muscle, Smooth—physiology—congresses. 2. Muscle Contraction—physiology—congresses. 3. Calcium—physiology—congresses. WE 500 R3444 1995]
QP321.5.R44 1995
591.1′852—dc20
DNLM/DLC
for Library of Congress 95–22863

Printed on acid-free paper



Typesetting: Best-set Typesetter Ltd., Hong Kong

# Preface

Smooth muscle is vital to most physiological functions. In general, smooth muscle controls the volume or diameter of hollow organs and its role is well known in the vascular system and gastrointestinal tract. The contractile elements in the walls of these organs can function to oppose an applied pressure, as in the tonic behavior of the major arteries and the control of blood pressure, or they may respond phasically, as in the uterus in childbirth. The responses are varied, the physiological behavior complex, and the pharmacology of membrane events diverse. Thus the study of smooth muscle is challenging and at times overwhelming, but since such complexity reflects a wide spectrum of cellular mechanisms, continued research certainly is justified. We are all too familiar with the consequences of altered smooth muscle function; high blood pressure and asthma are common examples.

Initially it was believed that smooth muscle was a disorganized relative of skeletal muscle and that the same mechanisms caused contraction or relaxation. While this is partly true, it is now realized that many aspects of smooth muscle function are distinct from those of skeletal muscle and involve different pathways. The ultrastructural basis of tension development appears to be common, and it is assumed that contraction in both muscle types involves cross-bridge interactions and the interdigitation of thick and thin filaments. Both muscle types are also activated by an increase in intracellular $Ca^{2+}$ concentrations. Besides these common fundamentals, however, the two muscle types are quite different. Many of the distinctions have been appreciated only relatively recently and new discoveries are rapidly adding to our knowledge of the unique aspects of smooth muscle function. Some of the areas where major differences are evident include: the diversity of membrane receptors and channels found in smooth muscle; the realization that excitation–contraction coupling need not be linked to membrane depolarization (so-called pharmacomechanical coupling); the connection of the latter to the phosphatidylinositol cascade and ultimately the release of $Ca^{2+}$ from the sarcoplasmic reticulum; the discovery that the activation of the smooth muscle contractile apparatus involves myosin phosphorylation and the $Ca^{2+}$-binding protein calmodulin, rather than the thin-filament protein troponin; the characterization of the two enzymes involved in myosin phosphorylation, myosin light chain kinase and phosphatase; and the discovery of two possible regulatory proteins associated with the thin filaments of smooth muscle, namely calponin and caldesmon.

The speakers at the Mie International Symposium have made important contributions to various aspects of smooth muscle research, and the individual contributions will provide details, support, and criticism for the above synopsis. Many areas of smooth muscle research are included, ranging from the plasma membrane to the molecular biology of individual proteins, and many recent advances are documented. It is hoped that these articles will provide an up-to-date summary for smooth muscle research and will stimulate new studies in normal and abnormal smooth muscle function.

David J. Hartshorne
Takeshi Nakano

# Contents

# Dynamics of Ion Channels Modified by Mobilized Calcium in Vascular Smooth Muscle Cells

Kenji Kitamura[1] and Hirosi Kuriyama[2,3]

*Summary.* We herein discuss the roles of mobilized $Ca^{2+}$ in the cytosol in relation to the activities of ionic channels such as $Ca^{2+}$-dependent $K^+$ and $Cl^-$ channels activated by mobilized $Ca^{2+}$ in the cytosol. The increase in cytosolic $Ca^{2+}$ concentration in vascular smooth muscle cells seems not to be homogeneous, and increase $Ca^{2+}$ may act differently on ionic channels and contractile proteins, indicating temporal and spatial differences in the site of release of $Ca^{2+}$ into the cytosol; this phenomenon may be causally related to the distribution of sarcoplasmic reticulum vesicles. It is clear that the mechanisms of both $Ca^{2+}$-induced $Ca^{2+}$ release (CICR) and $InsP_3$-induced $Ca^{2+}$ release (IICR) play essential roles in regulating mechanical responses and $Ca^{2+}$-dependent ion channels; however, a mutual relation between CICR and IICR in regulating the activity of smooth muscle cells has yet to be proved. Furthermore, a causal relation between $Ca^{2+}$-dependent $Cl^-$ and $K^+$ channels or these channels and the cytosolic free $Ca^{2+}$ concentration ($[Ca^{2+}]_i$) is not yet clear; nor is it clear whether changes in ionic channels leading to spontaneous oscillatory outward currents and spontaneous $Ca^{2+}$-dependent inward $Cl^-$ currents induced by changes in $[Ca^{2+}]_i$ have a physiological significance. The resolution of such problems may lead us toward a new concept of the role of mobilized $Ca^{2+}$ in the function of vascular smooth muscle cells.

*Key Words.* Ion channels—Calcium mobilization—$Ca^{2+}$ store sites—Phosphorylation—E-C coupling

## Introduction

In vascular and visceral smooth muscle cells, changes in the cytosolic free $Ca^{2+}$ concentration ($[Ca^{2+}]_i$) play an essential role in the regulation of the contraction-relaxation cycle. However, increased $[Ca^{2+}]_i$ not only regulates mechanical responses through the $4Ca^{2+}$–calmodulin (CaM)–myosin light chain kinase (MLCK)–20 kDa

[1] Department of Dental Pharmacology, Fukuoka Dental College, 2-15-1 Tamura, Sawara-ku, Fukuoka, 814-01 Japan

[2] Seinan Jogakuin University, Kita-Kyushu, Fukuoka, 802 Japan

[3] Department of Pharmacology, Faculty of Medicine, Kyushu University, Fukuoka, 812 Japan

myosin light chain ($MLC_{20}$) phosphorylation process; increased $[Ca^{2+}]_i$ also regulates metabolic processes in the cytosol and the permeability of ion channels in the sarcolemmal membrane (SL).

Increases in $[Ca^{2+}]_i$ occur on release of $Ca^{2+}$ from intracellular stores distributed in the sarcoplasmic reticulum (SR) and on influx of extracellular $Ca^{2+}$ through the SL. The former, intracellularly stored $Ca^{2+}$, is mainly released via activation of the ryanodine- and inositol 1,4,5-trisphosphate ($InsP_3$)-sensitive $Ca^{2+}$ channels, though $Ca^{2+}$ is also released from mitochondria. The latter influx of $Ca^{2+}$ occurs mainly via activation of the voltage-dependent $Ca^{2+}$ channel in the SL and also may occur partly through passive influx of $Ca^{2+}$ through the SL due to the presence of a concentration gradient across the cell membrane (millimolar order in the extracellular space and 100 nanomolar order in the cytosol). Furthermore, activation of receptors for vasoconstrictor agonists induces an increase in $[Ca^{2+}]_i$ through activation of a non-selective cation channel.

In visceral and vascular smooth muscle cells, some subtypes of $K^+$ and $Cl^-$ channels are regulated by $[Ca^{2+}]_i$, and these channels are called $Ca^{2+}$-dependent $K^+$ or $Cl^-$ channels, respectively. These ion channels are activated by increased $[Ca^{2+}]_i$. However, some ion channels (e.g., the voltage-dependent L-type $Ca^{2+}$ channel) are known to be inhibited by increased $[Ca^{2+}]_i$, as $Ca^{2+}$ stimulates the inactivation process of this channel.

We need to better understand the dynamic relation between electrical activity and mechanical response in vascular smooth muscle cells under conditions of increased $[Ca^{2+}]_i$; e.g., how increased $[Ca^{2+}]_i$ quantitatively regulates contractile proteins and ion channels, as SR vesicles are not homogeneously distributed in the cells. We also need to known how nonhomogeneous distribution of SR vesicles acts to regulate cell activity, and how $Ca^{2+}$ acts simultaneously to regulate contractile proteins and ion channels, which are released from the SR via different releasing mechanisms. In this chapter we focus on the role of mobilized $[Ca^{2+}]_i$ in relation to ionic channels in vascular smooth muscle cells, especially the $Ca^{2+}$-dependent $K^+$ and $Cl^-$ channels.

## Mobilization of Calcium Ions in the Cytosol of Vascular Smooth Muscle Cells

### Concentration of $[Ca^{2+}]_i$ and Its Relation to Mchanical Responses

Using luminescent and fluorescent dyes such as aequorine, quin-2, fura-2, rhod 1, or indo 1 [1–5], $[Ca^{2+}]_i$ has been measured in vascular smooth muscle cells or tissues in resting and active states [6–35]. The basal level of $[Ca^{2+}]_i$ measured using the above procedures has varied by tissue and species and has depended on tissue conditions such as temperature, ionic milieu, stretch, and muscle tone; it is generally assumed, however, to be 80–150 nM (i.e., about 100 nM). For example, Bukoski et al. [36] reported it to be 79 nM in Wistar rat mesenteric artery, and Jensen et al. [37] reported it to be 114 nM at the resting membrane potential of −61 mV. Itoh et al. [38] measured $[Ca^{2+}]_i$ and reported it to be 100–120 nM in physiological saline, whereas in nominal $Ca^{2+}$-free solution containing 0.1 mM EGTA the basal level was reduced to 50–60 nM. Ganitkevich and Isenberg [39] reported it to be 155 nM in coronary artery.

Many vascular smooth muscle tissues possess a resting muscle tone, and after application of EGTA the muscle tone declines to nearly zero. Moreover, an active tension developed in parallel with this decline when $[Ca^{2+}]_i$ was increased to 200 nM in intact tissues, using vascular tissues permeabilized by either saponin or β-escin [10, 40–42]. Increasing $[Ca^{2+}]_i$ in intact tissues or cells or applying $Ca^{2+}$ in skinned muscle tissues enhances the amplitude of the tension in a concentration-dependent manner; in skinned muscle tissues 1 μM $Ca^{2+}$ was required to develop the maximum amplitude of contraction. This amplitude of tension recorded in skinned muscle tissues was larger than that evoked by high $K^+$ or agonists in intact tissues (nonpermeabilized membrane) [43]. Presumably, many $Ca^{2+}$-buffering mechanisms, including uptake and release of $Ca^{2+}$, contribute to: (1) the regulation of $[Ca^{2+}]_i$ during the $[Ca^{2+}]_i$ increase by agonists and high concentration of $K^+$; (2) changes in the $Ca^{2+}$-sensitizing and desensitizing mechanisms; and (3) the various modulations of contractile proteins (actin and myosin systems) through synthesized second messengers, such as cyclic nucleotides, diacylglycerol (protein kinase C activator), and GTP-binding proteins [44–46]. Figure 1 shows the effects of high concentrations of $K^+$, noradrenaline (NAd), and caffeine on the $Ca^{2+}$-transient (measured using fura-2) and tension development recorded simultaneously from the rabbit mesenteric artery. Using the same preparation, the relation between the amplitude of the $Ca^{2+}$-transient and contraction differed depending on the $Ca^{2+}$ in the bath.

As shown in Figure 1, the amplitude of contraction induced by vasoconstrictor agonists or high $K^+$ solution in intact vascular smooth muscle tissues is not proportional to the amount of $Ca^{2+}$ released, as estimated from the $Ca^{2+}$–contraction relation in skinned muscle tissues prepared by saponin, α-toxin or β-escin. The mechanical responses were consistently larger for the agonist-induced contraction than for $K^+$-induced ones, even when both stimuli produced the same amplitudes of the $Ca^{2+}$-transients measured using fura-2. The development of a larger contraction than that evoked by the same amount of $Ca^{2+}$ is defined phenomenologically as an "increase in the $Ca^{2+}$ sensitivity of contractile proteins" and the converse as "$Ca^{2+}$ desensitization" [10, 13, 24, 26, 27, 48–69]. The amplitude of contraction is more closely related to the degree of phosphorylation of $MLC_{20}$ than to $[Ca^{2+}]_i$. This is because activation of contractile proteins by $Ca^{2+}$ may be modified by many factors during the steps from $Ca^{2+}$/CaM binding to the $MLC_{20}$ phosphorylation process; for example, modulation of $Ca^{2+}$/CaM kinase II, caldesmon, calponin, leiotonin, or myosin phosphatase regulate the amount of phosphorylation of $MLC_{20}$ and the activity of action [54, 67, 70–77]. Figure 2 shows arrangements for the $Ca^{2+}$-activated contractions observed in cardiac and skeletal muscle cells.

Furthermore, GTP and GTPγS enhance mechanical responses and $MLC_{20}$ phosphorylation with no modification of $[Ca^{2+}]_i$ or the activity of MLC phosphatase [50, 57, 78, 79]. Hirata et al. [71] reported that this GTPγS-induced increase in $Ca^{2+}$ sensitivity in vascular smooth muscle cells is mediated by rho p21, a low-molecular-weight soluble GTP-binding protein. The existence of $Ca^{2+}$-independent contraction has also been postulated in smooth muscle tissues [47, 52]. It is thought that second messengers such as cyclic AMP, cyclic GMP, or diacylglycerol and arachidonic acid (synthesized from phosphatidic acid, a co-product from phosphatidylinositol bisphosphate, PI-$P_2$) also regulate $Ca^{2+}$ sensitivity [54, 80]. Figure 3 shows the arrangement of contractile proteins in smooth muscle cells. As examples, specific actin and myosin (MLC) interactions are shown, such as 10S-6S conformational changes and the actions of $MLC_{20}$ in chicken gizzard smooth muscle.

Moreover, a dissociation of $[Ca^{2+}]_i$ from the mechanical response has been reported for the development of sustained (tonic) contraction. In many tonic vascular smooth

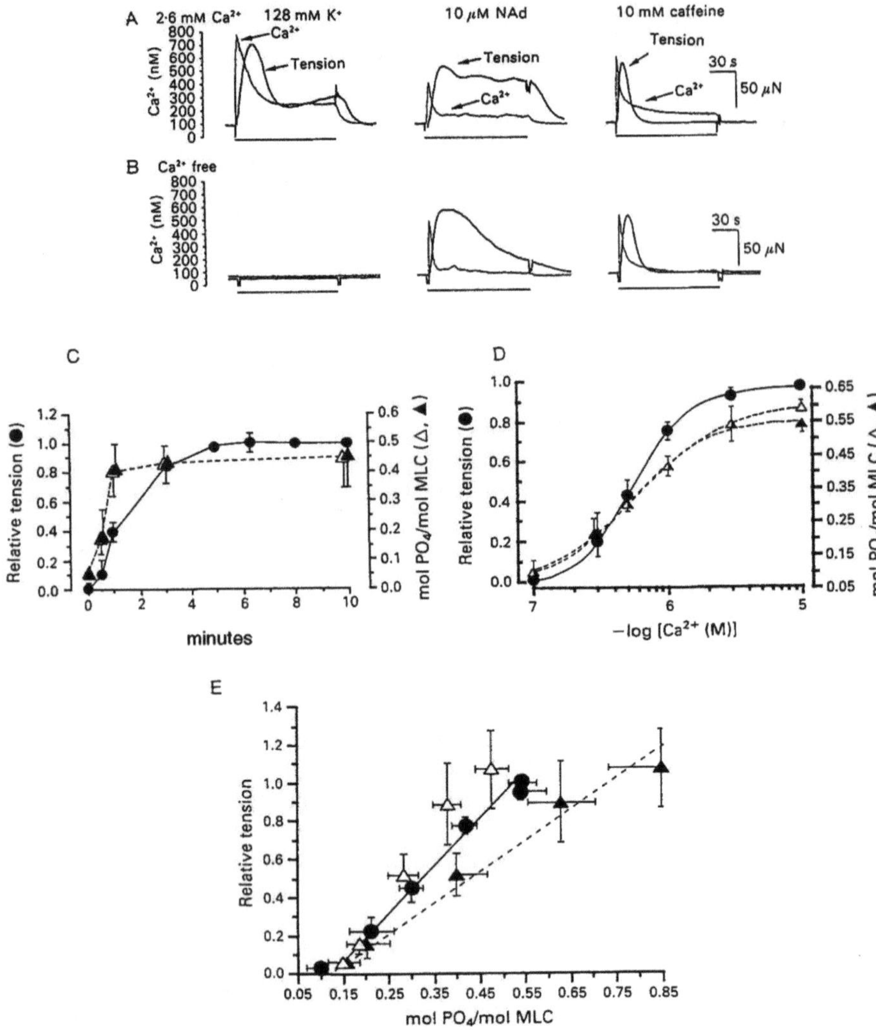


**Fig. 1.** **A, B.** Effects of 128 mM $K^+$, 10 μM noradrenaline (*NAd*), or 10 mM caffeine on cytosolic free $Ca^{2+}$ concentration ($[Ca^{2+}]_i$) and tension in a smooth muscle strip of the rabbit mesenteric artery in the presence or absence of extracellular $Ca^{2+}$. **A**, in the presence of 2.6 mM $Ca^{2+}$; **B**, in $Ca^{2+}$-free solution containing 2 mM EGTA. Individual stimuli were applied for 2 min at 20-min intervals. Traces indicate $[Ca^{2+}]_i$ and tension, respectively. The stimulus was applied where indicated by the *bar*. In **B**, the stimuli were applied 2 min after application of $Ca^{2+}$-free solution containing 2 mM EGTA [16]. **C.** Time-dependent changes in tension (*closed circles*) and monophosphorylated (*open triangles*) or all phosphorylated (mono + di, *closed triangles*) forms of myosin light chain (*MLC*) induced by 10 μM $Ca^{2+}$ in β-escin-treated skinned smooth muscle strips. The maximum amplitude of contraction induced by 10 μM $Ca^{2+}$ was normalized as a relative tension of 1.0 for each strip. $Ca^{2+}$ (10 μM) was applied at time zero in a relaxing solution containing 4 mM EGTA. **D.** Concentration-dependent effects of $Ca^{2+}$ on increase in tension and monophosphorylated or all phosphorylated (mono + di) forms of MLC in β-escin-treated skinned smooth muscle strips. $Ca^{2+}$ at any given concentration was applied for 10 min and 20 min in a relaxing solution containing 4 mM EGTA. The maximum amplitude of contraction induced by 10 μM $Ca^{2+}$ was normalized as a relative tension of 1.0 for each strip. **E.** Relation between tension and monophosphorylated or all phosphorylated (mono + di) forms of MLC in the presence of various concentrations of $Ca^{2+}$ or calyculin A in β-escin-treated skinned

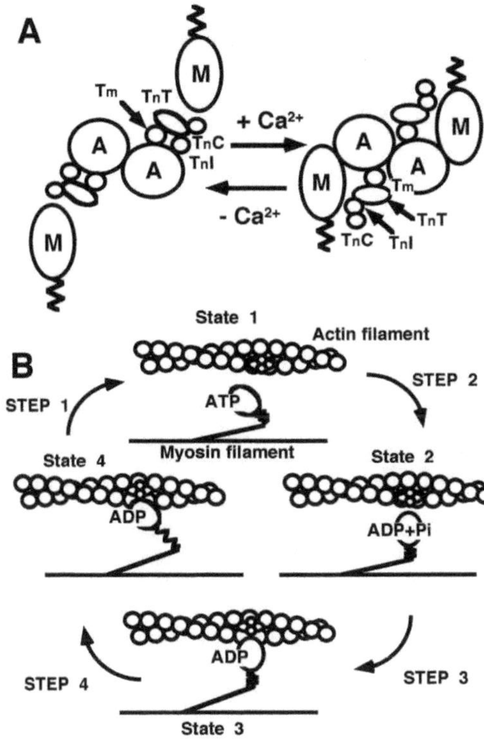


**Fig. 2.** Arrangement of contractile proteins in skeletal muscle. **A.** *M*, myosin filament; *A*, actin filament; *Tm*, tropomyosin; *TnT*, troponin T; *TnI*, troponin I; *TnC*, troponin C. In the absence of Ca, the tropomyosin–troponin complex covers the myosin-binding site on the actin filament. Binding of Ca ion to troponin C dislocates troponin I and tropomyosin, and the myosin-binding site is cropped out. **B.** ATP hydrolysis and the hinge movement of myosin head. STEP 1: ATP binding releases the myosin head from actin. STEP 2: Hydrolysis of ATP sets up the myosin head. STEP 3: Release of Pi from the myosin head and binding of Ca to troponin bind the myosin head to actin tightly. STEP 4: Conformational change of the hinge produces a "power stroke"

muscle tissues, both the phasically developed $Ca^{2+}$-transient and the degree of $MLC_{20}$ phosphorylation declined to near the basal level (tonic phase); however, the amplitude of contraction was sustained at a higher level than that expected from the $[Ca^{2+}]_i$. This dissociation between $[Ca^{2+}]_i$ and $MLC_{20}$ phosphorylation and the mechanical response represent the "latch" state [27, 81–90]. This latch state is postulated to be due to a slow detachment of actin-myosin cross-bridges in the presence of low $Ca^{2+}$ and

**Fig. 1.** (*Continued*) smooth muscle strips; *closed circles* monophosphorylated form of MLC in the presence of $Ca^{2+}$; *open triangles*, monophosphorylated form of MLC in the presence of calyculin A; *closed triangles*, all phosphorylated (mono + di) forms of MLC in calyculin A. The line was fitted by linear regression analysis. The slope of the line for the relation between tension and the monophosphorylated form of MLC in the presence of $Ca^{2+}$ or calyculin A was 2.35 ($r^2 = 1$) and that for the relation between tension and all phosphorylated forms of MLC in the presence of calyculin A was 1.53 ($r^2 = 0.999$). (**C**, **D**: From [47], with permission)

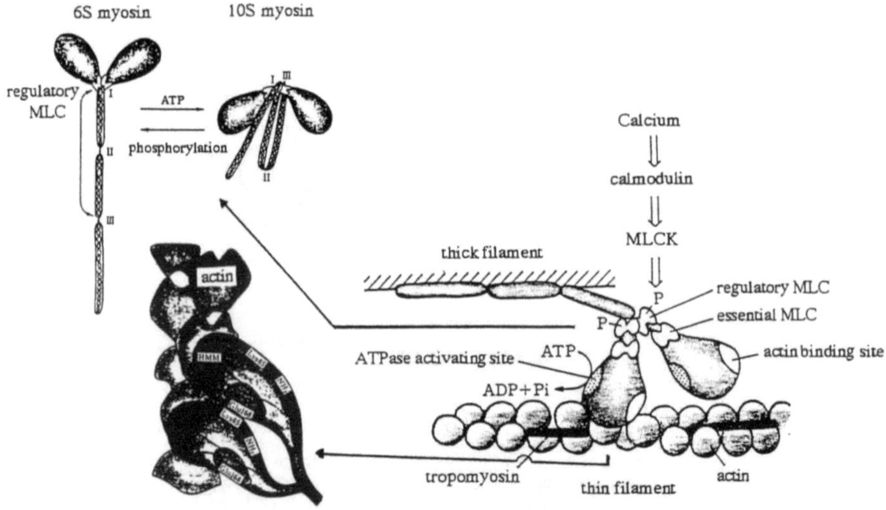


**Fig. 3.** Arrangement of contractile proteins in smooth muscle. Four Ca ions bind on the specific binding sites in the calmodulin. The Ca-calmodulin complex activates myosin light chain kinase (*MLCK*), which phosphorylates $MLC_{20}$. In the dephosphorylated condition MLC is folded (10S myosin) in the presence of ATP, but when $MLC_{20}$ is phosphorylated by MLCK the tail of MLC is stretched (6S myosin) even in the presence of ATP. Such conformational change of myosin is not seen on the myosin of cardiac and skeletal muscle. In 6S myosin, the activity of Mg-ATPase (72K peptide in the myosin head) is increased and modulates the actin-binding site in the 24K peptide

low phosphorylation following actin-myosin cross-bridge cycling at the high rate triggered by myosin phosphorylation. However, no consensus has yet developed concerning the details of this process [90–94].

## Influx of $Ca^{2+}$

Influx of $Ca^{2+}$ into vascular smooth muscle cells occurs mainly through activation of a voltage-dependent $Ca^{2+}$ channel. This channel is classified as long lasting (L; high voltage-activated), transient (T; low voltage-activated), neither L nor T (N), or P; different features of the individual channels have been described elsewhere [95–97]. In most vascular smooth muscle cells from different regions, the presence of both L- and T-type $Ca^{2+}$ channels has been detected, but in some vascular tissues only the L-type $Ca^{2+}$ channel is reported to be present. The L-type $Ca^{2+}$ channel has large conductance compared to that of the T-type, and it is sensitive to organic $Ca^{2+}$ channel blockers ($Ca^{2+}$ antagonists), such as the dihydropyridine (DHP) derivatives (nifedipine, nicardipine, nimodipine, nitrendipine, and 30 or more others), benzothiazepine derivatives (diltiazem and others), and phenylalkylamine derivatives (verapamil and gallopamil). Some examples of the voltage-dependent $Ca^{2+}$ channels distributed in vascular smooth muscle cells are guinea pig aorta conductance, 12pS [98]; rabbit mesenteric artery, 15–20 pS and 8 pS [99]; rabbit ear artery, 25 pS and 8 pS [100]; dog saphenous artery, 18–24 pS and 7–9 pS [101]; rabbit ear artery, 20 pS and 8 pS [102]; rabbit coronary artery, 21 pS [103]; rat small cerebral artery, 19 pS [104]; rabbit basilar artery, 23 pS and 12 pS [105] and 25 pS and 12 pS [106]; guinea pig portal vein, 18 pS [107]; guinea pig coronary artery, 28 pS and 7.5 pS [108]; and rabbit poste-

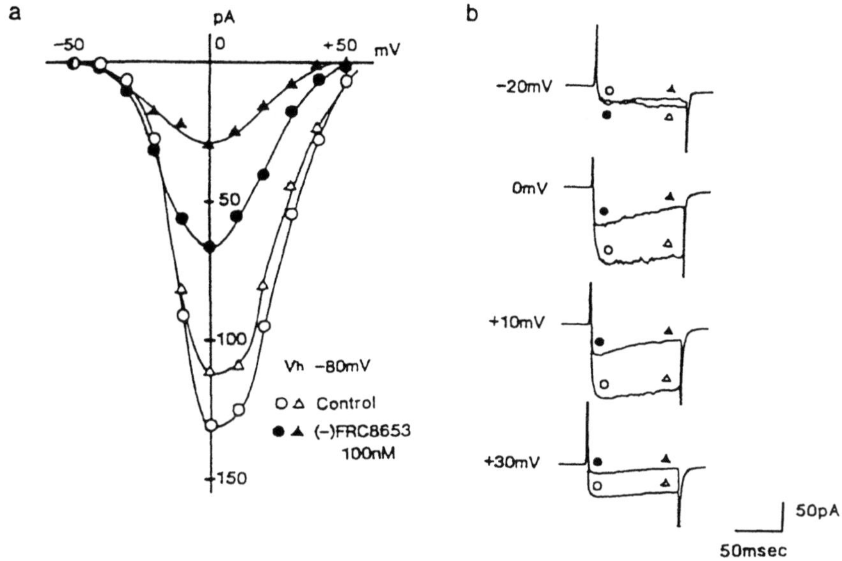


**Fig. 4.** Effects of 100 nM FRC8653, a DHP derivative, on the $Ba^{2+}$ inward current recorded from smooth muscle cells (rabbit basilar artery). **b** Two records of the inward currents obtained at four depolarizing pulses before and after application of a DHP derivative were superimposed. The membrane was held at −80 mV. **a** Current–voltage relations of the inward current in the presence or absence of 100 nM FRC8653. *Open symbols* represent the current amplitudes at peak of the pulse (*circles*) and at the end of the pulse (*triangles*). *Closed symbols* were those in the presence of the DHP derivative. The $Cs^+$-$TEA^+$ solution was in the pipette, and the 10 mM Ba-containing solution was in the bath. (From [105], with permission)

rior cerebral artery, 19 pS [104]. In cultured vascular smooth muscle cells, the features of the voltage-dependent $Ca^{2+}$ channels (especially the L type) differed from those recorded from freshly dispersed cells [109–112]. Figures 4 and 5 show the effects of DHP $Ca^{2+}$ antagonists on voltage-dependent $Ca^{2+}$ channels measured using whole-cell voltage-clamp (Fig. 4) and cell-attached patch-clamp (Fig. 5) procedures, respectively. When one type of the voltage-dependent $Ca^{2+}$ channel was distributed in the vascular smooth muscle cell, DHPs consistently inhibited this $Ca^{2+}$ channel, but when two types of the $Ca^{2+}$ channel were present, the large conductance $Ca^{2+}$ unitary current was inhibited by DHPs; DHPs reduced the open probability of the L-type $Ca^{2+}$ channel with no change in the unitary current amplitude (Fig. 5B). On the other hand, DHPs did not reduce the open probability of the small conductance $Ca^{2+}$ channel (presumably the T type) in the rabbit basilar artery (Fig. 5A). However, in guinea pig portal vein the T-type channel was also inhibited by DHPs, but it required more than 10 times higher concentration than that needed to inhibit the large conductance of $Ca^{2+}$ current (Fig. 5C,D).

Concerning $Ca^{2+}$ mobilization through the SL in vascular smooth muscle cells, Nelson et al. [114] calculated $Ca^{2+}$ influx through a voltage-dependent $Ca^{2+}$ channel using various assumptions, for example that the unitary current of 0.07 pA [115] corresponds to a $Ca^{2+}$ influx rate of $2 \times 10^5$ ions/s through a single open channel. Thus just one channel opening in a single cell would deliver $Ca^{2+}$ at a rate sufficient to raise $[Ca^{2+}]_i$ by 0.3 nM/s at 22°C [116] in the absence of buffering and $Ca^{2+}$ extrusion. If each

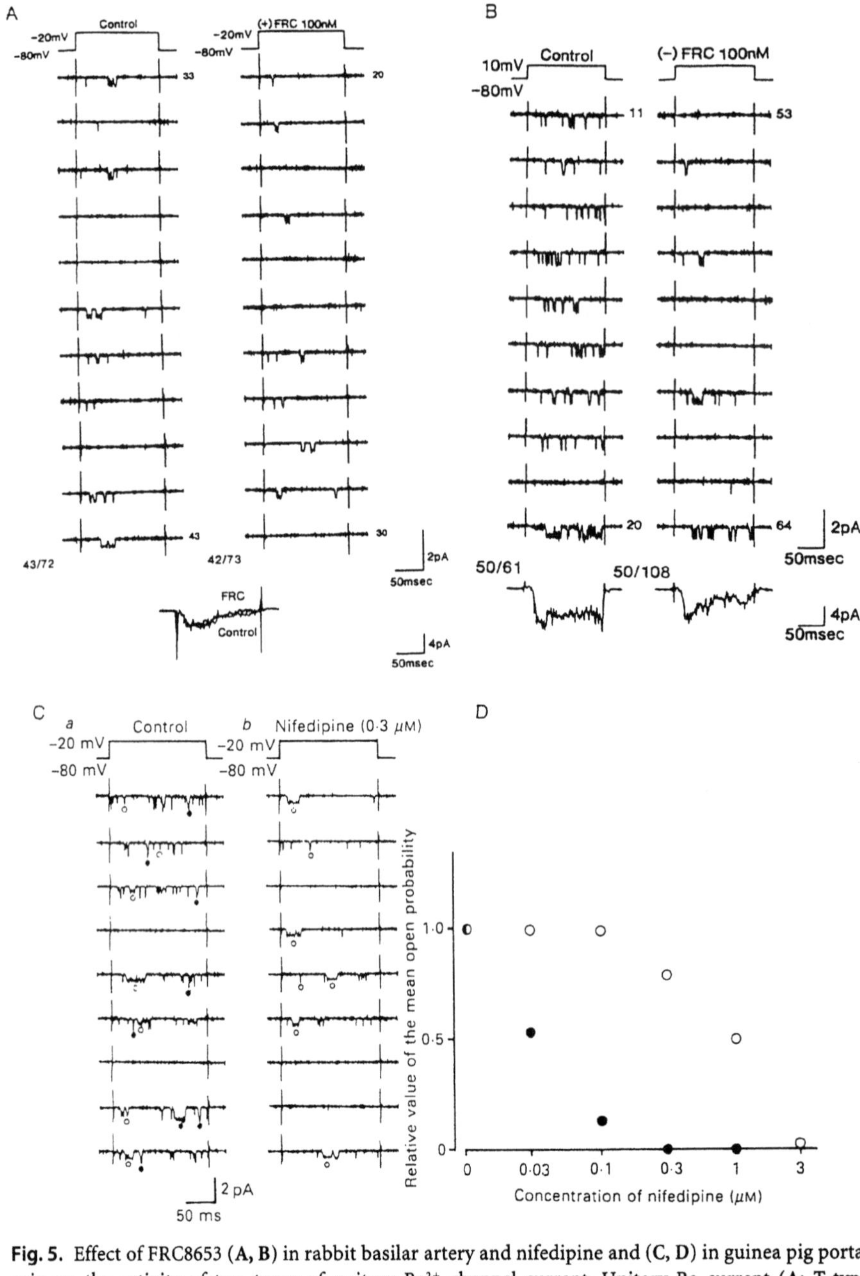


**Fig. 5.** Effect of FRC8653 (**A, B**) in rabbit basilar artery and nifedipine and (**C, D**) in guinea pig portal vein on the activity of two types of unitary $Ba^{2+}$ channel current. Unitary Ba current (**A**: T-type channel; **B**: L-type channel) evoked by depolarizing pulses to −20 mV and +10 mV from the holding potential of −80 mV in the smooth muscle cells of rabbit saphenous artery. A total of 72 (A, control), 73 (A, FRC8653), 61 (B, control), and 108 (B, FRC8653) pulses were applied before and during FRC8653 application, respectively. Tracings shown were obtained from samples 33 to 43 for the controls (A, T-type channel) and from samples 11 to 20 (B, L-type channel), 20 to 30 (A, T-type channel), and 53 to 64 (B, L-type channel) in the presence of FRC8653. Bottom tracings were summated currents in the absence and presence of FRC8653. *Fractions* at the bottom of the tracings

smooth muscle cell has more than 1000 such $Ca^{2+}$ channels, the cell would experience substantial $Ca^{2+}$ influx even when the open probability is as low as 0.001 [114]. Matsuda et al. [103] calculated the density of L-type $Ca^{2+}$ channels in the rabbit coronary artery using various assumptions (open probability = 0.13, average cell surface = $1.87 \times 10^3\,\mu m^2$, single $Ba^{2+}$ channel current = 0.53 pA, and average peak amplitude of the whole-cell $Ba^{2+}$ current = 217 pA in 110 mM $Ba^{2+}$) and concluded that the estimated channel density is 1.71 channels/$\mu m^2$. Sumimoto et al. [35] calculated the density of the nifedipine-sensitive $Ca^{2+}$ channel (L type) using labeled nifedipine or nitrendipine and estimated that more than six sites per square micrometer may be present. If these figures are roughly correct, it is plausible that depolarization of the membrane may lead to an influx of $Ca^{2+}$ that is sufficient to cause a significant increase in $[Ca^{2+}]_i$.

In general, the voltage-dependent $Ca^{2+}$ channel is selective to $Ca^{2+}$, but under extreme conditions this channel can be permeable to $Na^+$ and $K^+$, especially to $Na^+$; with low bath concentrations of $Ca^{2+}$ (less than 1 µM, but not $Ca^{2+}$-free solution), depolarization produced a large inward current that was blocked by DHPs but not by tetrodotoxin (a fast $Na^+$ channel blocker). Thus the action potential in $Ca^{2+}$-deficient $Na^+$-rich solution exhibited a long plateau phase with a slow onset and slow rate of rise. Presumably, under physiological conditions $Ca^{2+}$ may block the entrance of $Na^+$ through the $Ca^{2+}$ channel competitively at the channel's outer mouth. Because this $Na^+$ current generated through $Ca^{2+}$ channels ceased in the absence of $Ca^{2+}$ in the bath, at least a minimal concentration of $Ca^{2+}$ would be required to preserve the channel's activity [117].

Activation of vasoconstrictive agonist receptors can induce influx of $Ca^{2+}$ through the receptor-operated nonselective cation channel. Benham and Tsien [118] recorded the $Ca^{2+}$ influx evoked by activation of purinergic receptors by the application of ATP and concluded that one-tenth of the total receptor-activated current may be carried by $Ca^{2+}$ [119]. Xiong et al. [120] has reported in the rabbit portal vein that ATP activates the $P_{2X}$ and $P_{2Y}$ receptors. The former generated a fast, transient inward current that was inhibited by application of α,β-methylene ATP and the latter a slow, sustained current that was insensitive to the above agent. The latter activations of the $P_{2Y}$ receptor (activation of a nonselective cation channel) induce almost no activation of $Ca^{2+}$ influx. Figure 6 shows the effects of ATP on the rabbit portal vein as measured using the whole-cell voltage-clamp procedure.

On application of either phenylephrine ($\alpha_1$-adrenoceptor agonist) or acetylcholine (ACh; muscarinic receptor stimulation), receptor-induced responses with two components were evoked, that is, a transient, fast, large $Ca^{2+}$-dependent $Cl^-$ current (described in detail later) and a subsequent slow sustained current that was activated

**Fig. 5.** (*Continued*) represent the number of nonblank tracings/total number of tracings. **C.** Unitary $Ba^{2+}$ current evoked by a depolarizing pulse to −20 mV from the holding potential of −80 mV in the absence (a) and presence (b) of 0.3 µM nifedipine. Depolarizing pulses were applied every 10 s, and sequential traces were recorded. *Open and closed circles* indicate the typical openings of the T- and L-type channels, respectively. **D.** Relation between the open probability of two types of unitary $Ba^{2+}$ channel and the concentration of nifedipine. The open probability at each concentration of nifedipine was calculated from the total sweeps. Data were obtained only from the cells that showed activity of both types of unitary $Ba^{2+}$ channel during a depolarizing pulse of −20 mV at the holding potential of −80 mV in the control condition. (**A, B:** From [105]; **C:** from [113], with permission)

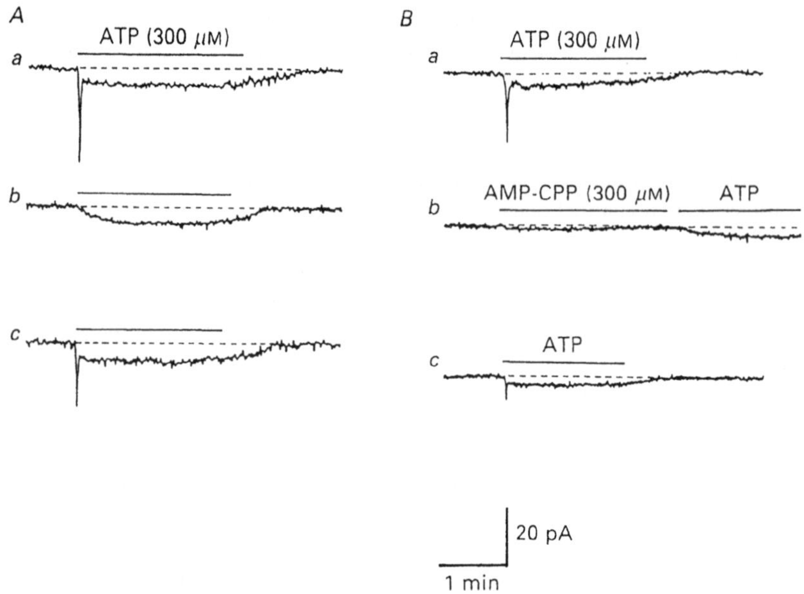


**Fig. 6.** Effects of 300 μM ATP on the smooth muscle cells of rabbit portal vein. The holding potential was −60 mV. **A.** ATP was continuously applied by pressure ejection during the period indicated by bars. Trace *b* was recorded 2 min after the end of trace *a*, and trace *c* was obtained 6 min after trace *b*. Traces *a* through *c* were obtained from the same cell. **B.** Effects of pretreatment with α,β-methylene ATP (AMP-CPP) (300 μM) on the ATP-induced inward current. Traces *a* through *c* were recorded at intervals of 10 min. For trace *b* ATP was applied 10 s after pretreatment with AMP-CPP. (From [120], with permission)

via the nonselective cation channel. The effects of prazosin (an $\alpha_1$-adrenoceptor blocker) and atropine (a muscarinic receptor blocker) were to inhibit the currents evoked by phenylephrine and ACh, respectively. When the two agonists were applied simultaneously, the generated currents were additively superimposed. The former current, $Cl^-$ current, is activated via increased $[Ca^{2+}]_i$ resulting from release of $Ca^{2+}$ from the SR following synthesis of $InsP_3$, and this current was blocked by $Cl^-$ channel blockers (disulfonic stilbene derivatives and others). The latter current is generated via the same mechanism as that observed on activation of the $P_{2Y}$ receptor (via activation of the nonselective cation channel). Thus influx of $Ca^{2+}$ seems to play only a minor role following the activation of these agonist receptors. Figure 7 shows an example of the effects of ACh and phenylephrine on rabbit portal vein smooth muscle cells recorded using the whole-cell patch-clamp procedure. Application of either of these agents evoked a transient $Ca^{2+}$-dependent $Cl^-$ current and a subsequently generated sustained small-amplitude current via the receptor-activated nonselective cation channel.

## Release of $Ca^{2+}$ from the SR

In vascular and visceral smooth muscles, the SR is poorly developed (perhaps 2%–5% of the total SL area) in comparison with that in skeletal muscle cells (5%–10% of the total membrane fraction). Many fragmented SR vesicles are distributed just underneath the SL, and contain a high concentration of $Ca^{2+}$ as estimated using x-ray

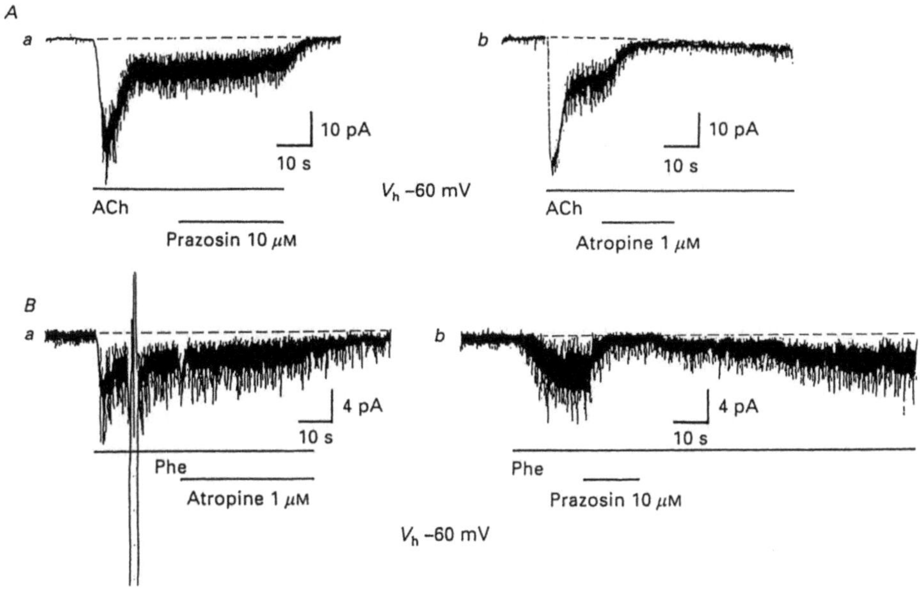


**Fig. 7.** Pharmacologic properties of the phenylephrine (*Phe*)- and acetylcholine (*ACh*)-induced inward currents (rabbit portal vein). At a holding potential of −60 mV, the inward currents were activated by either ACh or Phe. Two antagonists, prazosin and atropine, were adding to the bath (*bars*). **A.** ACh 100 μM. The effect of atropine 1 μM (b) was not completely reversible. **B.** Phe 100 μM. The vertical deflection represents an artifact induced by a ramp potential. (From [121], with permission)

analysis [122, 123], and some SR vesicles are distributed deep in the central region of the vascular smooth muscle cell. In some smooth muscle cells the presence of a foot-like structure between the SL and the SR vesicles has been recognized just underneath the SL, as also observed in the triad structure in skeletal muscle (A.P. Somlyo, personal communication).

Two $Ca^{2+}$-releasing mechanisms have been identified in vascular and visceral smooth muscle cells: (1) $Ca^{2+}$-induced $Ca^{2+}$ release mechanism (CICR): an increase in the $Ca^{2+}$ in the cytosol is capable of releasing $Ca^{2+}$ from the store site provided sufficient $Ca^{2-}$ in the store site; and (2) $InsP_3$-induced $Ca^{2+}$ release (IICR): $InsP_3$ is a product of the hydrolysis of PI-$P_2$ by phospholipase C (PLC) and activates the $InsP_3$-sensitive $Ca^{2+}$ channel in the SR.

A typical example of the first type, the ryanodine-sensitive $Ca^{2+}$ channel (ryanodine receptor), is composed of a homotetramer and is localized in the foot structure of the triad [124–126]. Cloning of the ryanodine receptor cDNA has been successfully carried out for skeletal, cardiac, and brain tissues. The receptor was classified into skeletal (RyR1), cardiac (RyR2), and nonmuscle (RyR3) types. In smooth muscle cells, the ryanodine receptors are mostly of the brain type (nonmuscle), though some are of the cardiac type [126–130]. After finding that the $InsP_3$ receptor was the same protein as that called P400 in the brain, this field of research advanced rapidly. This receptor protein is also composed of a homotetramer [131–135]. The $InsP_3$ receptor has been classified into $InsP_3$ R1 (a and b), R2, R3, and R4. The two $Ca^{2+}$ channels in the SR have some homology in their primary amino acid sequences (especially the C-terminal region [136, 137], but they have different features. Thus the CICR channel is sensitive to caffeine and ryanodine; and their effect is blocked by procaine [33, 41, 138–140]. In contrast, the IICR channel is blocked by heparin and thimerosal [141–147].

The CICR mechanism was first observed in skeletal and cardiac muscle [148, 149] and then in smooth muscle [138]. Activation of the IICR mechanism by $InsP_3$ occurs in many tissues [44, 45, 150–153], and this mechanism was first elucidated in smooth muscle cells by Suematsu et al. [154], Somlyo et al. [155], Yamamoto and Van Breemen [156], and Hashimoto et al. [157]. In 1968 Somlyo and Somlyo [158] suggested that in many vascular smooth muscle tissues agonist-induced contraction can occur with no detectable depolarization of the membrane, which they therefore termed "pharmacomechanical coupling." As described previously, depolarization of the membrane is required to activate the voltage-dependent $Ca^{2+}$ channel and thus induce an influx of $Ca^{2+}$. Therefore a mechanism other than $Ca^{2+}$ mobilization in the cytosol was postulated. In rabbit pulmonary and mesenteric arteries [159–161], NAd or ACh produce contraction without a marked change in the membrane potential. It is clear that activation of the agonist receptor stimulates PLC through a GTP-binding protein and synthesizes $InsP_3$ by hydrolysis of PI-$P_2$; the synthesized $InsP_3$ then activates release of $Ca^{2+}$ via the $InsP_3$-sensitive $Ca^{2+}$ channel in the SR. Activation of the agonist receptor also stimulates receptor-operated ion channels through GTP-binding proteins, which differs from that involved in the PLC activation. Presumably, the density and potency of GTP-binding proteins and the subsequent mechanisms may differ by tissue. At present, GTP-binding proteins are classified into large molecules (more than several subtypes, such as $G_i$, $G_S$, $G_O$, $G_P$, $G_Q$, $G_K$, $G_{PLC}$, and $G_{ACh}$) and small molecules (numerous isoforms such as rho p21 and smg p21). Each high-molecular-weight GTP-binding protein is composed of $\alpha$- and $\beta\gamma$-subunits, and their actions depend on the guanine nucleotides (GDP or GTP) [71, 162–173].

It is clear that the ryanodine- and $InsP_3$-sensitive $Ca^{2+}$ channels are distributed in the SR and release $Ca^{2+}$ under different stimulus conditions (CICR and IICR, respectively). Figure 8 shows the effects of caffeine and $InsP_3$ on the mechanical response evoked in saponin-treated skinned vascular smooth muscle cells.

It is not clear how densely distributed each receptor is in smooth muscle SR and how closely these two receptors interact. Iino [174, 175] postulated nonhomogeneous distributions of the SR and termed the compartments responsible for the release of $Ca^{2+}$ $S\alpha$ stores (contains CICR and IICR) and $S\beta$ stores (contains IICR alone) from their mode of $Ca^{2+}$ release and the contraction observed in intact and skinned muscle tissues. Iino [174] estimated that $S\alpha$ makes up 40% of the total $Ca^{2+}$ compartment in the guinea pig taenia coli, about 50% in pulmonary artery, less than 20% in portal vein, and less than 10% in myometrium (caffeine did not produce contraction in this tissue). Furthermore, Iino and Endo [176] found that IICR was enhanced, as is CICR, by way of a positive feedback control mechanism, whereas when $Ca^{2+}$ concentration was increased to more than 300 nM the IICR mechanism was inhibited by way of a negative feedback control mechanism. In relation to IICR and CICR, receptor-specific $Ca^{2+}$ sequestration-related substances, such as calsequesterine (in relation to the ryanodine receptor-related store site) or calreticulum and calciosome ($InsP_3$ receptor-related store) have been described [177, 178].

Refilling the store site (SR) with $Ca^{2+}$ requires the activation $Ca^{2+}$-ATPase (more details are given in the following section), but functional connections between SL and SR have been proposed, mostly to explain the SR refilling from the extracellular space

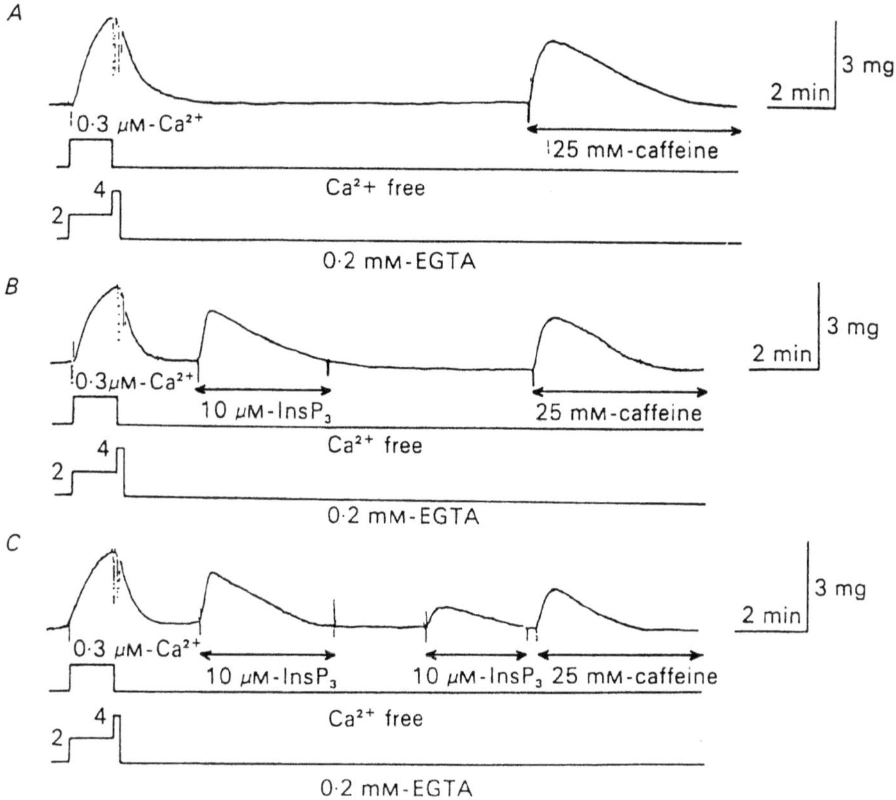


**Fig. 8.** Effects of 10 μM inisotol 1,4,5-trisphosphate ($InsP_3$) and 25 mM caffeine on the tension of the saponin-treated skinned muscle in the Ca-free solution after a Ca load. Ca was loaded by application of 0.3 μM Ca with 2 mM EGTA for 1 min; then the preparation was rinsed with Ca-free relaxing solution (with 4 mM EGTA) for 20 s. Finally, the preparation was rinsed with Ca-free test solution (0.2 mM EGTA). **A.** Caffeine 25 mM was applied 12 min after superfusion of Ca-free test solution. Caffeine produced transient contraction, but a second application of caffeine had no effect. **B.** $InsP_3$ was first applied 2 min after superfusion of Ca-free test solution for 5 min; then caffeine was applied at the same time as for the tracing in *A*. **C.** $InsP_3$ was applied twice before application of caffeine. *Bars* indicate Ca and EGTA concentrations. (From [157], with permission)

[30, 179–184]. Thus if a direct pathway between the extracellular space and SR is present, this process may bypass the SR $Ca^{2+}$-ATPase. On the other hand, it is also possible that the SR may be refilled by virtue of the SR $Ca^{2+}$ pump removing $Ca^{2+}$ from a restricted cytoplasmic space adjacent to the SL; this explanation would be consistent with the buffer-barrier hypothesis [181]. Using a preparation from the rabbit inferior vena cava, Chen and Van Breemen [183] examined the above two alternatives using caffeine, ryanodine, and thapsigargin (an inhibitor of $Ca^{2+}$ accumulation in the SR) and found that the refilling must be mediated by SR $Ca^{2+}$-ATPase. Inhibition of $Ca^{2+}$ accumulation via SR $Ca^{2+}$-ATPase causes an increase in $Ca^{2+}$ influx. Alternatively, these results suggest the existence of a continuous vectoral release of $Ca^{2+}$ from the SR lumen to the extracellular space. Further detailed investigations would be required to clarify the buffering system for $Ca^{2+}$ in the cytosol.

## $Ca^{2+}$ Extrusion from Cytosol

### $Ca^{2+}$-ATPase in SL and SR

Reduction of an elevated level of $Ca^{2+}$ in vascular smooth muscle cells is mainly due to activation of a CaM-binding $Ca^{2+}$-stimulated and $Mg^{2+}$-dependent ATPase ($Ca^{2+}$ pump) in the SL and SR, and partly in concert with activation of an orthodromic $3Na^+$ (influx)–$1Ca^{2+}$ (efflux) exchange diffusion. Comprehensive reviews concerning $Ca^{2+}$-ATPase have appeared [185–189].

There are two $Ca^{2+}$ pump mechanisms in vascular smooth muscle cells (i.e., the SR and SL $Ca^{2+}$ pumps). $Ca^{2+}$-ATPases in the SL are encoded by at least four genes (PMCA1–4), whereas in the SR they are encoded by at least three cDNAs (SERCA1–3). Differences in the features of SL and SR $Ca^{2+}$-ATPase have been investigated using pharmacological and biochemical procedures. It has been found that vanadate has a higher affinity for SL $Ca^{2+}$-ATPase than for SR $Ca^{2+}$-ATPase. Thapsigargin and 2,5-di(tert-butyl)-1,4-benzohydroquinone (tBuBHQ) and cyclopiazonic acid selectively block SR $Ca^{2+}$-ATPase but have no effect on SL $Ca^{2+}$-ATPase; the SL $Ca^{2+}$ ATPase is activated by CaM, acidic phospholipid, or protein kinase, but such activation has not been observed for the SR $Ca^{2+}$-ATPase [190–192]. In cardiac muscle cells, it has been calculated that one ATP molecule may transport two $Ca^{2+}$ ions into the SR. Using a crude membrane fraction of the bovine main pulmonary artery, Eggermont et al. [193] calculated that about 43% of the $Ca^{2+}$ pump activity is due to the $Ca^{2+}$ pump in the SR and 57% to that in the SL. However, the SR/SL $Ca^{2+}$-ATPase ratio may differ widely among the different types of smooth muscle tissue. Eggermont et al. [193] compared the levels of 100-kDa SR and 130-kDa SL phosphoprotein intermediates in isolated membranes from different smooth muscle tissues and reported that in the pig stomach, bovine aorta, and main pulmonary artery the amount of 130-kDa phosphoprotein was much higher than that of the 100-kDa type [186, 193]. The presence of phospholambane as observed in cardiac muscle [194] has also been reported in vascular smooth muscle cells (bovine pulmonary artery [193], bovine aorta [195], rat aorta [196], and canine iliac and ileum arteries [197]) though not in pig aorta [198].

Cyclic nucleotides such as cyclic AMP and cyclic GMP modify the activity of $Ca^{2+}$-ATPase in the SR or SL. For example, cyclic AMP (via protein kinase A) reduces $[Ca^{2+}]_i$ through activation of SR $Ca^{2+}$-ATPase [6, 11, 199–203], whereas cyclic GMP acts mainly via SL $Ca^{2+}$-ATPase to reduce $[Ca^{2+}]_i$ [17, 199, 204–208]. However, this acceleration of the activity of $Ca^{2+}$-ATPase by cyclic GMP (via protein kinase G [PKG]) may not be a direct action [186, 206, 209]. It is also thought that protein kinase C activated either by diacylglycerol, a co-product with $InsP_3$ of the hydrolysis of PI-$P_2$ and other metabolic paths, or by phorbol esters can also accelerate SL $Ca^{2+}$-ATPase and so reduce $[Ca^{2+}]_i$ [205, 206, 210, 211].

### Orthodromic $Na^+$–$Ca^{2+}$ Exchange Diffusion at the SL

In cardiac muscle cells, $Na^+$–$Ca^{2+}$ exchange diffusion normally occurs as a process that reduces $[Ca^{2+}]_i$; but in pathological conditions, an increase in $[Ca^{2+}]_i$ reverses the exchange direction to $Na^+$ efflux–$Ca^{2+}$ influx, thus inducing a further increase in $[Ca^{2+}]_i$. Much the same phenomenon has been postulated in vascular smooth muscle cells. Two main $Na^+$–$Ca^{2+}$ exchange diffusion mechanisms are known, one of which has a stoichiometry of $4Na^+:1Ca^{2+} + 1K^+$ as observed in the rods of the retina [212–

214], and the other $3Na^+$:$1Ca^{2+}$ as in cardiac muscle cells and neurons [215–221]. In vascular and visceral smooth muscle cells the stoichiometry of this exchanger is also thought to be $3Na^+$:$1Ca^{2+}$ [222–228]. Primary amino acid sequences and their characterization have been determined for $Na^+$-$Ca^{2+}$ exchangers in the brain [229–233].

In single cardiac muscle cells, a membrane current generated by this $Na^+$-$Ca^{2+}$ exchange diffusion has been identified by several groups [219, 221, 234–238]. However, in vascular smooth muscle cells, the role of $Na^+$-$Ca^{2+}$ exchange diffusion has been an area of controversy and confusion even though many investigators have supported the presence of this mechanism. It is generally thought that the $Na^+$-$Ca^{2+}$ exchanger plays a more important role in extruding $Ca^{2+}$ in large elastic arteries than in small resistance arteries [223–226; 239–248]. Blaustein et al. [224], considering the role of $Na^+$-$Ca^{2+}$ exchange diffusion in vascular smooth muscle cells, postulated that the $Na^+$-$Ca^{2+}$ exchanger is voltage-sensitive and promotes entry of $Ca^{2+}$ into smooth muscle cells when the $Na^+$ gradient across the SL was reduced or the membrane depolarized. It would play an important role in the extrusion of $Ca^{2+}$ following an elevation of $[Ca^{2+}]_i$, and it most likely contributes to the prolonged elevation of $[Ca^{2+}]_i$ during the maintenance of tonic tension. The predominant effect of a small reduction in the $Na^+$ gradient in quiescent smooth muscle cells appears to be an increase in the $Ca^{2+}$ sequestered in the SR. These authors therefore concluded that the $Na^+$-$Ca^{2+}$ exchanger in vascular smooth muscle cells plays a key role in contributing to the amount of $Ca^{2+}$ stored in the SR and so regulates muscle contractility. Furukawa et al. [205] reported that in aortic muscle cells at the peak of a stimulus-induced $[Ca^{2+}]_i$ transient the $Ca^{2+}$ efflux in $Na^+$-free solution was about 60% that in $Na^+$-containing solution, despite a higher peak $[Ca^{2+}]_i$ in $Na^+$-free solution. If $Na^+$-dependent and $Na^+$-independent $Ca^{2+}$ efflux does occur, these cells may possess both efflux components at the same order of magnitude. Thus under physiological conditions this exchanger may play a minor role in regulating $[Ca^{2+}]_i$ compared to the $Ca^{2+}$ pump mechanisms in the SR and SL. However, when the $Na^+$-$K^+$ pump activity is high under certain pathophysiological conditions, the $Na^+$-$Ca^{2+}$ exchange mechanism may play a more dominant role than under resting conditions [249]. It has also been reported that cyclic GMP with PKG stimulates the $Na^+$-$Ca^{2+}$ pump in smooth muscle cells [250, 251].

# Calcium-Dependent Potassium and Chloride Currents in Vascular Smooth Muscle Cells

## General Features of $K^+$ Currents

### Macroscopic $K^+$ Currents

When the membrane in rabbit and guinea pig portal veins was depolarized to levels more positive than −40 mV from the holding potential of −60 to −80 mV, a transiently developed voltage-dependent $Ca^{2+}$-inward current ($I_{Ca}$) occurs, followed by a transient outward current ($I_{TO}$) within several tenths of a millisecond, and then a small-amplitude sustained outward current ($I_{SO}$) is generated. Furthermore, spontaneously generated oscillatory outward currents (STOCs or $I_{OO}$) also occurred [117, 252–258]. In this tissue, both the $I_{TO}$ and $I_{OO}$ are $Ca^{2+}$-dependent $K^+$ currents, but it was reported that in general $I_{TO}$ is composed of $Ca^{2+}$-dependent and $Ca^{2+}$-independent $K^+$ currents. Furthermore, in some visceral smooth muscle cells, $I_{TO}$ is generated by activation of a $Ca^{2+}$-independent $K^+$ channel. Figure 9 shows depolarization-activated currents recorded from guinea pig portal vein.

In general, in many excitable and nonexcitable cells, $K^+$ currents are classified as voltage dependent or voltage independent; and the former is subclassified into $Ca^{2+}$-dependent and $Ca^{2+}$-independent $K^+$ currents. The $K^+$ channels in excitable and nonexcitable cells are different in terms of structure (amino acid sequences); see also details of short transmembrane spannings for inward rectifying and ATP-sensitive $K^+$ channels [259, 260] compared to those of the transient outward (A-)$K^+$ and delayed rectifing outward current channels [261]—and thresholds and rates for activation and inactivation. These currents have been called the delayed rectifying $K^+$ current, tran-

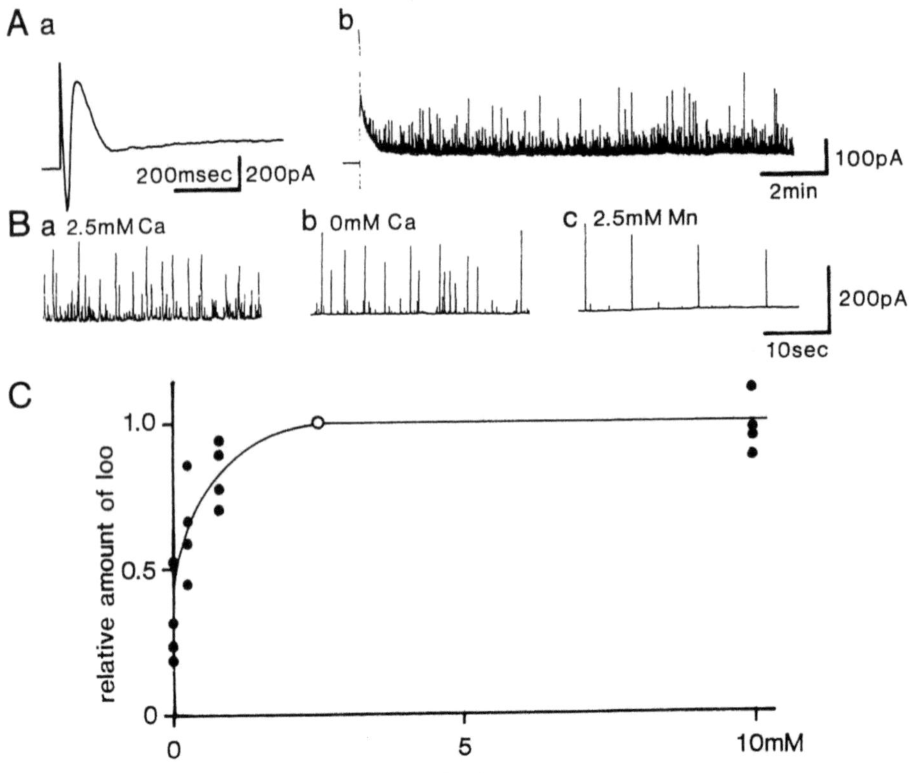


**Fig. 9.** Membrane currents recorded in response to a depolarization pulse to 0 mV (15 min) from the holding potential of −60 mV and effects of $[Ca^{2+}]_o$ on oscillatory outward currents ($I_{oo}$) in the rabbit portal vein. **A.** Tracings a of the membrane currents were measured at fast (*a*) and slow (*b*) recording speeds. Tracing *a* was reporoduced from the early part of tracing *b*. In tracing *b*, a recording of $I_{oo}$ at fast speed is also shown. **B.** The membrane potential was kept at 0 mV throughout the experiment. PSS (*a*), 0 mV $Ca^{2+}$ (*b*), and 2.5 mM $Mn^{2+}$ (*c*) solutions were successively superfused at 5-min intervals. Tracings *a*, *b*, and *c* were recorded after 4 min exposure to each solution. **C.** Relation between $[Ca^{2+}]_o$ and the amount of $I_{oo}$. The relative amount of $I_{oo}$ was calculated from the area of $I_{oo}$ generation for 1 min, divided by the value obtained from the same cell in the physiological salt solution. $I_{oo}$ was evoked by a depolarization to 0 mV, and the curve was drawn by eye. (From [107], with permission)

sient (A-)$K^+$ current and inward rectifying (anomalous-) $K^+$ current [262]. Delayed rectifying $K^+$ current is generated through the delayed rectifier $K^+$ channel and is turned on with a brief delay following the onset of membrane depolarization and persists while the depolarization is maintained. Distributions of this channel have been reported [254, 256, 263–269]. This delayed rectifying $K^+$ current was less dependent on $Ca^{2+}$ and was mostly blocked by tetraethylammonium (TEA) or 4-aminopyridine (4-AP). However, some delayed rectifying $K^+$ currents are sensitive to $Ca^{2+}$ [100, 252, 254, 256, 263–265, 269–272].

The transient (A-, anomalous) $K^+$ current is also activated by depolarization but decays automatically and rapidly while the depolarization is maintained. The channel activation by a given depolarizing voltage step is maximum when the step is imposed on a hyperpolarization and decreases steeply when a less negative conditioning pulse is applied prior to the voltage step. Distributions of this (A-)$K^+$ channel have been reported by several investigators [263, 273–276]. This (A-)$K^+$ current is developed in $Ca^{2+}$-free solution and is voltage- and time-dependent. This channel is sensitive to 4-AP and less sensitive to TEA.

The inward rectifying $K^+$ current mirrors the characteristics of the delayed rectifier outward $K^+$ current, and it can pass a larger $K^+$ current in the inward than in the outward direction. This rectification is due to a time-dependent gating process or to a time-independent rectification inherent to the permeability of the open pore (or both). Distributions of this channel in vascular smooth muscle cells have been reported [277–281]. Furthermore, agonists such as ACh have been found to inhibit a specific $K^+$ channel in smooth muscle cells (toad stomach) and induce depolarization of the membrane; this channel has been called the $K_M$ channel (M current) [282, 283]. An increase in cytosolic ATP inhibits a specific $K^+$ channel, which has been called an ATP-sensitive $K^+$ channel; conversely, a decrease in cytosolic ATP activates this channel [284–287]. Activation of the ATP-sensitive $K^+$ channel is inhibited by glyburide (glybenclamide) and activated by a series of $K^+$ channel openers, such as nicorandil, cromakalim, levcromakalim (lemakalim), pinacidil, and minoxidil (see reviews of Edwards and Weston [288, 289], Quast and Cook [290, 291], Kajioka et al. [286], and Kitamura et al. [292]. Some of the channels have been reported to be sensitive or not sensitive to $Ca^{2+}$. Moreover, in relation to the pacemaker activity in smooth muscle cells of the portal vein, so-called funny current ($I_f$) that is nonselective to $K^+$ but permeable to $Na^+$ and $K^+$, has been described [293].

The most common $Ca^{2+}$-dependent $K^+$ channel is voltage-dependent and sensitive to $[Ca^{2+}]_i$. In some cells, however, such as the rabbit portal vein and pig coronary artery, the $Ca^{2+}$-dependent $K^+$ channel was sensitive to $[Ca^{2+}]_o$ [294, 295].

## Unitary $K^+$ Currents

Using the patch-clamp procedures (cell-attached, inside-out and outside-out membrane patch configurations), the $Ca^{2+}$-sensitive and voltage-dependent $K^+$ channel has been classified into four types: large conductance (150–250 pS in 140 mM symmetrical high $K^+$; maxi-$K^+$ or B-$K^+$); intermediate conductance (about 30–100 pS; I-$K^+$); and small conductance (3–15 pS; $SK^+$). The maxi-$K^+$ and $SK^+$ channels are sensitive to $[Ca^{2+}]_i$; the fourth type of channel is a $[Ca^{2+}]_o$-sensitive 15–30 pS $K^+$ channel [113, 294, 296–312]. The maxi-$K^+$ channel is sensitive to charybdotoxin [313–316] with some exceptions. Extracellularly applied TEA, quinidine, and $Ba^{2+}$ also block this channel; but apamin (so-called S-$K^+$ channel blocker) and 4-AP are less effective [297, 299, 317]. The maxi-$K^+$ channel is activated by application of isoprenaline, adrenaline, or dibutyryl cyclic AMP [301] as well as GMP and cyclic GMP [302, 310, 318]. Distributions of at least two maxi-$K^+$ channels have been postulated from differences in their $Ca^{2+}$ sensitivity [319, 320].

The third subtype, the $SK^+$, was recorded from the rabbit coronary artery [321]. This current represents the delayed rectifier $K^+$ current. Noack et al. [267–269] classified the $Ca^{2+}$-independent $K^+$ current recorded from the guinea pig antrum into a transient component showing voltage-dependent activation and inactivation and another noninactivating component. The former had a conductance of about 2.5 pS and was insensitive to 4-AP, slightly sensitive to TEA, and blocked by caffeine. The latter was blocked by TEA but unaffected by 4-AP. Therefore the delayed rectifying $K^+$ current is composed of $Ca^{2+}$-dependent and $Ca^{2+}$-independent $K^+$ currents, and the latter can be subclassified into two $K^+$ channels that have different conductances and sensitivities to chemical agents. In the rabbit portal vein the presence of a $Ca^{2+}$-independent voltage-dependent transient (A-type; 5 pS) $K^+$ current has been reported [264–266, 276, 322]. This channel has been discussed in relation to the ATP-sensitive $K^+$ channel [264–269].

There is a conflict in results concerning the ATP-sensitive $K^+$ channel in that there are marked variations in the reported values of unitary current conductance (5–135 pS) and some of the channels were $[Ca^{2+}]_i$-and $[Ca^{2+}]_o$-sensitive and others were not. However, all these ATP-sensitive $K^+$ channels were blocked by glyburide. One example is the ATP-sensitive, $Ca^{2+}$-dependent $SK^+$ channel reported in rat portal vein (10 pS) [273, 287], which was less sensitive to TEA than to 4-AP and fairly insensitive to charybdotoxin and apamin and to voltage changes. Another reported channel is an ATP-sensitive, $Ca^{2+}$-independent $SK^+$ channel in rabbit portal vein (15–25 pS); this channel possessed much the same electrical properties as those observed for the ATP-sensitive, $Ca^{2+}$-dependent $K^+$ channel [286]. Figure 10 shows an example of the effects of LP805, a $K^+$ channel opener, on the maxi-$K^+$ and S-$K^+$ channels (small conductance, ATP- and glyburide-sensitive $K^+$ channel). Only the maxi-$K^+$ channel current was recorded in the control under the cell-attached patch-clamp procedure. LP805 did not have any effect on the large conductance maxi-$K^+$ channel but enhanced the open probability of the small conductance (S-)$K^+$ channel in rabbit portal veins. Pinacidil 0.1 mM, lemakalim 0.1 mM, and nicorandil 1 mM also opened the (S-)$K^+$ channel with the same values of the mean open time and mean burst duration (Fig. 11).

## General Features of $Cl^-$ Currents

In visceral and vascular smooth muscle cells, the $Cl^-$ equilibrium potential has been calculated from measurements of ionic concentrations and the value estimated to be −20 to −30 mV. This value is much higher than the resting membrane potential of −60 to −70 mV [325, 326]. Primary amino acid sequences of the voltage-dependent $Cl^-$ channel deduced from cDNA coding differ from those of $Na^+$, $K^+$, and $Ca^{2+}$ channels. The channel is composed of 13 transmembrane segments [327–331]. $Cl^-$ channels expressed in various cells are now classified as CIC-0, CIC-1, and CIC-2 [330–332]. They are distributed in different cells and show different electrical features. For example CIC-0 is distributed in *Torpedo* electric plaques and shows outward rectifying characteristics; CIC-0 and CIC-2, mainly distributed in brain and kidney, show a hyperpolarization-induced activation; CIC-1 is mainly distributed in skeletal

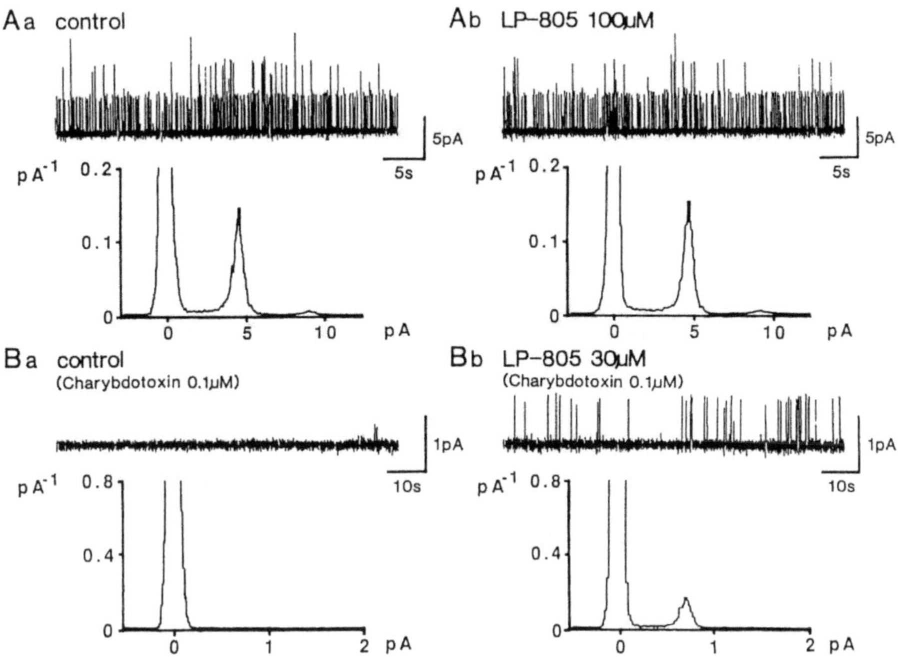


**Fig. 10.** Effect of LP805 on the unitary K channel current recorded in rabbit portal vein (cell attached condition). The physiological salt solution was filled in the pipette, and high K solution was superfused in the bath. The membrane potential was kept at 0 mV. LP805 was applied in the bath. **A.** Example of maxi-K channel current and amplitude histogram in the presence and absence of LP805. **B.** Example of small conductance K channel current and amplitude histogram. Charybdotoxin (100 nM) was placed in the pipette, and LP805 was superfused in the bath. Histograms were drawn from data obtained over 2 min with a sampling rate of 1 ms. (From [323], with permission)

muscle cells and shows hyperpolarization-induced inactivation. CIC-0 corresponds to the 9 and 18 pS $Cl^-$ channel recorded from intact cells and produces outward rectification; CIC-1 may correspond to the 45 pS $Cl^-$ channel recorded from skeletal muscle cells to judge from its hyperpolarization-induced inactivation and its sensitivity to anthracene-9-carboxylic acid (9AC). CIC-2 is thought to be related to the regulation of cell volumes [328, 329, 333, 334]. The features of cystic fibrosis transmembrane conductance regulator (CFTR) has been investigated in detail and shows much the same chemical structure as that of $Cl^-$ channels [335–343].

In general, $Cl^-$ currents are classified into voltage-dependent and $Ca^{2+}$-dependent $Cl^-$ currents. The former is subclassified into (1) depolarization-induced large conductance, or maxi-$Cl^-$ current (150–500 pS distributed in molluscan nerve cells and aorta), intermediate conductance (20–70 pS distributed in skeletal muscle cells and keratinocytes), and small conductance (1–15 pS distributed in aorta, keratinocytes, and *Torpedo* electric plaques) $Cl^-$ currents; (2) a hyperpolarization-induced $Cl^-$ current distributed in dendrites of *Aplysia* neurons; and (3) a bell-shaped voltage-dependent $Cl^-$ current distributed in rat skeletal muscle, Schwann cells, macrophages, secretory epithelial cells, alveoli, and the cardiac muscle of newborn rats [344–348]. Young et al. [349] found voltage-independent but $Ca^{2+}$-dependent $Cl^-$ currents in oocytes (380 pS). On the other hand, a $Ca^{2-}$-dependent $Cl^-$ channel with intermediate conductance (23 pS) was recorded in rat spinal cord [350].

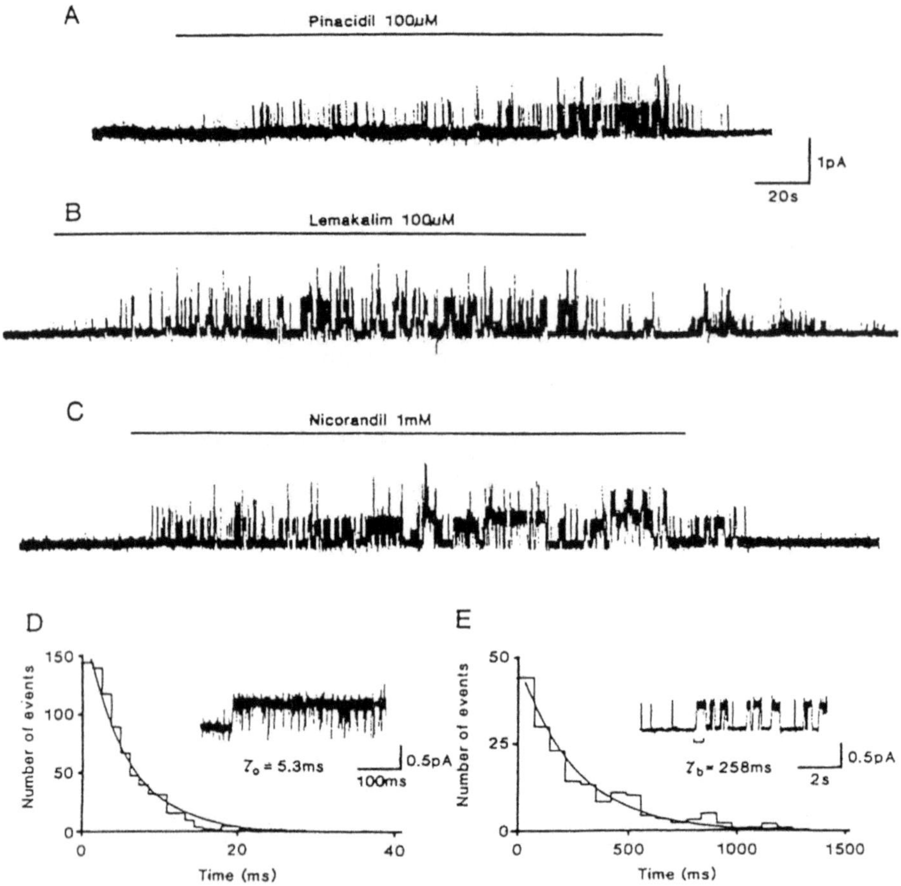


**Fig. 11.** Effects of pinacidil, lemakalim, and nicorandil on the unitary K channel current recorded from the rabbit portal vein using a cell-attached patch-clamp configuration. The patch membrane was kept at −30 mV. High-K solution was placed in the pipette, and high K (with 2 mM Mg and 1 mM EGTA) was superfused in the bath. All drugs were applied in the bath during the period indicated. **A–C.** Examples of actions of pinacidil (100 μM), lemakalim (100 μM), and nicorandil (1 mM). **D, E.** Histograms of open time and burst duration of the K channel current induced by 100 μM pinacidil. Open time and burst duration histograms were obtained using an active filter with cutoff frequencies of 100 Hz and 1 kHz, respectively. (From [324], with permission)

A 340 pS unitary $Cl^-$ conductance has been recorded in embryonic rat aorta [351]. This channel was activated in the potential range of +10 and −10 mV and exhibited excision excitation; it seems to exhibit a bell-shaped voltage-dependent $Cl^-$ current. In human mesenteric artery, a $Ca^{2+}$-dependent small-conductance (2.8 pS) $Cl^-$ channel has been identified [352]. This $Cl^-$ current, measured using the inside-out patch, ceased on reduction of the $Cl^-$ concentration from 150 mM to 5 mM and was activated by increasing the $Ca^{2+}$ in the bath. Therefore Klöckner [352] postulated that an elevation of $[Ca^{2+}]_i$ activates this $Ca^{2+}$-dependent $Cl^-$ channel current. In A7r5 cells prepared from fetal rat aorta, intracellularly applied disulfonic stilbene derivatives, blockers of $Cl^-$ channels, blocked the large conductance $Cl^-$ current [353].

## $Ca^{2+}$-Dependent $K^+$ and $Cl^-$ Currents Modified by Mobilized $Ca^{2+}$

### $Ca^{2+}$-Dependent $K^+$ Currents

*Transient Outward $K^+$ Current.* As already described, depolarizations of the vascular smooth muscle cell membrane produces $I_{Ca}$ and a subsequent $I_{TO}$. In the rabbit, rat, and guinea pig portal veins, this $I_{TO}$ occurred within 100 ms and depended on the voltage shift; it was blocked by reduction of $Ca^{2+}$ in the bath and by reduction of the $Ca^{2+}$ stored in the SR caused by application of ryanodine or caffeine ($Ca^{2+}$ releasers via the CICR channel). Furthermore, destruction of SR by a low concentration (100 nM) of the $Ca^{2+}$ ionophore A23187, which permeabilizes the SR membrane, also blocked the generation of $I_{TO}$. Figure 12 shows the effects of caffeine, charybdotoxin, and heparin on this $I_{TO}$. Caffeine and charybdotoxin blocked $I_{TO}$ generation, but heparin, an inhibitor of the $InsP_3$-induced $Ca^{2+}$-releasing channel in the SR, had no effect. The $I_{TO}$ was blocked by submillimolar concentrations of TEA but not by millimolar concentrations of 4-AP. Furthermore, generation of this transient outward current was required for the generation of the $I_{Ca}$.

From these results, Kitamura and Kuriyama [354] postulated that the $I_{TO}$ may be generated by activation of the maxi-$K^+$ channel by increased $[Ca^{2+}]_i$, which occurs through activation of the CICR mechanism via the ryanodine-sensitive $Ca^{2+}$ channel in the SR and that this process is terminated within 100 ms. Thus the influx of $Ca^{2+}$ induced by $I_{Ca}$ would not itself be enough to directly activate this $Ca^{2+}$-dependent $K^+$

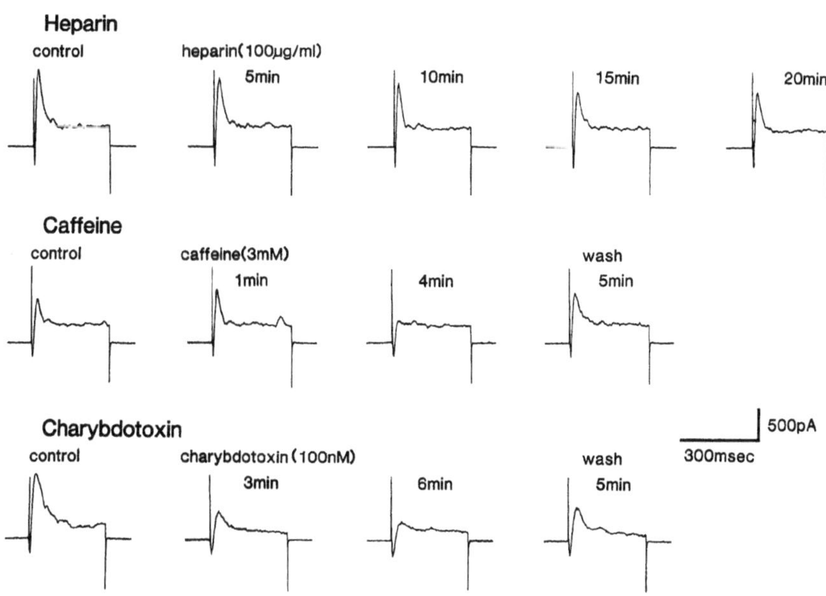


**Fig. 12.** Effects of caffeine (5 mM), charybdotoxin (100 nM), and heparin (100 µg/ml) on the membrane current evoked by a depolarizing pulse to 0 mV from the holding potential of −60 mV in the rabbit portal vein. Caffeine and charybdotoxin were applied to the bath, and heparin was perfused in the pipette. Depolarizing pulse produced a transient Ca inward current followed by a transient outward current ($I_{to}$) and a sustained outward current ($I_{so}$). Note that caffeine and charybdotoxin inhibits $I_{to}$ but not $I_{ca}$ or $I_{so}$

channel, but it would be enough to activate the $Ca^{2+}$ channel in the SR to induce CICR within a time course of 100 ms. It means that, for the generation of $I_{TO}$, influx of $Ca^{2+}$ induced by $I_{Ca}$ must trigger the $Ca^{2+}$ release mechanism from SR vesicles distributed just underneath the cell membrane to produce an increase in $[Ca^{2+}]_i$.

It must be mentioned here that Noack et al. [266] recorded a $Ca^{2+}$-independent $I_{TO}$ in guinea pig stomach fundus that showed both voltage-dependent activation and inactivation but also had another, noninactivating component. The former component had a conductance of 2.5 pS, was slightly sensitive to TEA, and was blocked by caffeine and $Cd^{2+}$. The latter component, a 60 pS channel, was blocked by TEA and $Cd^{2+}$ but was unaffected by 4-AP. This transient outward current had much the same properties as the A-current [263–265, 276, 355], though the noninactivating component seems to be correlated with the $Ca^{2+}$-insensitive component ($I_{SO}$). Therefore the transient outward currents recorded by Sakai et al. [257] and Noack et al. [266–268] may differ in their features; moreover, the $I_{TO}$ may differ in its features in different tissues and species. It can be said, however, that, at least in the portal vein, the $I_{TO}$ is generated by a $Ca^{2+}$-dependent $K^+$ current after influx and release of $Ca^{2+}$. Clapp and Gurney [356, 357] subdivided the outward current induced by depolarization into an A-current, as slowly inactivating ($Ca^{2+}$-activated $K^+$), and time-independent background $K^+$ currents.

*Spontaneous Transient Outward Current (STOCs; Oscillatory Outward Current, $I_{OO}$).* With the whole-cell voltage-clamp procedure, long electrical depolarization of the membrane or long application of vasoconstrictor agonists produce oscillatory outward currents (STOCs or $I_{OO}$). The amplitude and frequency of these currents are distributed randomly and they are modified in a voltage-dependent manner and depend on changes in $[Ca^{2+}]_o$. When the membrane is depolarized for more than several seconds and to less than −40 mV from the holding potential of −60 mV, an $I_{OO}$ occurs on the sustained outward current ($I_{SO}$). To block these currents more than 10 min is required after exposure to $Ca^{2+}$-free EGTA-containing solution, and this time is much longer than that needed to block $I_{TO}$ (immediate) and $I_{Ca}$. Caffeine and ryanodine transiently enhance the frequency and amplitude of $I_{OO}$ and then inhibit them. Procaine, a blocker of CICR, also blocks the $I_{OO}$ [117, 252, 253, 257, 258, 358–361]. Ryanodine blocks both the $I_{TO}$ and the $I_{OO}$, but heparin blocks only $I_{OO}$ [258]. $InsP_3$ generated the $I_{OO}$ but not the $I_{TO}$, so the $I_{TO}$ may have no causal relation to $InsP_3$-induced $Ca^{2+}$ release, IICR. The $I_{OO}$ can be generated by depolarization to −40 mV as described above, but $InsP_3$ could generate $I_{OO}$ at the resting membrane potential of −60 mV. This means the $I_{OO}$ can be evoked without depolarization from the membrane potential level of −60 mV in the case of $InsP_3$ application, though of course depolarization of the membrane accelerated the frequency and amplitude of $I_{OO}$. It has proved difficult, however, to demonstrate that such enhanced $I_{OO}$ is due to activation of IICR alone or in combination with an enhanced CICR mechanism. This situation implies that $I_{OO}$ can be generated by either depolarization or application of $InsP_3$, each in a different manner involving activation of either the CICR or the IICR mechanism via activation of $Ca^{2+}$ channels at the SR, as has been postulated for mechanical responses. Figures 13 and 14 show the effects of caffeine, noradrenaline, and $InsP_3$ on $I_{OO}$ evoked by depolarization of the membrane to 0 mV (Fig. 13) and at −60 mV and −40 mV (Fig. 14).

As shown in Figs. 13 and 14, unitary maxi-$K^+$ currents can be recorded from vascular smooth muscle cells using the cell-attached patch-clamp or inside-out patch-

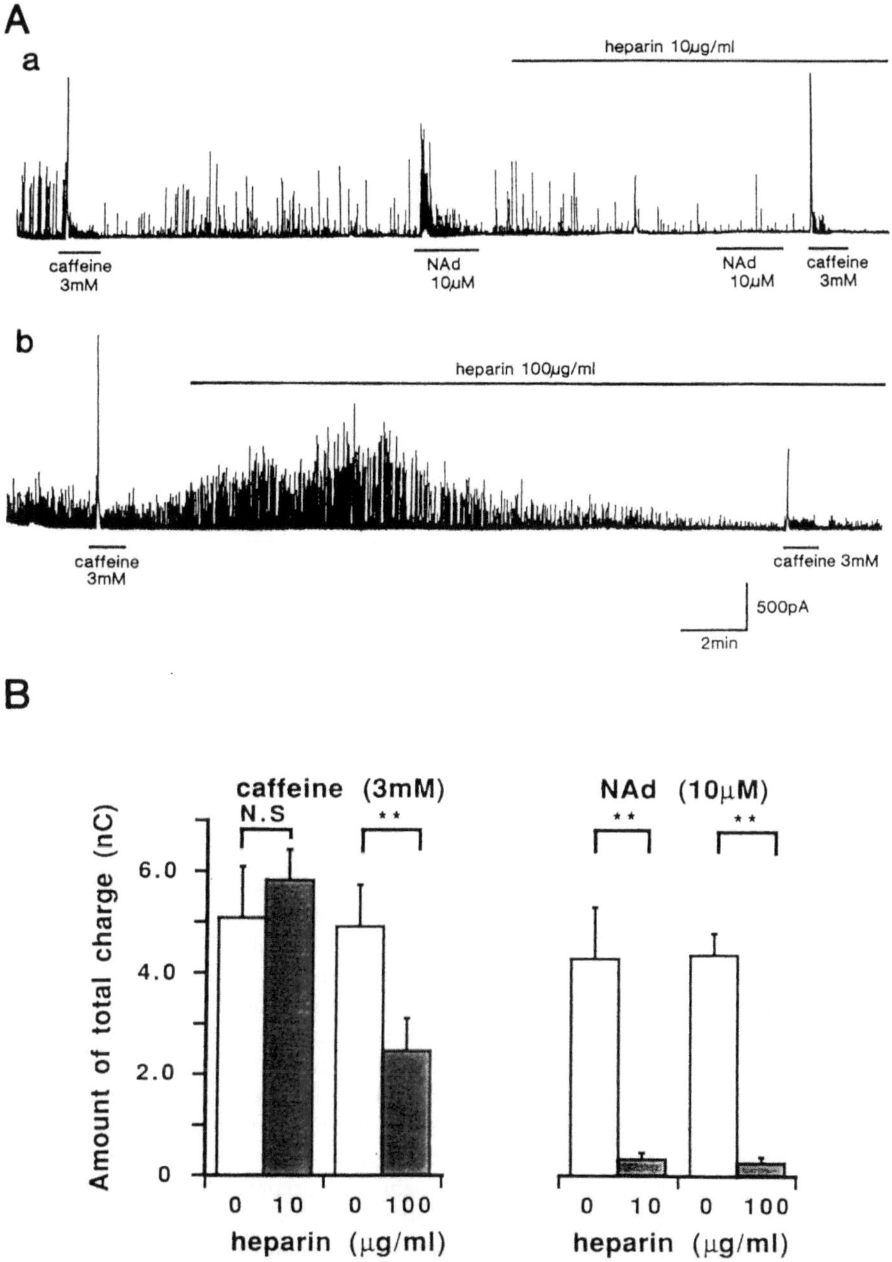


**Fig. 13.** Effects of heparin on the 3 mM caffeine- and 10 μM norepinephrine-induced responses in the rabbit portal vein. The membrane was kept at 0 mV. A. Effects of heparin, 10 μg/ml (*a*) and 100 μg/ml (*b*), on the oscillatory outward current modulated by 3 mM caffeine and 10 μM norepinephrine. B. Effects of heparin on the total charge produced by caffeine and norepinephrine. Each column indicates the mean with SD ($n$ = 6–7). $^{**}P < 0.01$; $^{*}P < 0.05$. (From [360], with permission)

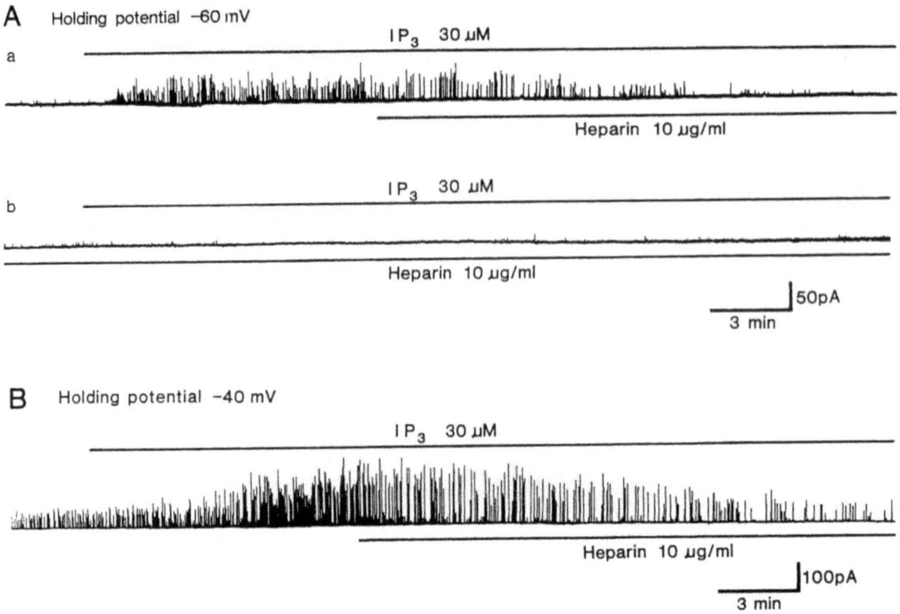


**Fig. 14.** **A, B.** Effects of heparin (10 μg/ml) on 30 μM $InsP_3$-induced generation of $I_{oo}$ at two holding potentials (−60 and −40 mV) (rabbit portal vein). $InsP_3$ and heparin were applied intracellularly where indicated by *bars*. **A.** The membrane was held at −60 mV. Heparin was applied 11 min after (*a*) or 15 min before (*b*) application of $InsP_3$. **B.** The membrane was held at −40 mV. Heparin was applied 10 min after application of $InsP_3$. Each trace was obtained from a different cell. (From [360], with permission)

clamp procedures. This unitary maxi-$K^+$ current is not sensitive to either heparin or caffeine (ryanodine), and more than 100 nM $Ca^{2+}$ in the medium (inside-out patch) or in the pipette is enough to activate this channel. Xiong et al. [258], using the rabbit portal vein, directly proved that in most of the preparations the maxi-$K^+$ unitary current (insensitive to caffeine or heparin) could be recorded, but in some cells or membrane fragments sporadic and simultaneous openings of dozens of unitary maxi-$K^+$ currents activated during depolarization of the membrane could be recorded. These currents may correspond to the $I_{OO}$ recorded using the whole-cell patch-clamp procedure, and they were consistently blocked by caffeine or ryanodine; in some cases heparin too blocked these sporadic burst discharges of the maxi-$K^+$ current. Therefore they postulated that the fragmented membrane prepared for the patch-clamp procedure may contain, or may have attached, SR vesicles, and that these SR vesicles possess a heterogeneous population of $InsP_3$-sensitive and ryanodine-sensitive $Ca^{2+}$ channels. The density or activity of the former may be less than that of the latter. Figures 15 and 16 show the effects of caffeine and heparin on the unitary currents and on sporadic burst discharges of the unitary currents recorded from excised patches (inside-out). When either ryanodine or heparin was applied, the sporadic burst discharges recorded from the fragmented membrane ceased, and only a uniform amplitude maxi-$K^+$ current remained. However, in some fragments heparin did not modify the appearance of sporadic burst discharges, but subsequently applied ryanodine abolished them.

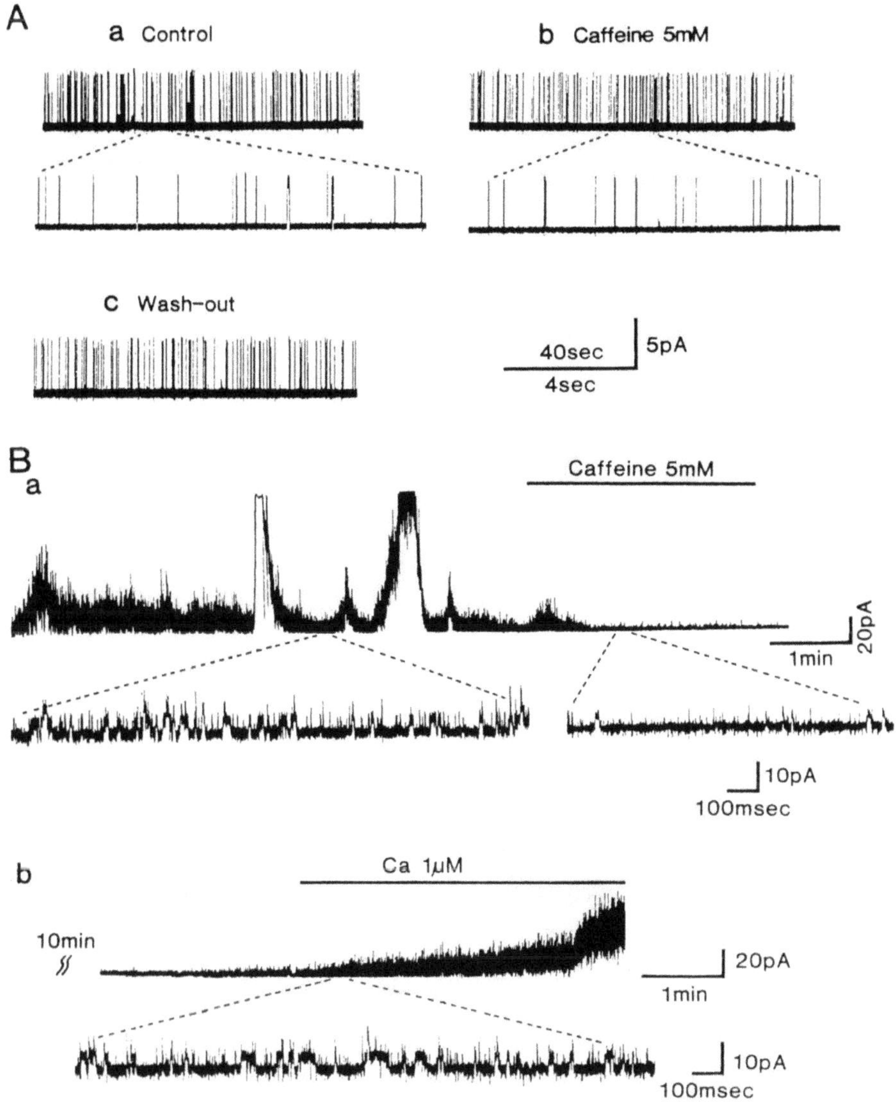


**Fig. 15.** Effects of 5 mM caffeine on the maxi K channel recorded from the inside-out membrane patch of the rabbit portal vein at a holding potential of −10 mV. The pipette was filled with PSS, and the bath was superfused with high-$K^+$ solution. **A.** Bath containing 4 mM EGTA and with a patch electrode of 6 MΩ resistance. Traces obtained just before (*a*, control), with caffeine in the bath (*b*, 5 min after), and 5 min after caffeine removal (*c*). Traces with an expanded scale are also shown. **B.** Bath containing 0.05 mM EGTA and with a patch electrode of 2.5 MΩ resistance. Caffeine 5 mM (*a*) and 1 μM free $Ca^{2+}$ (*b*) were added to the bath (intracellular side) at the time indicated by bars. Tracings *a* and *b* were obtained from the same membrane patch at a 10-min interval. Traces with an expanded scale before and during application of caffeine or $Ca^{2+}$ are shown. (From [120], with permission)

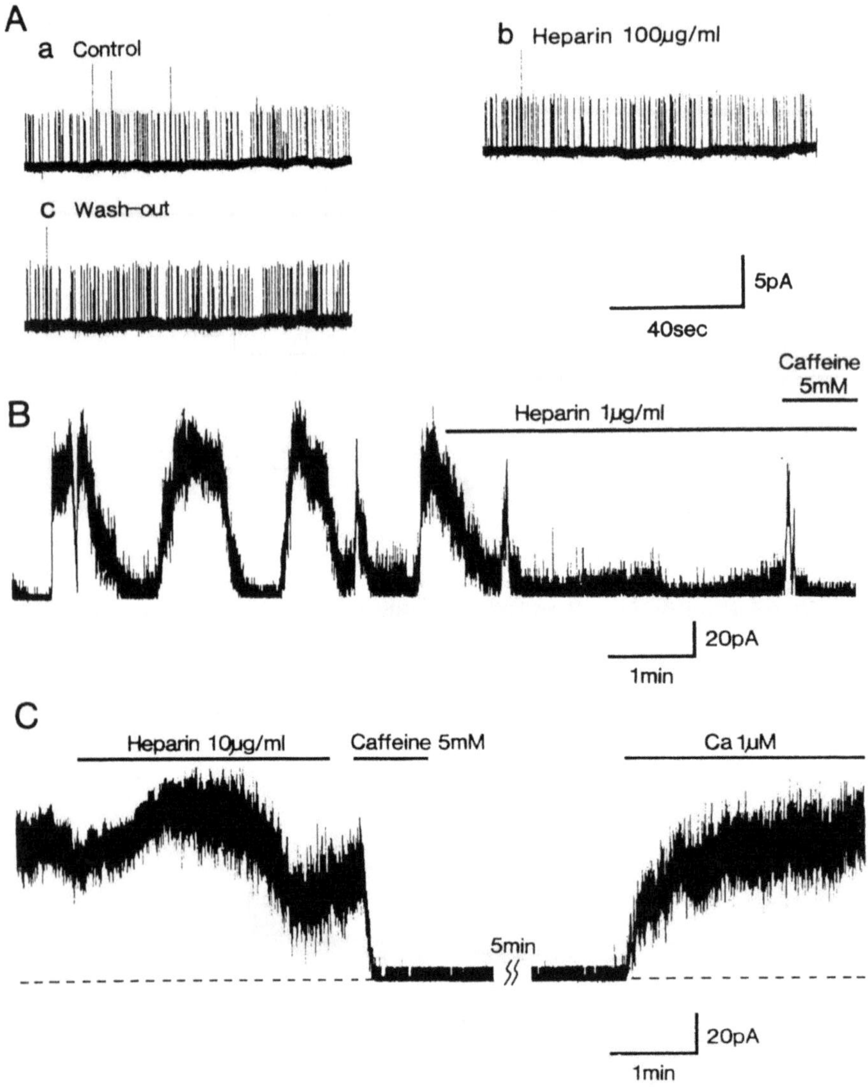


**Fig. 16.** Effects of heparin on the maxi K channel current recorded from the inside-out membrane patch of the rabbit portal vein at a holding potential of −10 mV. The pipette was filled with PSS, and the bath was superfused with the high-K solution containing 4 mM (**A**) and 0.05 mM (**B, C**) EGTA. **A.** Traces obtained just before (*a*, control) and during (*b*) bath application of heparin and 5 min after removal (*c*) of heparin (100 µg/ml). **B.** Low concentration heparin (1 µg/ml) was added to the bath at the time indicated by the *bar*, and then caffeine (5 mM) was applied. **C:** High concentration heparin (10 µg/ml) was added to the bath at the time indicated by the *bar*, and then the same membrane was exposed to 5 mM caffeine. After removal of caffeine, 1 µM free $Ca^{2+}$ was added to the bath. Note that caffeine but not heparin effectively inhibited $K^+$ channel currents. (From [120], with permission)

There are many unresloved problems related to the phenomena shown in Figures 13–16: (1) Whether the $I_{OO}$ plays a physiological role because it occurred only when the membrane was depolarized to more than −40 mV for a long period, except in the case of the $InsP_3$-induced $I_{OO}$. (2) Whether $Ca^{2+}$ release has a physiological function in these sporadic burst discharges because they occur more frequently before deterioration of the cell-attached patch condition, presumably while the SR vesicles are in an unphysiological condition, resulting from "excised excitation" or "excision activation" of the $Ca^{2+}$-dependent maxi-$Cl^-$ current. This maxi-$Cl^-$ current is rarely recorded in a cell-attached patch but is commonly recorded in cell-free patch conditions. The underlying mechanism of excision activation is postulated to involve the leakage away of an inhibitor, a 1-kDa substance thought to be an unsaturated fatty acid such as an arachidonic acid-like substance, which normally prevents this channel's opening [362, 363]. It is also plausible that, in the excised-patch condition, leakage of stored $Ca^{2+}$ in the SR vesicle may be caused by removal of some unknown substance. (3) How does the $Ca^{2+}$ release induced via the $InsP_3$ (heparin)-sensitive channel occur from the SR when the membrane is merely depolarized? One clue might be found in the reports of Itoh et al. [16] and Ito et al. [10] that changes in the membrane potential regulate the synthesis of $InsP_3$ (depolarization accelerated and hyperpolarization inhibited the synthesis of $InsP_3$). It may be possible that some unknown biochemical mechanism is triggered by depolarization of the membrane. (4) If the fragmented SL membrane and the SR vesicles are in an electrically independent state under patch conditions, how does release of $Ca^{2+}$ occur from the SR vesicle during depolarization of the membrane? Do the SR vesicles possess a functional connection with the fragmented SL membrane, such as in the triad structure of skeletal muscle? These questions remain unanswered.

## $Ca^{2+}$-Dependent Inward $Cl^-$ Current: Spontaneously Generated Transient Inward Current

In smooth muscle cells of the rabbit portal vein, the presence of spontaneously generated $Ca^{2+}$-dependent inward $Cl^-$ currents (STICs) has been detected during depolarization and agonist stimulation, and this channel is thought to be activated by $Ca^{2+}$ released from the SR via increased $InsP_3$ on depolarization or receptor activation [311, 364–368]. However, this current occurred only in $K^+$-free conditions. Activation of $Cl^-$ channels by released $Ca^{2+}$ has been reported not only in the portal vein but also in other vascular and visceral tissues using different agonists [364, 369–371]. Whether there is a mutual relation between $I_{OO}$ (STOCs) and STICs is not yet clear, but STICs are generated only in the absence of $K^+$ from the cytosol; one possibility is that STICs may be masked by the generation of $I_{OO}$, as postulated by Byrne and Large [364]. This explanation seems unlikely because the density and amplitude of STICs and $I_{OO}$ were much the same, and both phenomena occurred in a random manner. Therefore a relatively low $Ca^{2+}$ sensitivity of this $Ca^{2+}$-dependent $Cl^-$ channel or, alternatively, an inhibitory action of $K^+$ on the $Cl^-$ channel should be considered. When STICs occurred transiently with a large amplitude, the generation of these currents was blocked by application of either NAd or caffeine, as also observed for the generation of STOCs. Therefore this channel could be activated by $Ca^{2+}$ released from $InsP_3$- and ryanodine-induced $Ca^{2+}$ channels in the SR. These currents have been observed not only in vascular tissues but also in visceral smooth muscle tissues (intestine and stomach) [372, 373].

The unitary current of the $Ca^{2+}$-dependent $Cl^-$ channel was first described by Young et al. [349], who studied *Xenopus* oocytes; This channel has a conductance of 380 pS (maxi-$Cl^-$) and it was activated in the presence of a micromolar order of $Ca^{2+}$. However, this channel was more permeable to $I^-$ or $Br^-$ than to $Cl^-$. In contrast, in the rat spinal cord a low-conductance $Ca^{2+}$-dependent $Cl^-$ current has been recorded; this channel was almost closed at 10 nM $Ca^{2+}$ and activated in the presence of 500 nM $Ca^{2+}$ [350].

## Receptor-Activated Nonselective Cation Channel and $Cl^-$ Channels

*Receptor-Activated Nonselective Cation Channels.* Many vasoconstrictor agonists, such as NAd, histamine, ACh, and ATP depolarize the membrane and increase its ionic conductance. Using the whole-cell patch-clamp procedure, with measurement of $Ca^{2+}$ transients, the actions of ATP in the rabbit coronary artery have been studied in detail by Benham and Tsien [374] and Benham [119, 375, 376]. They found that ATP activates the $P_2$ receptor (α,β-methylene ATP-sensitive) and produces an inward current that has $Na^+$, $K^+$, and $Ca^{2+}$ components; and they discussed the possible role of $Ca^{2+}$ in mechanical responses (in relation to $Ca^{2+}$ transients). Xiong et al. [120] examined the rabbit portal vein and found that ATP activated both the α,β-methylene ATP-sensitive (a phasic component; presumably $P_{2x}$) and α,β-methylene ATP-insensive (a tonic component; presumably $P_{2Y}$) $P_2$ receptors. This means that activation of the two receptors induced activation of two different nonselective cation channels. The features of the former were the same as those described by Benham and Tsien [374]; and for the latter, sustained opening of the channel activated by the $P_{2Y}$ receptor induced mainly $Na^+$ and $K^+$ permeation and some $Ca^{2+}$ permeation. Thus an increase in $Ca^{2+}$ influx induced by ATP receptor activation may be due to activation of the $P_{2X}$ receptor, and the $Ca^{2+}$ influx may be sufficient to account for one-tenth of the total inward current [373]. Thus the receptor activation may release $Ca^{2+}$ from the SR via the synthesis of $InsP_3$ and also partly via activation of the CICR mechanism by the influx of $Ca^{2+}$ that occurs during generation of the fast-activated phasic current (nonselective cationic current). Concerning the tonic nonselective cationic current evoked by ATP stimulation in vascular tissues, much the same responses could also be recorded on the applications of either NAd (or phenylephrine) or ACh (see Fig. 7). Oike et al. [377] also reported that simultaneous application of GTPγS and PDBu mimicked a PTX-sensitive nonselective cation channel in rabbit portal vein.

Inoue and Kuriyama [121] concluded that at least two inward currents could be recorded following agonist stimulation, the transient large and sustained small inward currents. The former was a $Ca^{2+}$-dependent $Cl^-$ current, and the latter occurred by activation of the nonselective cation channel. Phenylephrine was more potent in activating these two channels than ACh, but both agents activated the same nonselective cationic channel as judged by an assessment of their biophysical properties and by drug actions on these channels (unitary conductances were estimated to be 23–25 pS). Therefore at least two types of receptor-activated nonselective cation channel may be distributed on vascular smooth muscle tissues, and the large transient component may be related to the influx of $Ca^{2+}$.

*$Ca^{2+}$-Dependent $Cl^-$ Current Generated by Agonist Stimulation.* As described earlier, STICs are generated by increased $[Ca^{2+}]_i$ resulting from release of $Ca^{2+}$ from the SR through activation of $InsP_3$ receptors. These responses occur with a latency longer than a few seconds. Moreover, it has also been reported in gastrointestinal and vascular smooth muscle cells that the transient phasic inward current generated by agonists is due to increased $[Ca^{2+}]_i$ via the synthesis of $InsP_3$ [121, 364, 367–369]. In vascular smooth muscle cells, the $Cl^-$ equilibrium potential is at a more positive level (more positive than −20 to −25 mV) than the resting membrane potential (−60 to

$-70\,mV$), and the $Ca^{2+}$-dependent $Cl^-$ channel has been found in porcine coronary artery [378], rabbit portal vein [121, 364, 367], rabbit ear artery [369], and pulmonary artery [368]. NAd and ACh each evoked an inward current with contributions from a $Ca^{2+}$-dependent $Cl^-$ current as well as from the nonselective cation channel. However, at a negative potential, the current elicited from the nonselective cation channel is much smaller than the transient inward current via the $Ca^{2+}$-dependent $Cl^-$ channel [368]. Because this $Ca^{2+}$-dependent $Cl^-$ channel current failed to generate an inward current after depletion of stored $Ca^{2+}$ in the SR, it may be generated by activation of the $InsP_3$-sensitive $Ca^{2+}$ channel in the SR following hydrolysis of $PI\text{-}P_2$ through activation of a G protein and phospholipase C. It implies that the synthesis of $InsP_3$ occurs within few seconds and that the ins$P_3$-sensitive $Ca^{2+}$ channel in the SR vesicles distributed just underneath the SL membrane may contribute to generation of the $Ca^{2+}$-dependent $Cl^-$ current.

However, it is not yet certain whether this transient inward current is the same as the current responsible for evoking the excitatory junction potential (EJP) in visceral and vascular smooth muscle. This is because, in these smooth muscle cells, the latency for generation of the EJP was within 100 ms, and iontophoretically applied NAd or ATP generated depolarization within a similar short period [379]. It is not yet certain whether 100 ms is enough time for the multiple steps involved in the metabolic path from agonist-receptor binding through GTP binding protein translocation—phospholipase C activation with the consequent hydrolysis of $PI\text{-}P_2$ to synthesize $InsP_3$—and the subsequent release of $Ca^{2+}$ from the SR, which finally leads to $Ca^{2+}$-induced activation of the $Cl^-$ channel. In vascular smooth muscle cells, EJPs generated by nerve stimulation do not directly correlate with the generation of a mechanical response; only the generation of an action potential superimposed on summated EJPs shows a relation to the generation of mechanical responses (except for bronchial and tracheal smooth muscles).

## References


1. Blinks JR, Wier WG, Hess P, Prendergast FG (1982) Measurement of $Ca^{2+}$ concentrations in living cells. Prog Biophys Mol Biol 40:1-114
2. Grynkiewicz G, Poenie M, Tsien RY (1985) A new generation of $Ca^{2+}$ indicators with greatly improved fluorescence properties. J Biol Chem 260:3440-3450
3. Tsien RW, Tsien RY (1990) Calcium channels, stores, and oscillations. Annu Rev Cell Biol 6:715-760
4. Tsien RY (1980) New calcium indicators and buffers with high selectivity against magnesium and protons: design, synthesis, and properties of prototype structures. Biochemistry 19:2396-2404
5. Tsien RY, Pozzan T, Rink TJ (1982) Calcium homeostasis in intact lymphocytes: cytoplasmic free calcium monitored with a new, intracellularly trapped fluorescent indicator. J Cell Biol 94:325-334
6. Abe A, Karaki H (1989) Effect of forskolin in cytosolic $Ca^{++}$ level and contraction in vascular smooth muscle. J Pharmacol Exp Ther 249:895-900
7. Abe A, Karaki H (1992) Mechanisms underlying the inhibitory effect of dibutyryl cyclic AMP in vascular smooth muscle. Eur J Pharmacol 211:305-311
8. Defeo TT, Briggs GM, Morgan KG (1987) $Ca^{2+}$ signals obtained with multiple indicators in mammalian vascular muscle cells. Am J Physiol 253:H1456-H1461
9. Defeo TT, Morgan KG (1985) Calcium-force relationships as detected with aequorin in two different vascular smooth muscles of the ferret. J Physiol (Lond) 369:269-282
10. Ito S, Kajikuri J, Itoh T, Kuriyama H (1991) Effects of lemakalim on changes in $Ca^{2+}$ concentration and mechanical activity induced by noradrenaline in the rabbit mesenteric artery. Br J Pharmacol 104:227-233

11. Kai H, Kanaide H, Matsumoto T, Nakamura M (1987) 8-Bromoguanosine 3′:5′-cyclic monophosphate decreases intracellular free calcium concentrations in cultured vascular smooth muscle cells from rat aorta. FEBS Lett 221:284–288
12. Kanaide H (1990) Calcium transients in primary culture rat aortic muscle cells. Asia Pac J Pharmacol 5:177–184
13. Karaki H (1989) $Ca^{2+}$ localization and sensitivity in vascular smooth muscle. Trends Pharmacol Sci 10:320–325
14. Karaki H (1990) The intracellular calcium-force relationship in vascular smooth muscle. Am J Hypertens 3:253S–256S
15. Katsuyama H, Ito S, Itoh T, Kuriyama H (1991) Effects of ryanodine on acetylcholine-induced $Ca^{2+}$ mobilization in single smooth muscle cells of the porcine coronary artery. Pflügers Arch 419:460–466
16. Itoh T, Kajikuri J, Kuriyama H (1992) Characteristic features of noradrenaline-induced $Ca^{2+}$ mobilization and tension in arterial smooth muscle of the rabbit. J Physiol (Lond) 457:297–314
17. Kobayashi S, Kanaide H, Nakamura M (1985) Cytosolic-free calcium transients in cultured vascular smooth muscle cells: microfluorometric measurements. Science 229:553–556
18. Kobayashi S, Kanaide H, Nakamura M (1986) Complete overlap of caffeine- and $K^+$ depolarization-sensitive intracellular calcium storage site in cultured rat arterial smooth muscle cells. J Biol Chem 261:15709–15713
19. Morgan JP, Morgan KG (1982) Vascular smooth muscle: the first recorded $Ca^{2+}$ transients. Pflügers Arch 395:75–77
20. Morgan JP, Morgan KG (1984) Alteration of cytoplasmic ionized calcium levels in smooth muscle by vasodilators in the ferret. J Physiol (Lond) 357:539–551
21. Morgan JP, Morgan KG (1984) Stimulus-specific patterns of intracellular calcium levels in smooth muscle of ferret portal vein. J Physiol (Lond) 351:155–167
22. Morgan KG, Suematsu E (1990) Calcium and vascular smooth muscle tone. Am J Hypertens 3:2915–1985
23. Ozaki H, Blondfield DP, Hori M, Publicover NG, Kato I, Sanders KM (1992) Spontaneous release of nitric oxide inhibits electrical, $Ca^{2+}$ and mechanical transients in canine gastric smooth muscle. J Physiol (Lond) 445:234–247
24. Ozaki H, Kwon S-C, Tajimi M, Karaki H (1990) Changes in cytosolic $Ca^{2+}$ and contraction induced by various stimulants and relaxants in canine tracheal smooth muscle. Pflügers Arch 416:351–359
25. Ozaki H, Ohyama T, Sato K, Karaki H (1990) $Ca^{2+}$-dependent and independent mechanisms of sustained contraction in vascular smooth muscle of the rat aorta. Jpn Pharmacol 52:509–512
26. Rembold CM (1990) Modulation of the $[Ca^{2+}]$ sensitivity of myosin phosphorylation in intact swine arterial smooth muscle. J Physiol (Lond) 429:77–94
27. Rembold CM (1992) Regulation of contraction and relaxation in arterial smooth muscle. Hypertension 20:129–137
28. Rembold CM, Murphy RA (1988) Myoplasmic $[Ca^{2+}]$ determines myosin phosphorylation in agonist-stimulated swine arterial smooth muscle. Circ Res 63:593–603
29. Somlyo AP, Himpens B (1989) Cell calcium and its regulation in smooth muscle. FASEB J 3:2266–2276
30. Van Breemen C, Saida K (1989) Cellular mechanisms regulating $[Ca^{2+}]_i$ smooth muscle. Annu Rev Physiol 51:315–329
31. Williams DA, Becker PL, Fay F (1987) Regional changes in calcium underlying contraction of single smooth muscle cells. Science 235:1644–1648
32. Williams DA, Fay F (1990) Intracellular calibration of the fluorescent calcium indicator fura-2. Cell Calcium 11:75–88
33. Watanabe C, Yamamoto Y, Hirano K, Kobayashi S, Kanaide H (1992) Mechanisms of caffeine-induced contraction and relaxation of rat aortic smooth muscle. J Physiol (Lond) 456:193–213
34. Sumimoto K, Domae M, Yamanaka K, Nakao K, Hashimoto T, Kitamura K, Kuriyama H (1987) Actions of nicorandil on vascular smooth muscles. J Cardiovasc Res 10(suppl 8):S66–S75
35. Sumimoto K, Hirata M, Kuriyama H (1988) Characterization of [$^3$H]nifedipine binding to intact vascular smooth muscle cells. Am J Physiol 254:C45–C52

36. Bukoski RD, Bergmann C, Gairard A, Stoclet J-C (1989) Intracellular $Ca^{2+}$ and force determined stimultaneously in isolated resistance arteries. Am J Physiol 257:H1728–H1735
37. Jensen PE, Mulvany MJ, Aalkjær C (1992) Endogenous and exogenous agonist-induced changes in the coupling between $[Ca^{2+}]_i$ and force in rat resistance arteries. Pflügers Arch 420:536–543
38. Itoh T, Suzuki A, Watanabe Y (1994) Effect of a peptide inhibitor of protein kinase C on G-protein-mediated increase in myofilament $Ca^{2+}$-sensitivity in rabbit arterial skinned muscle. Br J Pharmacol 111:311–317
39. Ganitkevich VY, Isenberg G (1993) Membrane potential modulates inositol 1,4,5-trisphosphate-mediated $Ca^{2+}$ transients in guinea-pig coronary myocytes. J Physiol (Lond) 470:35–44
40. Itoh T, Suzuki S, Suzuki A, Nakamura F, Naka M, Tanaka T (1994) Effects of exogenously applied calponin on $Ca^{2+}$-regulated force in skinned smooth muscle of the rabbit mesenteric artery. Pflügers Arch 427:301–308
41. Iino M (1989) Calcium-induced calcium release mechanism in the guinea-pig taenia caeci. J Gen Physiol 94:363–383
42. Iino M (1991) Effects of adenine nucleotides on inositol 1,4,5-trisphosphate-induced calcium release in vascular smooth muscle cells. J Gen Physiol 98:681–698
43. Itoh T, Kuriyama H, Suzuki H (1981) Excitation-contraction coupling in smooth muscle cells of the guinea-pig mesenteric artery. J Physiol (Lond) 321:515–535
44. Abdel-Latiff AA (1989) Calcium-mobilizing receptors, polyphospho-inositides, generation of second messengers and contraction in the mammalian iris smooth muscle: historical perspectives and current status. Life Sci 45:757–786
45. Berridge MJ (1993) Inositol trisphosphate and calcium signalling. Nature 361:315–325
46. Nishizuka Y (1992) Intracellular signaling by hydrolysis of phospholipids and activation of protein kinase C. Science 258:607–614
47. Suzuki A, Itoh T (1993) Effects of calyculin A on tension and myosin phosphorylation in skinned smooth muscle of the rabbit mesenteric artery. Br J Pharmacol 109:703–712
48. Bradley AB, Morgan KG (1987) Alteration in cytoplasmic calcium sensitivity during porcine coronary artery contractions as detected by aequorin. J Physiol (Lond) 385:437–448
49. Collins EM, Walsh MP, Morgan KG (1992) Contraction of single vascular smooth muscle cells by phenylephrine at constant $[Ca^{2+}]_i$. Am J Physiol 262:H754–H762
50. Fujiwara T, Itoh T, Kubota Y, Kuriyama H (1989) Effects of guanosine nucleotides on skinned smooth muscle tissue of the rabbit mesenteric artery. J Physiol (Lond) 408:535–547
51. Jiang MJ, Morgan KG (1989) Agonist-specific myosin phosphorylation and intracellular calcium during isometric contractions of arterial smooth muscle. Pflügers Arch 413:637–643
52. Karaki H (1991) $Ca^{2+}$ regulation of vascular smooth muscle and release of endothelium-derived relaxing factor. In: Sperelakis N, Kuriyama H (eds) Ion channels of vascular smooth muscle cells and endothelial cells. Elsevier, New York, pp 297–315
53. Karaki H, Sato K, Ozaki H (1988) Different effects of norepinephrine and KCl on the cytosolic $Ca^{2+}$-tension relationship in vascular smooth muscle of rat aorta. Eur J Pharmacol 151:325–328
54. Karaki H, Sato K, Ozaki H (1991) Different effects of verapamil on cytosolic $Ca^{++}$ and contraction in norepinephrine stimulated vascular smooth muscle. Jpn J Pharmacol 55:35–42
55. Kajikuri J, Itoh T, Kuriyama H (1992) Does $Ca^{2+}$ release by acetylcholine enhance the synthesis of inositol 1,4,5-trisphosphate in smooth muscle of the porcine coronary artery? Jpn J Physiol 42:665–671
56. Kitazawa T, Gaylin BD, Somlyo AP (1991) G-protein-mediated inhibition of myosin light chain phosphatase in vascular smooth muscle. Proc Natl Acad Sci USA 88:9307–9310
57. Kitazawa T, Kobayashi S, Horiuti T, Somlyo AV, Somlyo AP (1989) Receptor-coupled permeabilized smooth muscle: role of the phosphatidylinositol cascade, G-proteins and modulation of the contractile response to $Ca^{2+}$. J Biol Chem 264:5339–5342
58. Kitazawa T, Somlyo AP (1990) Desensitization and muscarinic re-sensitization of force and myosin light chain phosphorylation to cytoplasmic $Ca^{2+}$ in smooth muscle. Biochem Biophys Res Commun 172:1291–1297
59. Himpens B, Missiaen L, Droogmans G, Casteels R (1991) $A1F_4^-$ induces $Ca^{2+}$ oscillations in guinea-pig ileal smooth muscle. Pflügers Arch 417:645–650

60. Nishimura H, Moreland S, Ahn HY, Kawase T, Moreland RS, Van Breemen C (1992) Endothelin increase myofilament $Ca^{2+}$ sensitivity α-toxin-permeabilized rabbit mesenteric artery. Circ Res 71:951–959
61. Nishimura J, Khalil RA, Drenth JP, Van Breemen C (1990) Evidence for increased myofilament $Ca^{2+}$ sensitivity in norepinephrine-activated vascular smooth muscle. Am J Physiol 259:H2–H8
62. Nishimura J, Khalil RA, Van Breemen C (1989) Agonist-induced vascular tone. Hypertension 13:835–844
63. Nishimura J, Kolber M, Van Breemen C (1988) Norepinephrine and GTPγS increase myofilament $Ca^{2+}$ sensitivity in alpha-toxin permeabilized arterial smooth muscle. Biochem Biophys Res Commun 157:677–683
64. Nishimura J, Moreland S, Moreland RS, Van Breemen C (1991) Regulation of the $Ca^{2+}$-force relationship in permeabilized arterial smooth muscle. In: Moreland RS (ed) Regulation of smooth muscle contraction. Plenum Press, New York, pp 111–128
65. Somlyo AP, Kitazawa T, Himpens B, Mathus G, Horiuti K, Kobayashi S, Goldman YE, Somlyo AV (1989) Modulation of $Ca^{2+}$ sensitivity and of the time course of contraction in smooth muscle: a major role of protein phosphatases? Adv Protein Phosphatases 5:181–195
66. Somlyo AP, Somlyo AV (1990) Flash photolysis studies of excitation-contraction coupling, regulation, and contraction in smooth muscle. Annu Rev Physiol 52:857–874
67. Somlyo AP, Somlyo AV (1992) Pharmacomechanical coupling: the role of G-proteins in $Ca^{2+}$-release and modulation of $Ca^{2+}$-sensitivity. Jpn J Pharmacol 58(suppl II):54P–59P
68. Somlyo AP, Walker JW, Goldman YE, Trentham DR, Kobayashi S (1988) Inositol trisphosphate, calcium and muscle contraction. Philos Trans Soc B320:399–414
69. Yagi S, Becker PL, Fay FS (1988) Relationship between force and $Ca^{2+}$ concentration in smooth muscle as revealed by measurements on single cells. Proc Natl Acad Sci USA 85:4109–4113
70. Himpens B, Mathijs G, Somlyo AB, Butler TM, Somlyo AP (1988) Cytoplasmic free calcium, myosin light chain phosphorylation and force in phasic and tonic smooth muscle. J Gen Physiol 92:713–729
71. Hirata K, Kikuchi A, Sasaki T, Kuroda S, Kaibuchi K, Matsuura Y, Seki H, Saida K, Takai Y (1992) Involvement of rho p21 GTP-enhanced calcium ion sensitivity of smooth muscle contraction. J Biol Chem 267:8719–8722
72. Stull JT, Hsu L-C, Tansey MG, Kamm KE (1990) Myosin light chain kinase phosphorylation in tracheal smooth muscle. J Biol Chem 265:16683–16690
73. Sobue K, Sellers JR (1991) Caldesmon, a novel regulatory protein in smooth muscle and nonmuscle actomyosin systems. J Biol Chem 266:12115–12118
74. Takahashi K, Abe M, Hiwada K, Kokubu T (1988) A novel troponin T-like protein (calponin) in vascular smooth muscle: interaction with tropomyosin paracrystals. J Hypertens 6:S40
75. Takahashi K, Abe M, Hiwada K, Kokubu T (1989) Molecular organization of caldesmon, calponin and tropomyosin in smooth muscle thin filaments. Jpn Circ J 53:906
76. Takahashi K, Hiwada K, Kokubu T (1988) Vascular smooth muscle calponin: novel troponin-T-like protein. Hypertension 11:620–626
77. Tansey MG, Word RA, Hidaka H, Singer HA, Schworer CM, Kamm KE, Stull JT (1992) Phosporylation of myosin light chain kinase by the multifunctional calmodulin-dependent protein kinase II in smooth muscle cells. J Biol Chem 267:12511–12516
78. Kobayashi S, Kitazawa T, Somlyo AV, Somlyo AP (1989) Cytosolic heparin inhibits muscarinic and α-adrenergic $Ca^{2+}$-release and $Ca^{2+}$-sensitization in smooth muscle. J Biol Chem 264:17997–18004
79. Kubota Y, Nomura M, Kamm KE, Mumby MC, Stull JT (1992) GTPγS-dependent regulation of smooth muscle contractile elements. Am J Physiol 262:C405–C410
80. Gong MC, Fuglsang A, Alessi D, Kobayashi S, Cohen P, Somlyo AV, Somlyo AP (1992) Arachidonic acid inhibits myosin light chain phosphatase and sensitizes smooth muscle to calcium. J Biol Chem 267:21492–21498
81. Dillon PF, Aksoy MO, Driska SP, Murphy RA (1981) Myosin phosphorylation and the cross-bridge cycle in arterial smooth muscle. Science 211:495–497
82. Dillon PF, Murphy RA (1982) Tonic force maintenance with reduced shortening velocity in arterial smooth muscle. Am J Physiol 242:C102–C108

83. Hai C-M, Murphy RA (1988) Cross-bridge phosphorylation and regulation of latch state in smooth muscle. Am J Physiol 254:C99–C106
84. Hai C-M, Murphy RA (1988) Regulation of shortening velocity by cross-bridge phosphorylation in smooth muscle. Am J Physiol 255:C86–C94
85. Hai C-M, Murphy RA (1989) $Ca^{2+}$, crossbridge phosphorylation and contraction. Annu Rev Physiol 51:285–298
86. Hai C-M, Murphy RA (1989) Cross-bridge dephosphorylation and relaxation of vascular smooth muscle. Am J Physiol 256:C282–C287
87. Hai C-M, Murphy RA (1992) Adenosine 5′-triphosphate consumption by smooth muscle as predicted by the coupled four-state crossbridge model. Biophys J 61:530–541
88. Murphy RA (1989) Special topic: contraction in smooth muscle cells, introduction. Annu Rev Physiol 51:275–283
89. Murphy RA (1992) Do the cytoplasmic and muscle-specific isoforms of actin and myosin heavy and light chains serve different functions in smooth muscle? Jpn J Pharmacol 58(suppl II):67P–74P
90. Rembold CM, Richard JL, Chen X-L (1992) $Na^+$-$Ca^{2+}$ exchange, myoplasmic $Ca^{2+}$ concentration, and contraction of arterial smooth muscle. Hypertension 19:308–313
91. Butler TM, Siegman MJ, Mooers SU, Narayan SR (1990) Myosin-product complex in the resting state and during relaxation of smooth muscle. Am J Physiol 258:C1092–C1099
92. Marston SB (1989) A tight-binding interaction between smooth-muscle native thin filaments and heavy meromyosin in the presence of MgATP. Biochem J 259:303–306
93. Siegman MJ, Vyas TB, Mooers SU, Butler TM (1992) Energetics and regulation of smooth muscle contraction. Jpn J Pharmacol 58(suppl):75p–80p
94. Siegman ML, Butler TM, Mooers SU, Michalek A (1984) $Ca^{2+}$ can affect $V_{max}$ without changes in myosin light chain phosphorylation in smooth muscle. Pflügers Arch 401:385–390
95. Llinás R, Sugimori M, Lin J-W, Cherksey B (1989) Blocking and isolation of a calcium channel from neurons in mammals and cephalopods utilizing a toxin fraction (FTX) from funnel-web spider poison. Proc Natl Acad Sci USA 86:1689–1693
96. Rosenberg RL, Hess P, Tsien RW (1988) L-Type channels and calcium-permeable channels open at negative membrane potentials. J Gen Physiol 92:27–54
97. Tsien RW, Lipscombe D, Madison DV, Bley KR, Fox AP (1988) Multiple types of neuronal calcium channels and their selective modulators. Trends Neurosci 11:431–438
98. Caffrey JM, Josephson IR, Brown AM (1986) Calcium channels of amphibian stomach and mammalian aorta smooth muscle cells. Biophys J 49:1237–1242
99. Worley JF, Deitmer JW, Nelson MT (1986) Single nisoldipine-sensitive calcium channels in smooth muscle cells isolated from rabbit mesenteric artery. Proc Natl Acad Sci USA 83:5746-5750
100. Benham CD, Bolton TB, Denbigh JS, Lang RI (1987) Inward rectification in freshly isolated single smooth muscle cells of the rabbit jejunum. J Physiol (Lond) 383:461–476
101. Yatani A, Seidel CL, Allen J, Brown AM (1987) Whole-cell and single-channel calcium currents of isolated smooth muscle cells from saphenous vein. Circ Res 60:523–533
102. Aaronson PI, Bolton TB, Lang RJ, Mackenzie I (1988) Calcium currents in single isolated smooth muscle cells from the rabbit ear artery in normal-calcium and high-barium solutions. J Physiol (Lond) 405:57–75
103. Matsuda JJ, Volk, KA, Shibata EF (1990) Calcium currents in isolated rabbit coronary arterial smooth muscle myocytes. J Physiol (Lond) 427:657–680
104. Quayle JM, McCarron JG, Asbury JR, Nelson MT (1993) Single calcium channels in resistance-sized cerebral arteries from rats. Am J Physiol 264:H470–H478
105. Oike M, Inoue Y, Kitamura K, Kuriyama H (1990) Dual action of FRC8653, a novel dihydropyridine derivative, on the $Ba^{2+}$ current recorded from the rabbit basilar artery. Circ Res 67:993–1006
106. Worley JF, Quayle JM, Standen NB, Nelson MT (1991) Regulation of single calcium channels in cerebral arteries by voltage, serotonin, and dihydropyridines. Am J Physiol 261:H1951–H1960
107. Ohya Y, Sperelakis N (1989) Modulation of single slow (L-type) calcium channels by intracellular ATP in vascular smooth muscle cells. Pflügers Arch 414:257–264

108. Ganitkevich VY, Isenberg G (1990) Contribution of two types of calcium channels to membrane conductance of single myocytes from guinea-pig coronary artery. J Physiol (Lond) 426:19–42
109. Akaike N, Kanaide H, Kuga T, Nakamura M, Sadoshima J, Tomoike J (1989) Low-voltage-activated calcium current in rat aorta smooth muscle cells in primary culture. J Physiol (Lond) 416:141–160
110. Bodin P, Richard S, Travo C, Berta P, Stoclet J-C, Papin S, Travo P (1991) Responses of subcultured rat aortic smooth muscle myocytes to vasoactive agents and KCl-induced depolarization. Am J Physiol 260:C151–C158
111. Kuga T, Sadoshima JI, Tomoike H, Kanaide H, Akaike N, Nakamura M (1990) Actions of $Ca^{2+}$ antagonists on two types of $Ca^{2+}$ channels in rat aorta smooth muscle cells in primary culture. Circ Res 67:469–480
112. Neveu D, Nargeot J, Richard S (1993) Two high-voltage-activated dihydropyridine-sensitive $Ca^{2+}$ channel currents with distinct electrophysiological and pharmacological properties in cultured rat aortic myocytes. Pflügers Arch 424:45–53
113. Inoue Y, Oike M, Nakao K, Kitamura K, Kuriyama H (1990) Endothelin augments unitary calcium channel currents on the smooth muscle cell membrane of guinea-pig portal vein. J Physiol (Lond) 423:171–191
114. Nelson MT, Paltlak JB, Worley JF, Standen NB (1990) Calcium channels, potassium channels, and voltage dependence of arterial smooth muscle tone. Am J Physiol 259:C3–C18
115. Hess P, Lansman JB, Tsien RW (1984) Different modes of Ca channel gating behaviour favoured by dihydropyridine Ca antagonists and antagonists. Nature 311:538–544
116. Cavalie A, McDonald TF, Pelzer D, Trautwein W (1985) Temperature-induced transitory and steady-state changes in the calcium current of guinea pig ventricular myocytes. Pflügers Arch 405:294–296
117. Ohya Y, Kitamura K, Kuriyama H (1987) Cellular calcium regulates outward currents in rabbit intestinal smooth muscle cell. Am J Physiol 252:C401–C410
118. Benham CD, Tsien WA (1988) Noradrenaline modulations of calcium channels in single smooth muscle cells from rabbit ear artery. J Physiol (Lond) 404:767–784
119. Benham CD (1990) ATP-gated channels in vascular smooth muscle cells. Ann NY Acad Sci 603:275–286
120. Xiong Z, Kitamura K, Kuriyama H (1991) ATP activates cationic currents and modulated the calcium current through GTP-binding protein in rabbit portal vein. J Physiol (Lond) 440:143–165
121. Inoue R, Kuriyama H (1993) Dual regulation of cation-selective channels by muscarinic and $\alpha_1$-adrenergic receptors in the rabbit portal vein. J Physiol (Lond) 465:427–448
122. Devine CE, Somlyo AV, Somlyo AP (1972) Sarcoplasmic reticulum and excitation-contraction coupling in mammalian smooth muscle. J Cell Biol 52:690–718
123. Kowarski D, Shuman H, Somlyo AP, Somlyo AV (1985) Calcium release by noradrenaline from central sarcoplasmic reticulum in rabbit main pulmonary artery smooth muscle. J Physiol (Lond) 366:153–175
124. Franzini-Armstrong C (1970) Structure of the junction in frog twitch fibers. J Cell Biol 47:488–499
125. Fleischer S, Inui M (1989) Biochemistry and biophysics of excitation-contraction coupling. Annu Rev Biophys Biophys Chem 18:333–364
126. Takeshima H, Nishimura S, Matsumoto T, Ishida H, Kangawa K, Minamino N, Matsuo H, Heda M, Hanaoka M, Hirose T, Numa S (1989) Primary structure and expression from complementary DNA of skeletal muscle ryanodine receptor. Nature 339:439–445
127. Zorzato F, Fujii J, Otsu K, Phillips M, Green NM, Lai FA, Meissner G, MacLennan DH (1990) Molecular cloning of cDNA encoding human and rabbit forms of the $Ca^{2+}$ release channel (ryanodine receptor) of skeletal muscle sarcoplasmic reticulum. J Biol Chem 265:2244–2256
128. Nakai J, Imagawa T, Hakamata Y, Shigekawa M, Takeshima H, Numa S (1990) Primary structure and functional expression from cDNA of the cardiac ryanodine receptor/calcium release channel. FEBS Lett 271:169–177
129. Fujii J, Otsu K, Zorzato F, Leon S, Khanna VK, Weiler JEO Brien P, MacLennan DH (1991) Identification of a mutation in porcine ryanodine receptor associated with malignant hyperthermia. Science 253:448–451

130. Giannini G, Clementi E, Ceci R, Marziali G, Sorrentino V (1992) Expression of a ryanodine receptor-$Ca^{2+}$ channel that is regulated by TGF-beta. Science 257:91-94
131. Mikoshiba K (1993) Inositol 1,4,5-trisphosphate receptor. Trends Pharmacol Sci 14:86-89
132. Maeda N, Kawasaki T, Nakade S, Yokota N, Taguchi T, Kasai M, Mikoshiba K (1991) Structural and functional characterization of inositol 1,4,5-trisphosphate receptor channel from mouse cerebellum. J Biol Chem 266:1109-1116
133. Maeda N, Niinobe M, Mikoshiba K (1990) A cerebellar purkinje cell marker $P_{400}$ protein is an inositol 1,4,5-trisphosphate ($InsP_3$) receptor protein: purification and characterization of $InsP_3$ receptor complex. EMBO J 9:61-67
134. Mignery GA, Newton CL, Archer BT III, Sudhof TC (1990) Structure and expression of the rat inositol 1,4,5-trisphosphate receptor. J Biol Chem 265:12679-12685
135. Südhof TC, Newton CL, Archer BT III, Ushukaryov YA, Mignery GA (1991) Structure of a novel $InsP_3$ receptor. EMBO J 10:3199-3206
136. Fadool DA, Ache BW (1992) Plasma membrane inositol 1,4,5-trisphosphate-activated channels mediate signal transduction in lobster olfactory receptor neurons. Neuron 9:907-918
137. Miyazaki S, Yuzaki M, Nakada K, Shirakawa H, Nakanishi S, Nakade S, Mikoshiba K (1992) Block of $Ca^{2+}$ wave and $Ca^{2+}$ oscillation by antibody to the inositol 1,4,5-trisphosphate receptor in fertilized hamster eggs. Science 257:251-255
138. Itoh T, Kajiwara M, Kitamura K, Kuriyama H (1982) Roles of stored calcium on the mechanical response evoked in smooth muscle cells of the porcine coronary artery. J Physiol (Lond) 322:107-123
139. Lesh RE, Marks AR, Somlyo AV, Fleischer S, Somlyo AP (1993) Anti-ryanodine receptor antibody binding sites in vascular and endocardial endothelium. Circ Res 72:481-488
140. Saida K (1982) Intracellular Ca release in skinned smooth muscle. J Gen Physiol 80:191-202
141. Bootman MD, Taylor CW, Berridge MJ (1992) The thiol reagent, thimerosal, evokes $Ca^{2+}$ spikes in HeLa cells by sensitizing the inositol 1,4,5-trisphosphate receptor. J Biol Chem 267:25113-25119
142. Carroll J, Swann K (1992) Spontaneous cytosolic calcium oscillations driven by inositol trisphosphate occur during in vitro maturation of mouse oocytes. J Biol Chem 267:11196-11201
143. Hatzelmann A, Haurand M, Ulrich V (1990) Involvement of calcium in the thimerosal-stimulated formation of leukotriene by fMLP in human polymorphonuclear leukocytes. Biochem Pharmacol 39:559-567
144. Hecker M, Siegle I, MacArthur H, Sessa WC, Vane JR (1992) Role of intracellular thiols in the release of endothelium-derived relaxing factor from cultured endothelial cells. Am J Physiol 262:H888-H896
145. Missiaen L, De Smedt H, Droogmans G, Casteels R (1992) $Ca^{2+}$ release induced by inositol 1,4,5-trisphosphate is a steady-state phenomenon controlled by luminal $Ca^{2+}$ in permeabilized cells. Nature 357:599-602
146. Missiaen L, Taylor CW, Berridge MJ (1992) Luminal $Ca^{2+}$ promoting spontaneous $Ca^{2+}$ release from inositol trisphosphate-sensitive stores in rat hepatocytes. J Physiol (Lond) 455: 623-640
147. Mourey RJ, Verma A, Supattapone S, Snyder SH (1990) Purification and characterization of the inositol 1,4,5-trisphosphate receptor protein from rat vas deferens. Biochem J 272:383-389
148. Ford LE, Podolsky RJ (1972) Intracellular calcium movements in skinned muscle fibres. J Physiol (Lond) 223:21-33
149. Endo M, Tanaka M, Ogawa Y (1970) Calcium-induced release of calcium from the sarcoplasmic reticulum of skinned skeletal muscle fibres. Nature 228:34-36
150. Berridge MJ, Irvine RF (1984) Inositol trisphosphate, novel second-messenger in cellular signal transduction. Nature 312:315-321
151. Berridge MJ (1984) Inositol trisphosphate and diacylglycerol as second messengers. Biochem J 220:345-360
152. Shears SB (1989) Metabolism of the inositol phosphates produced upon receptor activation. Biochem J 260:313-324
153. Streb H, Irvine RF, Berridge MJ, Schulz I (1983) Release of $Ca^{2+}$ from a nonmitochondrial intracellular store in pancreatic acinar cells by inositol-1,4,5-trisphosphate. Nature 306:67-69

154. Suematsu E, Hirata M, Hashimoto T, Kuriyama H (1984) Inositol 1,4,5-trisphosphate releases $Ca^{2+}$ from intracellular store sites in skinned cells of porcine coronary artery. Biochem Biophys Res Commun 120:481–485
155. Somlyo AV, Bond M, Somlyo AP, Scarpa A (1985) Inositol trisphosphate-induced calcium release and contraction in vascular smooth muscle. Proc Natl Acad Sci USA 82:5231–5235
156. Yamamoto H, Van Breemen C (1985) Inositol 1,4,5-trisphosphate releases calcium from skinned cultured smooth muscle cells. Biochem Biophys Res Commun 130:270–274
157. Hashimoto T, Hirata M, Itoh T, Kanmura Y, Kuriyama H (1986) Inositol 1,4,5-trisphosphate activates pharmaco-mechanical coupling in smooth muscle of the rabbit mesenteric artery. J Physiol (Lond) 370:605–618
158. Somlyo AP, Somlyo AV (1968) Vascular smooth muscle. I. Normal structure, pathology, biochemistry, and biophysics. Pharmacol Rev 20:197–272
159. Suzuki H (1981) Effects of endogenous and exogenous noradrenaline on the smooth muscle of guinea-pig mesenteric vein. J Physiol (Lond) 321:495–512
160. Casteels R, Kitamura K, Kuriyama H, Suzuki H (1977) The membrane properties of the smooth muscle cells of the rabbit main pulmonary artery. J Physiol (Lond) 271:41–61
161. Casteels R, Kitamura K, Kuriyama H, Suzuki H (1977) Excitation-contraction coupling in the smooth muscle cells of the rabbit main pulmonary artery. J Physiol (Lond) 271:63–79
162. Abe K, Kusakabe Y, Tanemura K, Emori Y, Arai S (1993) Multiple genes for G protein-coupled receptors and their expression in lingual epithelia. FEBS Lett 316:253–256
163. Bourne HR, Sanders DA, McCormick F (1991) The GTPase superfamily: conserved structure and molecular mechanism. Nature 349:117–127
164. Gilman AG (1987) G proteins: transducers of receptor-generated signals. Annu Rev Biochem 56:615–649
165. Hall A (1990) The cellular functions of small GTP-binding proteins. Science 249:635–640
166. Jones DT, Reed RR (1989) Golf: an olfactory neuron specific-G protein involved in odorant signal transduction. Science 244:790–795
167. Katada T, Oinuma M, Ui M (1986) Mechanisms for inhibition of the catalytic activity of adenylate cyclase by the guanine nucleotide-binding protein serving as the substrate of islet-activating protein, pertussis toxin. J Biol Chem 261:5215–5221
168. Kishi K, Sasaki T, Kuroda S, Itoh T, Takai Y (1993) Regulation of cytoplasmic division of Xenopus embryo by rho p21 and its inhibitory GDP/GTP exchange protein (rho GDI). J Cell Biol 120:1187–1195
169. Miura Y, Kikuchi A, Musha T, Kuroda S, Yaku H, Sasaki T, Takai Y (1993) Regulation of morphology by rho p21 and its inhibitory GDP/GTP exchange protein (rho GDI) in Swiss 3T3 cells. J Biol Chem 268:510–515
170. Takai Y, Kaibuchi K, Kikuchi A, Kawata M (1992) Small GTP-binding proteins. Int Rev Cytol 133:187–230
171. Takaishi K, Kikuchi A, Kuroda S, Kotani K, Sasaki T, Takai Y (1993) Involvement of rho p21 and its inhibitory GDP/GTP exchange protein (rho GDI) in cell motility. Mol Cell Biol 13:72–79
172. Ui M (1984) Islet-activating protein, perussis toxin: a probe for functions of the inhibitory guanine nucleotide regulatory component of adenylate cyclase. Trends Pharmacol Sci 5:277–279
173. Ui M, Katada T, Murayama T, Kurose H, Yajima M, Tamura M, Nakamura T, Nogomori K (1984) Islet-activating protein, pertussis toxin: a specific uncoupler of receptor mediated inhibition of adenylate cyclase. Adv Cyclic Nucleotide Protein Phosphorylation Res 17:145–151
174. Iino M (1987) Calcium-dependent inositol trisphosphate-induced calcium release in the guinea-pig taenia caeci. Biochem Biophys Res Commun 142:47–52
175. Iino M (1990) Biphasic $Ca^{2+}$ dependence of inositol 1,4,5-trisphosphate-induced Ca release in smooth muscle cells of the guinea pig taenia caeci. J Gen Physiol 95:1103–1122
176. Iino M, Endo M (1992) Calcium-dependent immediate feedback control of inositol 1,4,5-trisphosphate-induced $Ca^{2+}$ release. Nature 360:76–78
177. Malgaroli A, Fesce R, Meldolesi J (1990) Spontaneous $[Ca^{2+}]_i$ fluctuations in rat chromaffin cells do not require inositol 1,4,5-trisphosphate elevations but are generated by a caffeine- and ryanodine-sensitive intracellular $Ca^{2+}$ store. J Biol Chem 263:3005–3008
178. Volpe P, Villa A, Damiani E, Sharp AH, Podini P, Snyder SH, Meldolesi J (1991) Heterogeneity of microsomal $Ca^{2+}$ stores in chicken purkinje neurons. EMBO J 10:3183–3189

179. Bourreau J-P, Abela AP, Kwan CY, Daniel EE (1991) Acetylcholine $Ca^{2+}$ stores refillig directly involves a dihydropride-sensitive channel in dog trachea. Am J Physiol 261:C497–C505
180. Brading AF, Burnett M, Sneddon P (1980) The effect of sodium removal on the contractile responses of the guinea-pig taenia coli to carbachol. J Physiol (Lond) 306:411–429
181. Casteels R, Droogmans G (1981) Exchange characteristics of the noradrenaline-sensitive $Ca^{2+}$ store in vascular smooth muscle cells of rabbit ear artery. J Physiol (Lond) 317:263–279
182. Chen Q, Cannel M, Van Breemen C (1992) The superficial buffer barrier in vascular smooth muscle. Can J Physiol Pharmacol 70:509–514
183. Chen Q, Van Breemen C (1993) The superficial buffer barrier in venous smooth muscle: sarcoplasmic reticulum refilling and unloading. Br J Pharmacol 109:336–343
184. Putney JW Jr (1986) A model for receptor-regulated $Ca^{2+}$ entry. Cell Calcium 7:1–12
185. Carafoli E (1991) Calcium pump of the plasma membrane. Physiol Rev 71:129–153
186. Raeymaekers L, Wuytack F (1993) $Ca^{2+}$ pumps in smooth muscle cells. J Muscle Res Cell Motil 14:141–157
187. Schatzmann HJ (1989) The calcium pump of the surface membrane and of the sarcoplasmic reticulum. Annu Rev Physiol 51:473–485
188. Strehler EE (1991) Recent advances in the molecular characterization of the plasma membrane $Ca^{2+}$ pumps. J Membr Biol 120:1–15
189. Wuytack F, Raeymaekers L (1992) The $Ca^{2+}$-transport ATPases from the plasma membrane. J Bioenerg Biomembr 24:285–300
190. Campbell AM, Kessler PD, Sagara Y, Inesi G, Fambrough DM (1991) Nucleotide sequences of avian cardiac and brain SR/ER $Ca^{2+}$ ATPase and functional comparison with fast twitch $Ca^{2+}$ ATPase. J Biol Chem 266:16050–16055
191. Lytton J, Westlin M, Burk SE, Shull GE, MacLennan DH (1992) Functional comparisons between isoforms of the sarcoplasmic or endoplasmic reticulum family of calcium pumps. J Biol Chem 267:14483–14489
192. Lytton J, Zarain-Herzberg A, Periasamy M, MacLennan DH (1989) Molecular cloning of the mammalian smooth muscle sarco(endo)plasmic reticulum $Ca^{2+}$-ATPase. J Biol Chem 264:7059–7065
193. Eggermont JA, Vrolix M, Raeymaeders L, Wuytack F, Casteels R (1988) $Ca^{2+}$-transport ATPases of vascular smooth muscle. Circ Res 62:266–278
194. Tada M, Kodama M, Fujii J, Kimura Y, Kijima Y (1989) Molecular structure and function of phospholamban: the regulatory protein of calcium pump in cardiac sarcoplasmic reticulum. Adv Exp Med Biol 255:79–89
195. Watras J (1988) Regulation of calcium uptake in bovine aortic sarcoplasmic reticulum by cyclic AMP-dependent protein kinase. J Mol Cell Cardiol 20:711–723
196. Sarcebvic B, Brookes V, Martin TJ, Kemp BE, Robinson PJ (1989) Atrial natriuretic peptide-dependent phosphorylation of smooth muscle cell particulate fraction proteins is mediated by cGMP-dependent protein kinase. J Biol Chem 264:20648–20654
197. Ferguson DG, Young EF, Raeymaekers L, Kranias EG (1988) Localization of phospholamban in smooth muscle using immunogold electron microscopy. J Cell Biol 107:555–562
198. Raeymaekers L, Jones LR (1986) Evidence for the presence of phospholamban in the endoplasmic reticulum of smooth muscle. Biochim Biophys Acta 882:258–265
199. Itoh T, Kanmura Y, Kuriyama H, Sasaguri T (1985) Nitroglycerine- and isoprenaline-induced vasodilatation: assessment from the actions of cyclic nucleotides. Br J Pharmacol 84:393–406
200. Parker I, Ito Y, Kuriyama H, Miledi R (1987) β-Adrenergic agonists and cyclic AMP decrease intracellular resting free-calcium concentration in ileum smooth muscle. Proc R Soc Lond [Biol] 230:207–214
201. Cornwell TL, Lincoln TM (1989) Regulation of intracellular $Ca^{2+}$ levels in cultured vascular smooth muscle cells: reduction of $Ca^{2+}$ by atriopeptin and 8-bromo-cyclic GMP is mediated by cyclic GMP-dependent protein kinase. J Biol Chem 264:1146–1155
202. Felbel J, Trockur B, Ecker T, Landgraf W, Hofmann F (1988) Regulation of cytosolic calcium by cAMP and cGMP in freshly isolated smooth muscle cells from bovine trachea. J Biol Chem 263:16764–16771

203. McDaniel NL, Rembold CM, Richard HM, Murphy RA (1991) Cyclic AMP relaxes swine arterial smooth muscle predominantly by decreasing cell $Ca^{2+}$ concentration. J Physiol (Lond) 439:147–160
204. Furukawa K, Nakamura H (1984) Characterization of the ($Ca^{2+}$-$Mg^{2+}$)ATPase purified by calmodulin-affinity chromatography from bovine aortic smooth muscle. J Biochem 96:1343–1350
205. Furukawa KI, Tawada Y, Shigekawa M (1988) Regulation of the plasma membrane $Ca^{2+}$ pump by cyclic nucleotides in cultured vascular smooth muscle cells. J Biol Chem 263:8058–8065.
206. Imai S, Yoshida Y, Sun H-T (1990) Sarcolemmal effects of protein kinases thereupon. J Biochem 107:755–761
207. Rashatwar SS, Cornwell TL, Lincoln TM (1987) Effects of 8-bromo-cGMP on $Ca^{2+}$ levels in vascular smooth muscle cells: possible regulation of $Ca^{2+}$-ATPase by cGMP-dependent protein kinase. Proc Natl Acad Sci USA 84:6586–6589
208. Vrolix M, Raeymaekers L, Wuytack F, Hofmann F, Casteels R (1988) Cyclic GMP-dependent protein kinase stimulates the plasmalemmal $Ca^{2+}$ pump of smooth muscle via phosphorylation of phosphatidylinositol. Biochem J 255:855–863
209. Yoshida Y, Sun H-T, Cai J-Q, Imai S (1991) Cyclic GMP-dependent protein kinase stimulates the plasma membrane $Ca^{2+}$ pump ATPase of vascular smooth muscle via phosphorylation of a 240-KDa protein. J Biol Chem 266:19819–19825
210. Fukuda T, Ogurusu T, Furukawa K-I, Shigekawa M (1990) Protein kinase C-dependent phosphorylation of sarcolemmal $Ca^{2+}$-ATPase isolated from bovine aortic smooth muscle. J Biochem 108:629–634
211. Popescu LM, Hinescu ME, Musat S, Ionescu M, Pistritzu F (1986): Inositol trisphosphate and the contraction of vascular smooth muscle cells. Eur J Pharmacol 123:167–169
212. Yau K-W, Nakatani K, Tamura T (1991) Sodium-calcium exchange and phototransduction in retinal photoreceptors. Ann NY Acad Sci 639:275–284
213. Hodgkin AL, McNaughton PA, Nunn BJ (1987) Measurement of sodium-calcium exchanges in salamander rods. J Physiol (Lond) 391:347–370
214. Lagnado L, McNaughton PA (1990) The electrogenic properties of the Na:Ca exchange. J Membr Biol 113:177–191
215. Baker PF, Blaustein MP, Hodgkin AL, Steinhardt RD (1969) The influence of calcium ions on sodium efflux in squid axons. J Physiol (Lond) 200:431–458
216. Blaustein MP, Hodgkin AL (1969) The effect of cyanide on the efflux of calcium from squid axons. J Physiol (Lond) 200:497–527
217. Dahan D, Spanier R, Rahamimoff H (1991) The modulation of rat brain $Na^{+}$-$Ca^{2+}$ exchange by $K^{+}$. J Biol Chem 266:2067–2075
218. Ehara T, Matsuoka S, Noma A (1989) Measurement of reversal potential of $Na^{+}/Ca^{2+}$ exchange current in single guinea-pig ventricular cells. J Physiol (Lond) 410:227–249
219. Kimura J, Miyamae S, Noma A (1987) Identification of sodium-calcium exchange current in single ventricular cells of guinea-pig. J Physiol (Lond) 384:199–222
220. Nicoll DA, Longoni S, Philipson KD (1990) Molecular cloning and functional expression of the cardiac sarcolemmal $Na^{+}$-$Ca^{2+}$ exchanger. Science 250:562–565
221. Yasui K, Kimura J (1990) Is potassium co-transported by the cardiac Na-Ca exchanger? Pflügers Arch 415:513–515
222. Allen TJA, Noble D, Reuter H (1989) Sodium-calcium exchange. Oxford University Press, Oxford
223. Blaustein MP (1977) Sodium ions, calcium ions, blood pressure regulation and hypertension: a reassessment and a hypothesis. Am J Physiol 232:C165–C173
224. Blaustein MP, Ambsesi A, Bloch RJ, Goldman WF, Juhaszova M, Lindenmayer GE, Veiss DN (1992) Regulation of vascular smooth muscle contractility: roles of the sarcoplasmic reticulum (SR) and the sodium/calcium exchanger. Jpn J Pharmacol 58(suppl II):107P–114P
225. Blaustein MP, Goldman WF, Fontana G, Krueger B, Santiago EM, Steele TD, Weiss DN, Yarowsky PJ (1991) Physiological roles of the sodium-calcium exchanger in nerve and muscle. Ann NY Acad Sci 639:254–274
226. Sheu SS, Blaustein MP (1992) Sodium/calcium exchange and control of cell calcium and contractility in cardiac and vascular smooth muscles. In: Fozzard HA, Haber E, Jennings RB, Katz AM, and Morgan HE (eds) The heart and cardiovascular system Raven, New York, pp 903–943

227. Smith JB, Zheng T, Lyu R-M (1989) Ionomycin releases calcium from the sarcoplasmic reticulum and activates $Na^{+}$-$Ca^{2+}$ exchange in vascular smooth muscle cells. Cell Calcium 10:125–134
228. Vigne P, Breittmayer JP, Duval D, Frelin C, Lazdunski M (1988) The $Na^{+}/Ca^{2+}$ antiporter in aortic smooth muscle cells: characterization and demonstration of an activation by phorbol esters. J Biol Chem 263:8078–8083
229. Cook NJ, Kaupp UB (1988) Solubilization, purification, and reconstitution of the sodium-calcium exchanger from bovine reinal rod outer segments. J Biol Chem 263:11382–11388
230. Nicoll DA, Philipson KD (1991) Molecular studies of the cardiac sarcolemmal sodium-calcium exchanger. Ann NY Acad Sci 639:181–188
231. Philipson KD, Longoni S, Ward R (1988) Purification of the cardiac $Na^{+}$-$Ca^{2+}$ exchange protein. Biochim Biophys Acta 945:298–306
232. Porzig H (1991) Characterization of monoclonal antibodies cross-reacting with myocardial and retinal sodium-calcium exchanger poteints. Ann NY Acad Sci 639:202–209
233. Rahamimoff H, Dahan D, Furman I, Spanier R, Tessari M (1991) Molecular and mechanistic heterogeneity of the $Na^{+}$-$Ca^{2+}$ exchanger. Ann NY Acad Sci 639:210–221
234. Li J, Kimura J (1990) Translocation mechanism of Na:Ca exchange in single cardiac cells of guinea pig. J Gen Physiol 96:777–788
235. Li J, Kimura J (1991) Translocation mechanism of cardiac Na-Ca exchange. Ann NY Acad Sci 639:48–60
236. Miura Y, Kimura J (1989) Sodium-calcium exchange curren: dependence on internal Ca and Na and competitive binding of external Na and Ca. J Gen Physiol 93:1129–1145
237. Mechmann S, Pott L (1986) Identification of $Na^{+}$-$Ca^{2+}$ exchange current in single cardiac myocytes. Nature 319:597–599
238. Hume JR, Uehara A (1986) "Creep currents" in single frog strial cells may be generated by electrogenic Na/Ca exchange. J Gen Physiol 87:857–884
239. Aaronson PI, Poston L, Woolfson RG, Smirnov SV (1991) Evidence for Na-Ca exchange in human resistance arteries. Ann NY Acad Sci 639:521–530
240. Borin ML Goldman WF, Blaustein MP (1993) Intracellular free $Na^{+}$ in resting and activated A7r5 vascular smooth muscle cells. Am J Physiol 264:C1513–C1524
241. Droogmans G, Casteels R (1979) Sodium and calcium interactions in vascular smooth muscle cells of the rabbit ear artery. J Gen Physiol 74:57–70
242. Karaki H, Ozaki H, Urakawa N (1978) Effects of ouabain and potassium-free solution on the contraction of isolated blood vessels. Eur J Pharmacol 48:439–443
243. Matlib MA (1991) Role of sarcolemmal membrane sodium-calcium exchanger in vascular smooth muscle tension. Ann NY Acad Sci 639:531–542
244. Mulvany MJ, Aalkjaer C, Jensen PE (1991) Sodium-calcium exchange in vascular smooth muscle. Ann NY Acad Sci 639:498–504
245. Moore DW, Fogarty KE, Fay FS (1992) Role of $Na^{+}$-$Ca^{2+}$ exchange in β-adrenergic relaxation of single smooth muscle cells. Ann NY Acad Sci 639:543–549
246. Nabel EG, Berk BC, Brock TA, Smith TW (1988) $Na^{+}$-$Ca^{2+}$ exchange in vascular smooth muscle cells. Circ Res 62:486–493
247. Reuter H, Blaustein MP, Hausler G (1973) Na-Ca exchange and tension development in arterial smooth muscle. Philos Trans R Soc Lond [Biol] 265:87–94
248. Van Bremen C, Aaronson P, Loutzenhiser R (1979) Sodium-calcium interactions in mammalian smooth muscle. Pharmacol Rev 30:167–208
249. Motley ED, Paul RJ, Matlib MA (1993) Role of $Na^{+}$-$Ca^{2+}$ exchange in the regulation of vascular smooth muscle tension. Am J Physiol 33:H1028–H1040
250. Furukawa KI, Oshima N, Tawada Y, Shigekawa M (1991) Cyclic GMP stimulates $Na^{+}/Ca^{2+}$ exchange in vascular smooth muscle cells in primary culture. J Biol Chem 266:12337–12341
251. Ganitkevich VY, Isenberg G (1993) $Ca^{2+}$ entry through $Na^{+}$-$Ca^{2+}$ exchange can trigger $Ca^{2+}$ release from $Ca^{2+}$ stores in $Na^{+}$-loaded guinea-pig coronary myocytes. J Physiol (Lond) 468:225–243
252. Benham CD, Bolton TB (1986) Spontaneous transient outward currents in single visceral and vascular smooth muscle cells of the rabbit. J Physiol (Lond) 381:385–406
253. Bolton TB, Lim SP (1989) Properties of calcium stores and transient outward currents in single smooth muscle cells of rabbit intestine. J Physiol (Lond) 409:385–401

254. Hume JR, Leblanc N (1989) Macroscopic $K^+$ currents in single smooth muscle cells of the rabbit portal vein. J Physiol (Lond) 413:49–73
255. Komori S, Bolton TB (1990) Role of G-proteins in muscarinic receptor inward and outward currents in rabbit jejunal smooth muscle. J Physiol (Lond) 427:395–419
256. Muraki K, Imaizumi Y, Kojima T, Kawai T, Watanabe M (1990) Effects of tetraethylammonium and 4-aminopyridine on outward currents and excitability in canine tracheal smooth muscle cells. Br J Pharmacol 100:507–515
257. Sakai T, Terada K, Kitamura K, Kuriyama H (1988) Ryanodine inhibits the Ca-dependent $K_L$ current after depletion of Ca stored in smooth muscle cells of the rabbit ileal longitudinal muscle. Br J Pharmacol 95:1089–1100
258. Xiong Z, Kitamura K, Kuriyama H (1992) Evidence for contribution of $Ca^{2+}$ storage sites on unitary $K^+$ channel currents in inside-out membrane of rabbit portal vein. Pflügers Arch 420:112–114
259. Ho K, Nichols CG, Lederer WJ, Lytton J, Vassilev PM, Kanazirska MV, Hebert SC (1993) Cloning and expression of an inwardly rectifying ATP-regulated potassium channel. Nature 362:31–38
260. Kubo Y, Baldwin TJ, Jan YN, Jan LY (1993) Primary structure and functional expression of a mouse inward rectifier potassium channel. Nature 362:127–133
261. Jan LY, Jan YN (1992) Tracing the roots of ion channels. Cell 69:715–718
262. Adams DJ, Nonner W (1990) Voltage-dependent potassium channels: gating, ion permeation and block. In: Cook NS (ed) Potassium channels Ellis Horwood, Chichester, pp 40–69
263. Beech DJ, Bolton TB (1989) A voltage-dependent outward current with fast kinetics in single smooth muscle cells isolated from rabbit portal vein. J Physiol (Lond) 412:397–414
264. Beech DJ, Bolton TB (1989) Properties of the cromakalim-induced potassium conductance in smooth muscle cells isolated from the rabbit portal vein. Br J Pharmacol 98: 851–864
265. Beech DJ, Bolton TB (1989) Two components of potassium current activated by depolarization of single smooth muscle cells from the rabbit portal vein. J Physiol (Lond) 418:293–309
266. Noack T, Deitmer P, Lammel E (1992) Characterization of membrane currents in single smooth muscle cells from the guinea-pig gastric antrum. J Physiol (Lond) 451:387–417
267. Noack T, Deitmer P, Edwards G, Weston AH (1992) Characterization of potassium currents modulated by BRL 38227 in rat portal vein. Br J Pharmacol 106:717–726
268. Noack T, Edwards G, Deitmer P, Weston AH (1992) Potassium channel modulation in rat portal vein by ATP depletion: a comparison with the effects of levcromakalim (BRL 38227). Br J Pharmacol 107:945–955
269. Okabe K, Kitamura K, Kuriyama H (1987) Features of 4-aminopyridine sensitive outward current observed in single smooth muscle cells from the rabbit pulmonary artery. Pflügers Arch 409:561–568
270. Smirnov SV, Aaronson PI (1993) Alteration of the transmembrane $K^+$ gradient during development of delayed rectifier potassium current in single rat pulmonary artery cells. J Physiol (Lond) 467:310P
271. Langton, PD, Nelson MT, Huang Y, Standen NB (1991) Block of calcium-activated potassium channels in mammalian arterial myocytes by tetraethylammonium ions. Am J Physiol 260: H927–H934
272. Ohya Y, Terada K, Kitamura K, Kuriyama H (1986) Membrane currents recorded from a fragment of rabbit intestinal smooth muscle cells. Am J Physiol 251:C335–C346
273. Okabe K, Kajioka S, Nakao K, Kitamura K, Kuriyama H, Weston AH (1990) Actions of cromakalim on ionic currents recorded from single smooth muscle cells of the rat portal vein. J Pharmacol Exp Ther 250:832–839
274. Smirnov SV, Zholos AV, Shuba MF (1992) A potential-dependent fast outward current in single smooth muscle cells isolated from the newborn rat ileum. J Physiol (Lond) 454:573–589
275. Lang RJ (1989) Identification of the major membrane currents in freshly dispersed single smooth muscle cells of the guinea-pig ureter. J Physiol (Lond) 412:375–395
276. Imaizumi Y, Muraki K, Watanabe M (1990) Characteristics of transient outward currents in single smooth muscle cells from the ureter of the guinea-pig. J Physiol (Lond) 427:301–324
277. Edwards FR, Hirst GDS (1988) Inward rectification in submucosal arterioles of guinea pig ileum. J Physiol (Lond) 404:437–454

278. Edwards FR, Hirst GDS, Siverberg GD (1988) Inward rectification in rat cerebral arterioles; involvement of potassium ions in auto regulation. J Physiol (Lond) 404:455–466
279. Hirst GDS, Edwards FR (1989) Sympathetic neuroeffector transmission in arteries and arterioles. Physiol Rev 69:546–604
280. McCarron JG, Halpern W (1990) Potassium dilates rat cerebral arteries by two independent mechanisms. Am J Physiol 259:H902–H908
281. Quayle JM, McCarron JG, Brayden JE, Nelson MJ (1993) Inward rectifier potassium currents in smooth muscle cells from rat resistance-sized cerebral arteries. In: Proceedings of the Physiological Society UK, UCL meeting
282. Sims SM, Singer JJ, Walsh JV Jr (1985) Cholinergic agonists suppress a potassium current in freshly dissociated smooth muscle cells of the toad. J Physiol (Lond) 367:503–529
283. Sims SM, Singer JJ, Walsh JV Jr (1988) Antagonistic adrenergic-muscarinic regulation of M current in smooth muscle cells. Science 239:190–193
284. Standen NB, Quayle JM, Davies NW, Brayden JE, Huang Y, Nelson MT (1989) Hyperpolarizing vasodilators activates ATP-sensitive $K^+$ channels in arterial smooth muscle. Science 245:177–180
285. Kajioka S, Kitamura K, Kuriyama H (1991) Guanosine diphosphate activates an adenosine-5′-triphosphate-sensitive $K^+$ channel in the rabbit portal vein. J Physiol (Lond) 444:397–418
286. Kajioka S, Nakashima M, Kitamura K, Kuriyama H (1991) Mechanism of vasodilation induced by potassium-channel activators. Clin Sci 81:129–139
287. Kajioka S, Oike M, Kitamura K (1990) Nicorandil opens a calcium-dependent potassium channel in smooth muscle cells of the rat portal vein. J Pharmacol Exp Ther 254:905–913
288. Edwards G, Weston AH (1990) Structure activity relationships of $K^+$ channel openers. Trends Pharmacol Sci 11:417–422
289. Edwards G, Weston AH (1993) The pharmacology of ATP-sensitive potassium channels. Annu Rev Pharmacol Toxicol 33:597–637
290. Quast U (1992) Potassium channel openers: pharmacological and clinical aspects. Fundam Clin Pharmacol 6:279–293
291. Quast U, Cook NS (1989) Moving together: $K^+$ channel openers and ATP-sensitive $K^+$ channels. Trends Pharmacol Sci 10:431–435
292. Kitamura K, Kamouchi M, Kuriyama H (1994) Mechanisms of action of $K^+$ channel openers and closers in smooth muscle cells. In: Sperelakis N (ed) Physiology and pathophysiology of the heart. Kluwer Academic/Norwell, New York
293. Kamouchi M, Xiong Z, Teramoto N, Kajioka S, Okabe K, Kitamura K (1991) Ionic currents involved in vasodilating actions of E4080, a newly synthesized bradycardia-inducing agent, in dispersed smooth muscle cells of the rabbit portal vein. J Pharmacol Exp Ther 259:1396–1403
294. Inoue R, Okabe K, Kitamura K, Kuriyama H (1986) A newly identified $Ca^{2+}$-dependent $K^+$ channel in the smooth muscle membrane of single cells dispersed from the rabbit portal vein. Pflügers Arch 406:138–143
295. Inoue I, Nakaya Y, Nakaya S, Mori H (1989) Extracellular $Ca^{2+}$-activated $K^+$ channel in coronary artery smooth muscle cells and its role in vasodilation. FEBS Lett 255:281–284
296. Berger W, Grygorcyk R, Schwarz W (1984) Single $K^+$-channels in smooth muscles in membrane evaginations of smooth muscle cells. Pflügers Arch 402:18–23
297. Benham CD, Bolton TB, Lang RJ, Takewaki T (1985) The mechanism of action of $Ba^{2+}$ and TEA on single $Ca^{2+}$-activated $K^+$ channels in arterial and intestinal smooth muscle cell membrane. Pflügers Arch 403:120–127
298. Benham CD, Bolton TB, Lang RJ, Takewaki T (1986) Calcium-activated potassium channels in single smooth muscle cells of rabbit jejunum and guinea-pig mesenteric artery. J Physiol (Lond) 371:45–67
299. Inoue R, Kitamura K, Kuriyama H (1985) Two Ca-dependent K-channels classified by application of tetraethylammonium distribute to smooth muscle membranes of the rabbit portal vein. Pflügers Arch 405:173–179
300. Kihara M, Matsuzawa K, Tokuno H, Tomita T (1990) Effects of calmodulin antagonists on calcium-activated potassium channels in pregnant rat myometrium. Br J Pharmacol 100:353–359

301. Kume H, Takai A, Tokuno H, Tomita T (1989) Regulation of $Ca^{2+}$-dependent $K^+$ channel activity in tracheal myocytes by phosphorylation. Nature 341:152–154
302. Fujino K, Nakaya S, Wakatsuki T, Miyoshi Y, Nakaya Y, Mori H, Inoue I (1991) Effects of nitroglycerin on ATP-induced $Ca^{++}$-mobilization porcine coronary artery. J Pharmacol Exp Ther 256:371–377
303. Hu SL, Kim HS, Jeng AY (1991) Dual action of endothelin-1 on the $Ca^{2(+)}$-activated $K^+$ channel in smooth muscle cells of porcine coronary artery. Eur J Pharmacol 194:31–36
304. McCann JD, Welsh MJ (1986) Calcium-activated potassium channels in canine airway smooth muscle. J Physiol (Lond) 372:113–127
305. Singer JJ, Walsh JV Jr (1984) Large conductance $Ca^{2+}$-activated $K^+$-channels in smooth muscle cell membrane: reduction in unitary currents due to internal $Na^+$ ions. Biophys J 45:68–70
306. Singer JJ, Walsh JV Jr (1987) Characterization of calcium-activated potassium channels in single smooth muscle cells using the patch-clamp technique. Pflügers Arch 408:98–111
307. Silberberg SD, Van Breemen C (1990) An ATP, calcium and voltage sensitive potassium channel in porcine coronary artery smooth muscle cells. Biochem Biophys Res Commun 172:517–522
308. Toro L, Ramos-Franco J, Stefani E (1990) GTP-dependent regulation of myometrial KCa channels incorporated into lipid bilayers. J Gen Physiol 96:373–394
309. Toro L, Vaca L, Stefani E (1991) Calcium-activated potassium channels from coronary smooth muscle reconstituted in lipid bilayers. Am J Physiol 260:H1779–H1789
310. William DL, Katz GM, Roy-Contancin L, Reuben JP (1988) Guanosine 5′-monophosphate modulates gating of high-conductance $Ca^{2+}$-activated $K^+$ channels in vascular smooth muscle cells. Proc Natl Acad Sci USA 85:9360–9364
311. Wang Y, Mathers DA (1993) $Ca^{2+}$-dependent $K^+$ channels of high conductance in smooth muscle cells isolated from rat cerebral arteries. J Physiol (Lond) 462:529–545
312. Inoue I, Nakaya S, Nakaya Y (1990) An ATP-sensitive $K^+$ channel activated by extracellular $Ca^{2+}$ and $Mg^{2+}$ in primary cultured arterial smooth muscle cells. J Physiol (Lond) 430:132
313. Beech DJ, Bolton TB, Castle NA, Strong PN (1987) Characterization of a toxin from scorpion (Leiurus quinquestriatus) venom that blocks in vitro both large (BK) $K^+$ channels in rabbit vascular smooth muscle and intermediate (IK) conductance $Ca^{2+}$-activated $K^+$ channels in human red cells. J Physiol (Lond) 387:32
314. Cook NS (1989) The effect of some potassium channel blockers upon contractile responses of the rabbit aorta. J Cardiovasc Pharmacol 13:299–306
315. Cook NS, Quast U (1990) Potassium channel pharmacology. In: Cook NS (ed) Potassium channels, structure, classification, function and therapeutic potential. Ellis Horwood, Chichester, pp 181–255
316. Miller C, Moczydlowski E, Latorre R, Phillips M (1985) Charybdotoxin, a protein inhibitor of single $Ca^{2+}$-activated $K^+$ channels from mammalian skeletal muscle. Nature 313:316–318
317. Wong BS (1989) Quinidine blockade of calcium-activated potassium channels in dissociated gastric smooth muscle cells. Pflügers Arch 414:416–422
318. Thornbury KD, Ward SM, Dalziel HH, Carl A, Westfall DP, Sanders KM (1991) Nitric oxide and nitrosocysteine mimic nonadrenergic, noncholinergic hyperpolarization in canine proximal colon. Am J Physiol 261:G553–G557
319. Toro L, Silberberg SD, Brown AM, Stefani E (1989) $K^+$ and $Ca^{2+}$ currents from coronary single smooth muscle cells. Biophys J 55:543a
320. Wilde DW, Lee KS (1989) Outward potassium currents in freshly isolated smooth muscle cell of dog coronary arteries. Circ Res 65:1718–1734
321. Volk KA, Shibata EF (1993) Single delayed rectifier potassium channels from rabbit coronary artery myocytes. Am J Physiol 264:H1146–H1153
322. Imaizumi Y, Muraki K, Watanabe M (1989) Ionic currents in single smooth muscle cells from the ureter of the guinea-pig. J Physiol (Lond) 411:131–159
323. Kamouchi M, Kajioka S, Sakai T, Kitamura K, Kuriyama H (1993) A target $K^+$ channel for LP-805-induced hyperpolarization in smooth muscle cells of the rabbit portal vein. Naunyn Schmiedebergs Arch Pharmacol 347:329–335
324. Kamouchi M, Kitamura K (1994) Regulation of ATP-sensitive $K^+$ channels by ATP and nucleotide diphosphate in rabbit portal vein. Am J Physiol 266:H1687–H1698

325. Aickin CC, Brading AF (1983) Towards an estimate of chloride permeability in the smooth muscle of guinea-pig vas deferens. J Physiol (Physiol) 336:179–197
326. Casteels R (1981) Membrane potential in smooth muscle cells. In: Bulbring E, Brading AF, Jones AW, Tomita T (eds) Smooth muscle. Edward Arnold, London, pp 105–126
327. Collins FS: Cystic fibrosis: molecular biology and therapeutic implications. Science 256:774–779
328. Bauer CK, Steinmeyer K, Schwarz JR, Jentsch TJ (1991) Completely functional double-barreled chloride channel expressed from a single Torpedo cDNA. Proc Natl Acad Sci USA 88:11052–11056
329. Grunder S, Thienmann A, Pusch M, Jentsch TJ (1992) Regions involved in the opening of CIC-2 chloride channel by voltage and cell volume. Nature 360:759–762
330. Steinmeyer K, Ortland C, Jentsch TJ (1991) Primary structure and functional expression of a developmentally regulated skeletal muscle chloride channel. Nature 354:301–304
331. Thienmann A, Grunder S, Pusch M, Jentsch TJ (1992) A chloride channel widely expressed in epithelial and non-epithelial cells. Nature 356:57–60
332. Jentsch TJ, Steinmeyer K, Schwarz G (1990) Primary structure of Torpedo marmorata chloride channel isolated by expression cloning in Xenopus oocytes. Nature 348:510–514
333. Blatz AL (1985) Single chloride-selective channels active at resting membrane potentials in cultured rat skeletal muscle. Biophys J 47:119–123
334. Miller C, Richard EA (1990) The voltage-dependent chloride channel of Torpedo electroplax. In: Alvarez-Leefmans FJ, Russel JM (eds) Chloride channels and carrier in nerve, muscle and glia cells. Plenum, New York, pp 383–405
335. Anderson MP, Gregory RJ, Thompson S, Souza DW, Paul S, Mulligan RC, Smith AE, Welsh MJ (1991) Demonstration that CFTR is a chloride channel by alteration of its anion selectivity. Science 253:202–205
336. Anderson MP, Rich DP, Gregory RJ, Smith AE, Welsh MJ (1991) Generation of cAMP-activated chloride currents by expression of CFTR. Science 251:679–682
337. Gregory RJ, Cheng SH, Rich DP, Marshall J, Paul S, Hehir K, Ostedgaard L, Klinger KW, Welsh MJ, Smith A-E (1990) Expression and characterization of the cystic fibrosis transmembrane conductance regulator. Nature 347:382–386
338. Halley DJ, Bijman J, De Jonge HR, Sinaasappel M, Neijens HJ, Niermeijer MF (1990) The cystic fibrosis defect approached from different angles—new perspectives on the gene, the chloride channel, diagnosis and therapy. Eur J Pediatr 149:670–677
339. Higgins CF, Hyde SC (1991) Channeling our thoughts. Nature 352:194–195
340. Kartner N, Hanrahan JW, Jensen TJ, Naismith AL, Sun SZ, Ackerley CA, Reyes EF, Tsui LC, Rommens JM, Bear CE (1991) Expression of the cystic fibrosis gene in non-epithelial invertebrate cells produces a regulated anion conductance. Cell 64:681–691
341. Rich DP, Gregory RJ, Anderson MP, Manavalan P, Smith AE, Welsh MJ (1991) Effect of deleting the R domain on CFTR-generated chloride channels. Science 253:205–207
342. Riordan JR, Rommens JM, Kerem B, Alon N, Rozmahel R, Grzelczak Z, Zielenski J, Lok S, Plavsic N, Chou JL (1989) Identification of the cystic fibrosis gene: cloning and characterization of complementary DNA. Science 245:1066–1073
343. Welsh MJ (1991) Acidification indication. Nature 352:23–24
344. Blatz AL, Magleby KL (1983) Single voltage-dependent chloride-selective channels of large conductance in cultured rat muscle. Biophys J 43:237–241
345. Coulombe A, Duclohier H, Coraboeuf E, Touzet N (1987) Single chloride-permeable channels of large conductance in cultured cardiac cells of new-born rats. Eur Biophys J 14:155–162
346. Franciolini F, Petris A (1990) Chloride channels of biological membranes. Biochim Biophys Acta 1031:247–259
347. Gray PT, Bevan S, Ritchie JM (1984) High conductance anion-selective channels in rat cultured Schwann cells. Proc R Soc Lond B221:395–409
348. Li M, McCann J-D, Liedtke CM, Nairn AC, Greengard P, Welsh MJ (1988) Cyclic AMP-dependent protein kinase opens chloride channels in normal but not cystic fibrosis airway epithelium. Nature 331:358–360
349. Young GPH, Young JDE, Deshpande AK, Goldstein M, Koide SS, Cohn ZA (1984) A $Ca^{2+}$-activated channel from Xenopus laevis oocyte membranes reconstituted into planar bilayers. Proc Natl Acad Sci USA 81:5155–5159

350. Mayer M-K, Owen DG, Barker JL (1990) Calcium-dependent chloride currents in vertebrate central neurons. In: Alvarez-Leefmans FJ, Russel JM (eds) Chloride channels and carriers in nerve, muscle and Glial cells. Plenum, New York, pp 355–364
351. Soejima M, Kokubun S (1988) Single anion-selective channel and its ion selectivity in the vascular smooth muscle cell. Pflügers Arch 411:304–311
352. Klöckner U (1993) Intracellular calcium ions activate a low-conductance chloride channel in smooth-muscle cells isolated from human mesenteric artery. Pflügers Arch 424:231–237
353. Kokubun S, Saigusa A, Tamura T (1991) Blockade of Cl channels by organic and inorganic blockers in vascular smooth muscle cells. Pflügers Arch 418:204–213
354. Kitamura K, Kuriyama H (1989) Role of calcium released from the sarcoplasmic reticulum of smooth muscle cells as induced by inositol phosphatides. In: Hidaka H, Carafori E, Means AR, Tanaka T (eds) Calcium protein signalling. Plenum, New York, pp 49–60
355. Smirnov SV, Aaronson PI (1992) $Ca^{2+}$-activated and voltage-gated $K^+$ currents in smooth muscle cells isolated from human mesenteric arteries. J Physiol (Lond) 457:431–454
356. Clapp LH, Gurney AM (1992) ATP-sensitive $K^+$ channels regulate the resting potential of pulmonary arterial smooth muscle cells. Am J Physiol 262:H916–H920
357. Clapp LH, Gurney AM (1991) Outward currents in rabbit pulmonary artery cells dissociated with a new technique. Exp Physiol 76:677–693
358. Komori S, Bolton TB (1989) Actions of guanine nucleotide and cyclic nucleotides on calcium stores in single patch-clamped smooth muscle cells from rabbit portal vein. Br J Pharmacol 97:973–982
359. Komori S, Bolton TB (1991) Calcium release induced by inositol 1,4,5-trisphosphate in single rabbit intestinal smooth muscle cells. J Physiol (Lond) 433:495–517
360. Kitamura K, Xiong Z, Teramoto N, Kuriyama H (1992) Roles of inositol trisphosphate and protein kinase C in the spontaneous outward current modulated by calcium release in rabbit portal vein. Pflügers Arch 421:539–551
361. Ohya Y, Terada K, Yamaguchi K, Inoue R, Okabe K, Kitamura K, Hirata M, Kuriyama H (1988) Effects of inositol phosphates on the membrane activity of smooth muscle cells of the rabbit portal vein. Pflügers Arch 412:382–389
362. Krick W, Disser J, Hazama A, Burckhardt G, Frömter E (1991) Evidence for a cytosolic inhibitor of epithelial chloride channels. Pflügers Arch 418:491–499
363. Kunzelmann K, Tilmann M, Hansen CP, Greger R (1991) Inhibition of epithelial chloride channels by cytosol. Pflügers Arch 418:479–490
364. Byrne NG, Large WA (1988) Membrane ionic mechanisms activated by noradrenaline in cells isolated from rabbit portal vein. J Physiol (Lond) 404:557P–573P
365. Hogg RC, Wang Q, Large WA (1993) Time course of spontaneous calcium-activated chloride currents in smooth muscle cells from the rabbit portal vein. J Physiol (Lond) 464:15–31
366. Wang Q, Hogg RC, Large WA (1992) Properties of spontaneous inward currents recorded in smooth muscle cells isolated from the rabbit portal vein. J Physiol (Lond) 451:525–537
367. Wang Q, Large WA (1991) Noradrenaline-evoked cation conductance recorded with the nystatin whole-cell method in rabbit portal vein cells. J Physiol (Lond) 435:21–39
368. Wang Q, Large WA (1993) Action of histamine on single smooth muscle cells dispersed from the rabbit pulmonary artery. J Physiol (Lond) 468:125–139
369. Amédée T, Large WA, Wang Q (1990) Characteristics of chloride currents activated by noradrenaline in rabbit ear artery cells. J Physiol (Lond) 428:501–516
370. Droogmans G, Callewaert G, Declerck I, Casteels R (1991) ATP-induced $Ca^{2+}$ release and $Cl^-$ current in cultured smooth muscle cells form pig aorta. J Physiol (Lond) 440:623–634
371. Pacaud P, Loirand G, Bolton TB, Mironneau C, Mironneau J (1992) Intracellular cations modulate noradrenaline-stimulated calcium entry into smooth muscle cells of rat portal vein. J Physiol (Lond) 456:541–556
372. Ohta T, Ito S, Nakazato Y (1993) Chloride currents activated by caffeine in rat intestinal smooth muscle cells. J Physiol (Lond) 465:149–162
373. Ohta T, Ito S, Noto T, Tachibana R, Nakazato Y, Ohga A (1992) The inhibitory action of cyclic AMP on responses to carbachol dependent on calcium stores in rat gastric smooth muscle. J Physiol (Lond) 453:367–384

374. Benham CD, Tsien RW (1987) A novel receptor-operated $Ca^{2+}$-permeable channel activated by ATP in smooth muscle. Nature 328:275–278
375. Benham CD (1989) ATP-activated channels gate calcium entry in single smooth muscle cells dissociated from rabbit ear artery. J Physiol (Lond) 419:689–701
376. Benham CD (1992) ATP-gated cation channels in vascular smooth muscle cells. Jpn J Pharmacol 58(suppl II):179–184
377. Oike M, Kitamura K, Kuriyama H (1993) Protein kinase C activates non-selective cation channel in the rabbit portal vein. Pflügers Arch 424:159–164
378. Klöckner U, Isenberg G (1991) Endothelin depolarizes myocytes from porcine coronary and human mesenteric arteries through a Ca-activaed chloride current. Pflügers Arch 418:168–175
379. Suzuki H (1985) Electrical responses of smooth muscle cells of the rabbit ear artery to adenosine triphosphate. J Physiol (Lond) 359:401–415

# Dual Regulation of Smooth Muscle Contraction

Hideaki Karaki

*Summary.* Smooth muscle contraction is initiated by an increase in cytosolic $Ca^{2+}$. $Ca^{2+}$, together with calmodulin, activates myosin light chain kinase to phosphorylate myosin light chain. Phosphorylated myosin interacts with actin to form cycling cross-bridges and generates force. During maintained force, however, the amount of phosphorylated myosin gradually decreases, suggesting that not only phosphorylated myosin but also unphosphorylated myosin is involved in cross-bridge cycling. A latch model and the positive cooperativity model suggest that formation of unphosphorylated cycling cross-bridges is regulated by the amount of phosphorylated cross-bridges. In addition to the time-dependent dissociation between myosin phosphorylation and contraction, there is a different type of dissociation, a stimulus-dependent dissociation between cytosolic $Ca^{2+}$ level and contraction. Part of this dissociation is explained by the agonist-induced increase in $Ca^{2+}$ sensitivity of myosin phosphorylation. Our results suggest that agonists also activate the cycling of unphosphorylated cross-bridges in the absence of myosin phosphorylation or myosin light chain kinase activity. This mechanism is activated at lower $Ca^{2+}$ levels than myosin light chain kinase and may play an important role in agonist-induced $Ca^{2+}$ sensitization. Although the details of this mechanism are not clear, we know that it is not regulated by the activation of protein kinase C. Thus there seem to be two regulation mechanisms of smooth muscle contraction: the mechanism dependent on and the mechanism not dependent on the phosphorylation of myosin light chain.

*Key Words.* Smooth muscle—Myosin light chain kinase—Myosin phosphorylation—Cytosolic $Ca^{2+}$ level

## Regulation of Cross-Bridge Cycling: Time-Dependent Dissociation

Contraction of smooth muscle is initiated by an increase in cytosolic $Ca^{2+}$ levels ($[Ca^{2+}]_i$), as shown in Figure 1A. $Ca^{2+}$, together with calmodulin, activates myosin light chain kinase (MLCK) to phosphorylate myosin regulatory light chain (MLC). Phosphorylated myosin interacts with actin filament to form cycling cross-bridges and generates force. Removal of contractile stimulation decreases $[Ca^{2+}]_i$ followed

Department of Veterinary Pharmacology, Graduate School of Agriculture and Life Sciences, University of Tokyo, 1-1-1 Yayoi, Bunkyo-ku, Tokyo, 113 Japan

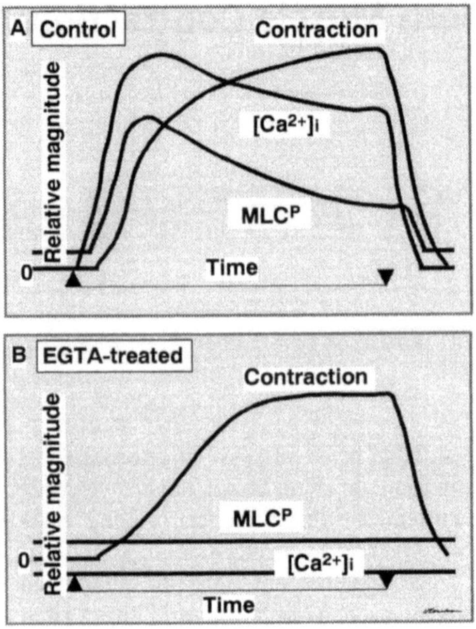


**Fig. 1.** Changes in cytosolic $Ca^{2+}$ level ($[Ca^{2+}]_i$), myosin light chain phosphorylation (*MLC^P*), and contraction in smooth muscle. **A.** In normal solution. Stimulation of the muscle increased $[Ca^{2+}]_i$, myosin light chain phosphorylation, and contractile force. During sustained contraction, both $[Ca^{2+}]_i$ and MLC phosphorylation decreased. **B.** In $Ca^{2+}$-free solution with EGTA, $Ca^{2+}$ stores were previously depleted. Agonists but not high $K^+$ induced contraction without changing $[Ca^{2+}]_i$ or MLC phosphorylation. Agonist was added at ▲ and removed at ▼

by a decrease in MLCK activity. MLCK/MLC phosphatase (MLCP) balance is changed toward dephosphorylation of MLC, and dephosphorylated myosin is detached from actin to relax the smooth muscle. This hypothesis seems to explain clearly the mechanism of smooth muscle contraction. However, the amount of phosphorylated myosin gradually decreases during sustained contraction (Fig. 1A). Hence there is no linear correlation between MLC phosphorylation and the force of the contraction measured during sustained contraction. Almost maximum force is developed when only 20%–40% MLC is phosphorylated during sustained contraction. In contrast, shortening velocity, a measure of the cycling rate, gradually decreases during sustained contraction. These results suggest that not only phosphorylated myosin but also unphosphorylated myosin is involved in cross-bridge cycling and that the rate of cycling of unphosphorylated cross-bridges is slower than that of the phosphorylated cross-bridges.

To explain the formation of unphosphorylated crossbridges, Murphy proposed a four-state (or latch) model [review: 1]. In this model, as shown in Figure 2A, not only detached cross-bridges ($K_2$) but also attached cross-bridges ($K_5$) are dephosphorylated by phosphatase. As a result, a part of the contraction is maintained by dephosphorylated and attached cross-bridges, or latch bridges. Thus the latch bridge is formed only when myosin is phosphorylated. Vyes et al. [2] found that thiophosphorylated myosin also formed latch bridges, although this myosin is not dephosphorylated by phosphatase and therefore is not expected to form the dephosphorylated and attached cross-bridges. This result suggests that $K_5$ is not the mechanism of latch bridge formation, so they proposed a positive cooperativity model (Fig. 2B). According to this model, a latch bridge is formed without previous phosphorylation ($K_8$), and therefore phosphatase activity ($K_5$) is not necessary to make the latch bridge. However, the rate of formation of unphosphorylated cross-bridges is regulated by the amount of phosphorylated MLC (*thick arrow*, Fig. 2B). The

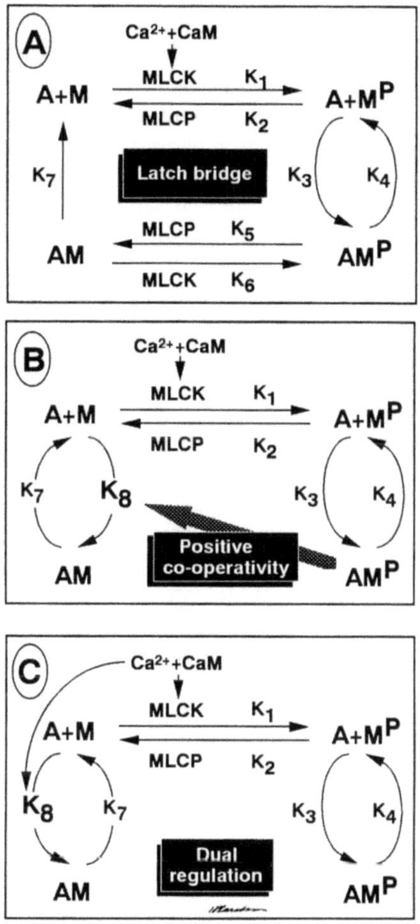


**Fig. 2.** Models of smooth muscle contraction. **A.** Four-state (or latch) model [1]. $Ca^{2+}$, together with calmodulin (*CaM*), activate myosin light chain kinase (*MLCK*) to phosphorylate MLC ($K_1$). Phosphorylated myosin ($M^P$) interacts with actin (*A*) to activate cross-bridge cycling ($K_3$ and $K_4$). During contraction, phosphatase (*MLCP*) dephosphorylates both the detached cross-bridge ($K_2$) and the attached cross-bridge ($K_5$). $K_5$ results in a dephosphorylated, attached cross-bridge or a latch bridge (*AM*). Because AM detaches to form A and M slowly ($K_7$), the latch bridge contributes to maintain force. **B.** Positive cooperativity model [2], suggesting that a dephosphorylated, cycling cross-bridge (AM) is formed without previous phosphorylation ($K_8$). However, $K_8$ is regulated by the amount of phosphorylated myosin ($M^P$), and therefore MLCK directly and indirectly regulates both mechanisms ($K_1$ and $K_8$). **C.** We propose a dual regulation model suggesting that a dephosphorylated, cycling cross-bridge (AM) is formed without previous phosphorylation ($K_8$), and that $K_8$ is not regulated by the amount of phosphorylated myosin ($M^P$) and therefore is independent of MLCK activation. $K_8$ is more sensitive to $Ca^{2+}$ than MLCK. All of these mechanisms may exist in smooth muscle

phosphorylated MLC somehow changes unphosphorylated myosin to produce an unphosphorylated cross-bridge in a positive cooperative manner.

In both the latch model and the positive cooperativity model, MLCK activity or phosphorylation of MLC is necessary to make the latch bridge. Recently, however, we found that a latch bridge is formed without previous phosphorylation of MLC, so we propose a new model (Fig. 1C), which is summarized in this chapter. Related subjects have been discussed elsewhere [3–7].

## Stimulus-Dependent Dissociation

Simultaneous measurements of $[Ca^{2+}]_i$ and contractile force show a stimulus-dependent dissociation between these two parameters. In the rat aorta, for example, cumulative addition of KCl or receptor agonists induced a graded increase in both $[Ca^{2+}]_i$ and contractile force. Comparing the effects of high concentrations of KCl, agonists induced greater contraction, although $[Ca^{2+}]_i$ stimulated by KCl was greater than that stimulated by agonists [3, 6, 8]. 12-Deoxyphorbol 13-isobutyrate (DPB) had effects similar to those of receptor agonists [7]. The resulting $[Ca^{2+}]_i$–force relation showed that the slope of the curve for agonists was steeper than that for KCl, as shown in Figure 3. These results were confirmed by experiments in permeabilized smooth muscle in which agonists increased the force of contraction in the presence of constant $Ca^{2+}$ concentration [9, 10]. GTPγS and phorbol ester showed effects similar to those of agonists in permeabilized smooth muscle. These results suggest that agonists activate receptor-coupled GTP-binding protein to increase $Ca^{2+}$ sensitivity of contractile elements, and that activation of protein kinase C (PKC) may activate a similar mechanism.

## $Ca^{2+}$ Sensitization of MLC Phosphorylation

Because an increase in $[Ca^{2+}]_i$ results in MLC phosphorylation followed by contraction (Fig. 2A), dissociation between $[Ca^{2+}]_i$ and contraction may be due to dissociation between $[Ca^{2+}]_i$ and MLC phosphorylation or MLC phosphorylation and contraction [4]. In the presence of maximally effective concentrations of stimulants, agonists induced greater contraction than KCl, although these stimulants increased $[Ca^{2+}]_i$ to similar levels. Measurement of MLC phosphorylation and contraction showed that there was no positive correlation between these two parameters. The efficiency of stimulants to increase MLC phosphorylation and contractile force at a given $[Ca^{2+}]_i$ was dependent on the type of stimulant. In the presence of maximally effective concentrations of KCl, MLC is phosphorylated by only about 20%. In contrast, in the presence of maximally effective concentrations of agonists, 20%–40% MLC was phosphorylated. Endothelin-1 induced the greatest MLC phosphorylation

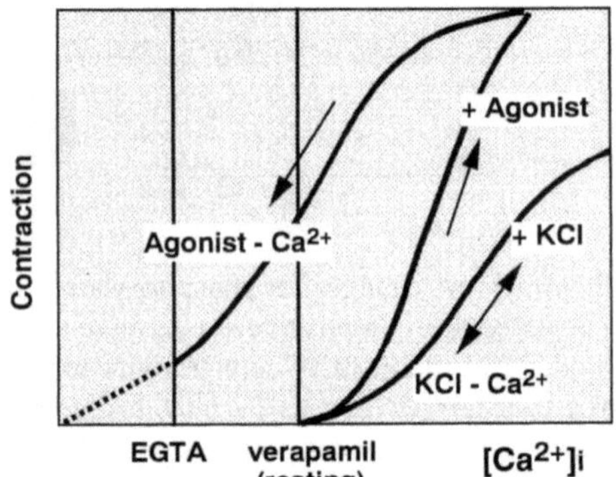


**Fig. 3.** Relation between $[Ca^{2+}]_i$ and contraction. Cumulative addition of KCl (*+KCl*) increased both $[Ca^{2+}]_i$ and contraction. Addition of verapamil in the presence of KCl (*KCl-Ca²⁺*) decreased both $[Ca^{2+}]_i$ and contraction to their respective resting level. The curve obtained by cumulative addition of KCl and that obtained by the addition of verapamil were superimposable. Cumulative addition of agonist (*+Agonist*) also increased both $[Ca^{2+}]_i$ and contraction. However, the slope of the curve was steeper than that due to KCl. Addition of verapamil in the presence of agonist (*Agonist-Ca²⁺*) decreased $[Ca^{2+}]_i$ to the resting level and only partially inhibited contraction. Addition of EGTA decreased $[Ca^{2+}]_i$ below the resting level and further inhibited contraction. However, a part of the agonist-induced contraction was not inhibited in the presence of verapamil and EGTA. This portion of contraction may be due to a small amount of $Ca^{2+}$ remaining in the cell. If $[Ca^{2+}]_i$ were further decreased, this contraction would be inhibited (*dotted line*). The resulting correlation curve located in parallel to the left of that due to KCl suggests that agonists increase $Ca^{2+}$ sensitivity of contractile elements

followed by prostaglandin $F_{2\alpha}$ > norepinephrine > high $K^+$[5–7]. In the presence of phorbol ester, about 80% of the MLC was phosphorylated. The maximum contractions induced by agonists and phorbol ester were greater than that induced by KCl, reflecting the larger amount of phosphorylated MLC [6, 7]. Thus the increase in $Ca^{2+}$ sensitivity of smooth muscle contraction is explained partly by the increase in $Ca^{2+}$ sensitivity of MLC phosphorylation. After measurement of the time course of MLC phosphorylation during agonist-induced $Ca^{2+}$ sensitization, however, it was reported that the increase in phosphorylation was only transient [11]. This result suggests that agonist-induced $Ca^{2+}$ sensitization is only partially explained by $Ca^{2+}$ sensitization of MLC phosphorylation.

Mechanisms coupled to GTP-binding protein have been suggested to be involved in the $Ca^{2+}$ sensitivity of MLC phosphorylation [9, 10]. Small GTP-binding protein may be involved in $Ca^{2+}$ sensitization in skinned smooth muscle [12]. There are two possible mechanisms for changes in $Ca^{2+}$ sensitivity of MLC phosphorylation (Fig. 4). The first possibility is the inhibition of MLC phosphatase, which results in more MLC phosphorylation and therefore more contraction at a given $[Ca^{2+}]_i$. It was recently suggested that activation of the receptor–GTP-binding protein–phospholipase $A_2$ system cleaves arachidonic acid from the cell membrane, which selectively inhibits MLC phosphatase [13, 14]. Because relative MLCK activity, and hence the amount of phosphorylated MLC, is determined by the difference between MLCK activity and MLC phosphatase activity, receptor-mediated inhibition of MLC phosphatase activity may shift the $[Ca^{2+}]_i$-MLC phosphorylation relation to the left. An alternative or additional mechanism is that an increase in $[Ca^{2+}]_i$ activates not only MLCK but also $Ca^{2+}$/calmodulin-dependent protein kinase II (CaMKII), and CaMKII phosphorylates MLCK to inhibit its activity [15, 16], as shown in Figure 4B. If agonists uncouple this negative feedback loop, phosphorylation of MLCK is prevented and $Ca^{2+}$ sensitivity of MLC phosphorylation is kept high. In vascular smooth muscle, it has been reported

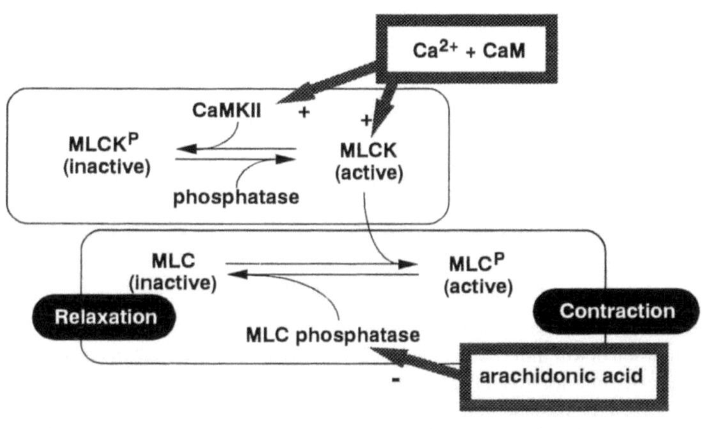


**Fig. 4.** Possible mechanisms to increase $Ca^{2+}$ sensitivity of MLC phosphorylation. **(Top)**: Stimulant-induced increase in $[Ca^{2+}]_i$ activates not only MLCK but also $Ca^{2+}$/calmodulin-dependent protein kinase II (*CaMKII*). Activated CaMKII phosphorylates MLCK to inhibit its activity. Agonists may somehow inhibit this negative feeback loop and phosphorylate more MLC at a give $[Ca^{2+}]_i$. **(Bottom)** Agonist activates the receptor–GTP-binding protein–phospholipase $A_2$ pathway to cleave arachidonic acid from the membrane. Arachidonic acid inhibits MLC phosphatase to increase the MLCK/phosphatase balance and phosphorylate more MLC at a given $[Ca^{2+}]_i$

that $Ca^{2+}$ sensitivity of MLCK extracted from KCl-treated tissues was lower than that extracted from unstimulated or histamine-treated tissues [17], supporting the above suggestion.

## Contraction Not Dependent on MLC Phosphorylation

In most of the experiments reported so far, various concentrations of stimulant were added to change $[Ca^{2+}]_i$, MLC phosphorylation, and contraction. Thus concentrations of stimulants are changed together with $[Ca^{2+}]_i$. Because the $Ca^{2+}$-sensitizing effect is dependent on the concentration of agonist, it is interesting to change $[Ca^{2+}]_i$ in the presence of a constant concentration of agonist. To change $[Ca^{2+}]_i$ without changing the agonist concentration, we used the $Ca^{2+}$ channnel blocker verapamil plus EGTA. In the presence of a maximally effective concentration of KCl, verapamil decreased $[Ca^{2+}]_i$ to the resting level. Contraction was also decreased to the resting level. The resulting $[Ca^{2+}]_i$-force relation was similar to that obtained by the addition of KCl

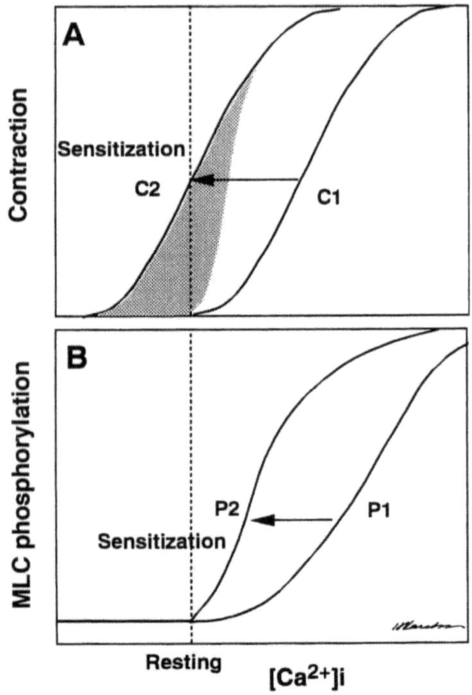


**Fig. 5.** Relation between $[Ca^{2+}]_i$ and contractile force and $[Ca^{2+}]_i$ and MLC phosphorylation. **A.** Cumulative addition of KCl induced a graded increase in both $[Ca^{2+}]_i$ and force, resulting in curve *C1*. Maximally effective concentration of agonist induced greater contraction than KCl, although agonists increased $[Ca^{2+}]_i$ to a level similar or lower than that due to high $K^+$. Addition of verapamil and EGTA decreased the $[Ca^{2+}]_i$ to and below the resting $[Ca^{2+}]_i$, respectively. Agonist-induced contraction was also inhibited but to a lesser extent than $[Ca^{2+}]_i$, resulting in curve *C2*. In the presence of agonists, therefore, force is maintained at or below the resting $[Ca^{2+}]_i$. **B.** KCl increased MLC phosphorylation, as shown by curve *P1*. Maximally effective concentration of agonist induced more MLC phosphorylation than KCl. Addition of verapamil decreased $[Ca^{2+}]_i$ to the resting level, creating curve *P2*. Addition of EGTA did not change the resting MLC phosphorylation. Thus part of the contraction (indicated by the *shading*) is not explained by MLC phosphorylation

(Fig. 3). In contrast, the $[Ca^{2+}]_i$–force relation obtained by decreasing $[Ca^{2+}]_i$ in the presence of agonist was different from that obtained by the cumulative addition of agonist. Verapamil decreased $[Ca^{2+}]_i$ to or near the resting level, whereas contraction was only partially inhibited. Addition of EGTA further decreased $[Ca^{2+}]_i$ below a resting level and inhibited the contraction, although some of the contraction remained. The resulting $[Ca^{2+}]_i$–force relation was located to the left of that obtained with KCl, indicating that contraction induced by agonist is maintained at a lower $[Ca^{2+}]_i$ than that needed to maintain the KCl-induced contraction.

The results obtained with intact preparations were confirmed by the results with permeabilized smooth muscle in which agonists decreased the threshold $Ca^{2+}$ concentration for contraction in the presence of GTP, shifting the $Ca^{2+}$–force curve to the lower $Ca^{2+}$ level [9, 10]. It was also found that GTPγS and phorbol esters show similar effects as agonists.

If smooth muscle contraction is fundamentally regulated by MLC phosphorylation, and if $Ca^{2+}$ sensitization of contraction is due to $Ca^{2+}$ sensitization of MLC phosphorylation, it is expected that the $[Ca^{2+}]_i$–MLC phosphorylation relation would shift to the left, as is the case with the $[Ca^{2+}]_i$–force relation. Changes in MLC phosphorylation are summarized in Figure 5. In the presence of agonists or phorbol ester, both $[Ca^{2+}]_i$ and MLC phosphorylation were higher than that in the presence of KCl, as described previously. In the muscle stimulated by KCl, addition of verapamil decreased $[Ca^{2+}]_i$ and MLC phosphorylation to their respective resting levels. In muscle stimulated by agonists or phorbol ester, addition of verapamil also decreased $[Ca^{2+}]_i$ and MLC phosphorylation to their respective resting levels, as shown in Figure 5. Thus neither the contractions nor the $Ca^{2+}$ sensitization of contraction observed in the presence of verapamil or EGTA is explained by the increase in MLC phosphorylation.

In this series of experiments, however, $[Ca^{2+}]_i$ was decreased after contraction was induced in normal solution. Thus the initial increase in $[Ca^{2+}]_i$ and MLC phosphorylation might have produced an unphosphorylated cycling cross-bridge, as suggested by the latch model (Fig. 2A). To examine this possibility, agonist was added in the presence of EGTA. A muscle strip was pretreated with $Ca^{2+}$-free solution (with EGTA) for a long enough time to deplete $[Ca^{2+}]_i$ in storage sites. As shown in Figure 1B, addition of agonist to the $Ca^{2+}$-depleted muscle changed neither $[Ca^{2+}]_i$ nor MLC phosphorylation. However, agonists still induced slow, sustained contraction [6]. These results suggest that agonist can induce contraction without an initial increase in $[Ca^{2+}]_i$ or MLC phosphorylation.

To confirm the possibility that some of the contraction is not dependent on MLC phosphorylation, a selecive inhibitor of MLCK, wortmannin [18], was used. This inhibitor is highly selective for MLCK and was expected to inhibit only the portion of the contraction that is dependent on MLC phosphorylation. It was observed that wortmannin completely inhibited the KCl-induced contraction. However, the same concentration of wortmannin only partially inhibited the contractions induced by agonists and phorbol ester [19]. In the absence of external $Ca^{2+}$, agonists and phorbol ester induced sustained contraction without changing $[Ca^{2+}]_i$ or MLC phosphorylation. Wortmannin did not inhibit these contractions. The magnitude of the contraction that is not inhibited by wortmannin in the presence of $Ca^{2+}$ was almost the same as the contraction obtained in the absence of external $Ca^{2+}$. These results support the suggestion that some of the agonist-induced contraction is due to an MLC phosphorylation-independent mechanism. It was also observed that contraction induced by phorbol ester, which was not dependent on MLC

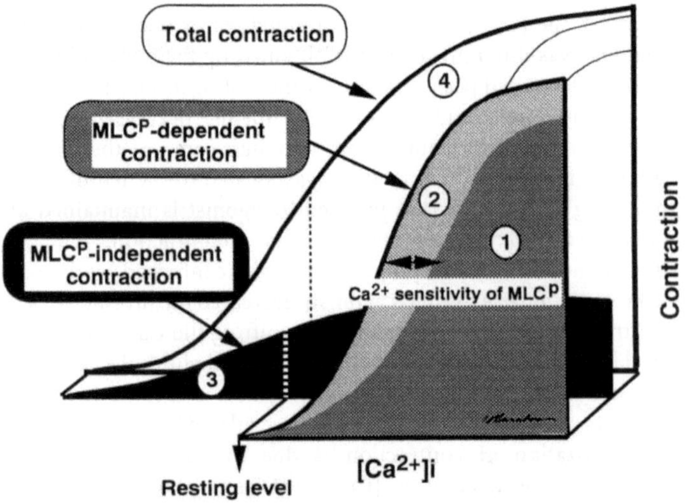


**Fig. 6.** Dual regulation of smooth muscle contraction. Agonists increase $[Ca^{2+}]_i$ above the resting level to phosphorylate MLC and induce contraction (curve *1*). Agonist also increase $Ca^{2+}$ sensitivity of MLC phosphorylation to sift the $[Ca^{2+}]_i$–force curve to the lower $[Ca^{2+}]_i$ levels and induce more contraction at a given $[Ca^{2+}]_i$ (curve *2*). The mechanism that depends on MLC phosphorylation (curves 1 and 2) is inhibited by a selective inhibitor of MLCK, wortmannin. At the same time, agonists activate the MLC phosphorylation-independent mechanism to induce contraction (curve *3*). The MLC phosphorylation-independent mechanism is more sensitive to $Ca^{2+}$ than the MLC phosphorylation-dependent mechanism and is almost fully activated in the resting $Ca^{2+}$ level. This contraction is not inhibited by the MLCK inhibitor wortmannin. Smooth muscle contraction is the sum of the contractions induced by these two mechanisms (curves 1 + 2 + 3 = curve *4*). The $[Ca^{2+}]_i$ above the resting level is the physiological $Ca^{2+}$ range. In the presence of agonists, a decrease in the $[Ca^{2+}]_i$ to a resting level by, for example, $Ca^{2+}$ channel blockers only partially inhibits muscle tension (down to the *dotted line*) because of mechanism 3

phosphorylation, was inhibited by neither $Ca^{2+}$-free solution nor wortmannin. Because the contractions induced by agonists and phorbol ester were completely inhibited by a nonselective kinase inhibitor, K252a, kinase other than MLCK may be involved in the MLC phosphorylation-independent contraction. Details of the MLC phosphorylation-independent contraction remain to be examined, although it is tempting to suggest that actin-binding proteins such as calponin and caldesmon are involved in this contraction.

Previously, we showed that complete inhibition of phosphatase activity by phosphatase inhibitors such as okadaic acid and calyculin A increased MLC phosphorylation and induced contraction even in the absence of $Ca^{2+}$ and suggested the presence of $Ca^{2+}$-independent MLCK activity [20, 21]. This possibility is supported by the fact that the basal level of MLC phosphorylation was not decreased in the absence of $Ca^{2+}$. When $[Ca^{2+}]_i$ is increased, MLCK activity is increased above the basal level in a concentration-dependent manner. Because agonists did not increase MLC phosphorylation in the absence of external $Ca^{2+}$, agonist-induced inhibition of phosphatase is not as strong as that seen with phosphatase inhibitors. Even so, there was basal MLC phosphorylation during the contraction induced by agonists in $Ca^{2+}$-free solution. Thus the possibility was not excluded that basal MLC phosphorylation or $Ca^{2+}$-independent MLCK activity plays some role in the "MLC phosphorylation-independent" contraction.

From these results, we suggest that smooth muscle contraction is composed of two mechanisms: MLC phosphorylation dependent and independent mechanisms. Because the MLCK-dependent mechanism requires high concentraions of $Ca^{2+}$, removal of external $Ca^{2+}$ preferentially inhibits this mechanism. Furthermore, this mechanism is dependent on MLCK activity, so wortmannin selectively inhibits this contraction.

## "Intact" Preparation and Permeabilized Preparation

In "intact" (not permeabilized) smooth muscle preparations, verapamil, EGTA, or both only partially inhibited the contractions induced by agonists and phorbol esters, although $[Ca^{2+}]_i$ was inhibited to or below the resting level [6]. Agonists and phorbol ester induced a sustained contraction in the presence of EGTA without changing $[Ca^{2+}]_i$ or MLC phosphorylaion (Fig. 1B). The resulting $[Ca^{2+}]_i$–force relation suggested that, although MLC-dependent contraction is regulated by $[Ca^{2+}]_i$, a part of the contraction obtained in the presence of EGTA (MLC phosphorylation-independent contraction) may not be dependent on $Ca^{2+}$ (Fig. 3). In the experiments with permeabilized smooth muscle, in contrast, removal of $Ca^{2+}$ completely inhibited the contractions. Thus the $Ca^{2+}$–force curve was shifted to the lower $Ca^{2+}$ level in parallel by agonists, GTPγS, or phorbol esters. The results with permeabilized muscle suggest that smooth muscle contraction, irrespective of whether it is MLC phosphorylation dependent or independent, is regulated by $Ca^{2+}$. This difference may be explained by the possibility that in intact muscle EGTA did no decrease $[Ca^{2+}]_i$ below the threshold level for contraction; the contraction observed in the presence of EGTA may be dependent on a small amount of $Ca^{2+}$ remaining in the cell even in the presence of EGTA (Fig. 3). It is also suggested that the MLC phosphorylation-independent mechanism is more sensitive to $Ca^{2+}$ than MLCK.

In intact muscle, contraction induced at or below resting $[Ca^{2+}]_i$ was not accompanied by an increase in MLC phosphorylation. In permeabilized muscle, however, the "resting" $Ca^{2+}$ level is not clear, and the agonist-induced shift of the $Ca^{2+}$–force curve is not as evident as that in intact muscle. From these reasons, it is not clarified if there is a contraction in permeabilized muscle that is not dependent on MLC phosphorylation. Our preliminary results indicated that an inhibitor of MLCK, wortmannin, at the concentration needed to abolish the $Ca^{2+}$–induced contraction, did not inhibit a part of the $Ca^{2+}$-induced contraction in permeabilized muscle in the presence of GTPγS, suggesting that some of the $Ca^{2+}$-sensitizing effect is not due to MLC phosphorylation and that GTP-binding protein is involved in this MLC phosphorylation-independent mechanism.

The $Ca^{2+}$-sensitizing effects of agonist and phorbol esters are more clearly observed in intact than in permeabilized muscle. In intact muscle, $Ca^{2+}$ sensitivity was increased by agonists and phorbol ester, in the order: norepinephrine < prostaglandin $F_{2\alpha}$ < endothelin-1 < 12-deoxyphorbol 13-isobutyrate [6, 7]. Such a clear difference was not observed in permeabilized muscle. Although the reason for this difference is not clear, some factor regulating $Ca^{2+}$ sensitivity may be lost from permeabilized smooth muscle. Because smooth muscle was permeabilized with bacterial α toxin, which makes only small pores, this factor may have molecular weight less than 1000 daltons.

## Relation Between MLC Phosphorylation and Contraction

Although smooth muscle contraction is fundamentally due to cycling of phosphorylated cross-bridges, it seems reasonable to assume that unphosphorylated cross-bridges also play an important role. Thus the time-dependent dissociation between MLC phosphorylation and contraction, which is observed irrespective of the type of stimulation, is explained by MLCK-dependent recruiting of unphosphorylated cross-bridges. In contrast, the stimulus-dependent dissociation between $[Ca^{2+}]_i$ and contraction is explained by an increase in MLC phosphorylation at a given $[Ca^{2+}]_i$ and by MLC phosphorylation-independent recruiting of unphosphorylated cross-bridges. Figure 7 shows the relation between MLC phosphorylation and contraction stimulated by KCl and agonists. KCl increases $[Ca^{2+}]_i$, activates MLCK, and activates cycling of phosphorylated cross-bridges. During sustained contraction, the amount of phosphorylated MLC decreases. However, force is maintained because unphosphorylated cross-bridges are recruited as suggested by the latch model and the positive cooperativity model. Agonist also increases $[Ca^{2+}]_i$ and the amount of MLC phosphorylation. In addition, agonist increases the MLCK/MLCP balance to phosphorylate more MLC than KCl at a given $[Ca^{2+}]_i$. During sustained contraction, part of the contraction is maintained by cycling of unphosphorylated cross-bridges, as is the case with KCl-induced contraction. At the same time, agonist activates the MLC phosphorylation-independent mechanism to further increase the number of unphosphorylated cross-bridges. In the absence of external $Ca^{2+}$, all the MLCK-dependent mechanisms do not work. However, the MLC phosphorylation-independ-

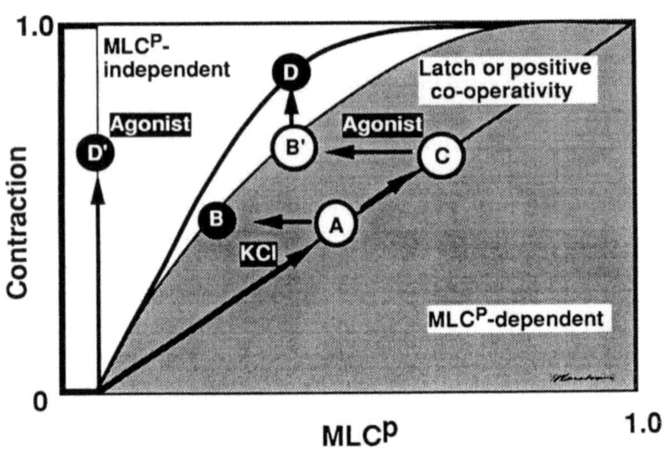


**Fig. 7.** Relation between MLC phosphorylation (*MLC*$^P$) and contraction. High $K^+$ increases MLC$^P$ followed by a proportional increase in the contractile force (*A*). During sustained contraction MLC$^P$ decreases, whereas force is maintained (*B*) by the formation of a latch bridge. Agonist also increases the MLC$^P$ and force to that at point A. In addition, agonist increases the MLCK/MLC phosphatase balance and further increases both MLC$^P$ and force to point *C*. During sustained contraction MLC$^P$ decreases, whereas force is maintained by the latch bridge (*B'*). Agonist also activates the MLC phosphorylation-independent mechanism to increase force without changing MLC$^P$ (*D*). Thus agonist induces more contraction than high $K^+$ at a given $[Ca^{2+}]_i$. In the absence of external $Ca^{2+}$, high $K^+$ is ineffective. However, agonist increases force without changing MLC$^P$ by activating the MLC phosphorylation-independent mechanism (*D'*)

ent mechanism is activated by agonist and induces contraction without changing MLC phosphorylation.

## Do Agonists Activate Protein Kinase C?

The activators of PKC, phorbol esters, showed effects similar to those of agonists in that they increased $Ca^{2+}$ sensitivity of MLC phosphorylation and activated a mechanism that is not dependent on MLC phosphorylation [7]. Because agonists acivate PKC through the formation of diacylglycerol, it has been widely accepted that $Ca^{2+}$ sensitization induced by agonists is attributable to this mechanism. To examine this possibility, rat aorta was treated with phorbol ester for 24 h to inhibit the PKC activity. In the preparations without PKC activity, agonists still increased $Ca^{2+}$ sensitivity [22]. In the permeabilized muscle without PKC activity, agonists also potentiated $Ca^{2+}$-induced contraction. In the absence of PKC activity, in contrast, the effects of phorbol ester to increase $Ca^{2+}$ sensitivity were completely abolished. Thus the effects of phorbol ester may be due to activation of PKC, whereas the effect of agonists are not (Fig. 8).

## Conclusion

The results of recent studies suggest that there are two regulatory mechanisms in smooth muscle. The first mechanism is MLC phosphorylation ($K_1$ to $K_4$ in Fig. 2C). Changes in MLCK/MLC phosphatase balance change the $Ca^{2+}$ sensitivity of phosphorylation between curves 1 and 2 in Figure 6. The mechanism dependent on MLC phosphorylation (curves 1 and 2) is inhibited by wortmannin. The second mechanism is MLC phosphorylation-independent, indicated by $K_8$ in Figure 2C and curve 3 in Figure 6. This mechanism is more sensitive to $Ca^{2+}$ than MLCK, and activation of this mechanism results in increased $Ca^{2-}$ sensitivity. This mechanism is

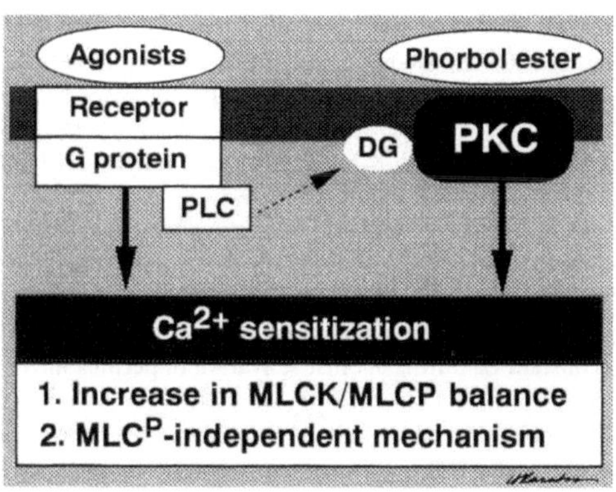


**Fig. 8.** Effect of agonist and phorbol ester on $Ca^{2+}$ sensitization of smooth muscle. Agonists increase $Ca^{2+}$ sensitivity of contractile elements by increasing the MLCK/MLC phosphatase balance and by activating the MLC phosphorylation-independent mechanism. The effects of agonists are mediated by GTP-binding protein. However, protein kinase C (*PKC*) is not involved in this pathway at least in rat aorta. Phorbol esters activate PKC but not GTP-binding protein and increase $Ca^{2-}$ sensitivity of contractile elements

not inhibited by wortmannin. High $K^+$ activates only the MLC phosphorylation (curve 1 in Fig. 6 or $K_1$ to $K_4$ in Fig. 2C). Agonists activate MLC phosphorylation (curve 1), $Ca^{2+}$ sensitization of MLC phosphorylation (curve 2), and the MLC phosphorylation-independent mechanism (curve 3 or $K_1$ to $K_8$ in Fig. 2C) and thus shift the $Ca^{2+}$–force curve to lower the $Ca^{2+}$ level, as shown by curve 4 in Figure 6. The MLC phosphorylation-independent mechanism is activated without previous myosin phosphorylation or activation of MLCK. This mechanism is regulated by $Ca^{2+}$ but not by protein kinase C. Although our results suggest that the MLC phosphorylation-independent mechanism is activated by agonists and phorbol ester, but not by high $K^+$ (Fig. 8), a latch bridge is formed during the high $K^+$-induced contraction as well, possibly through dephosphorylation of an attached cross-bridge (Fig. 1A) or positive cooperativity (Fig. 1B). Thus there may be more than one pathway to create the unphosphorylated attached cross-bridges.

## References


1. Murphy RA (1994) What is special about smooth muscle? The significance of covalent crossbridge regulation. FASEB J 8:311–318
2. Vyes TB, Mooers SU, Narayan SR, Witherell JC, Siegman MJ, Butler TM (1992) Cooperative activation of myosin by ligh chain phosphorylation in permeabilized smooth muscle. Am J Physiol 263:C210–C219
3. Karaki H (1989) $Ca^{2+}$ localization and sensitivity in vascular smooth muscle. Trends Pharmacol Sci 10:320–325
4. Karaki H (1990) The intracellular calcium-force relationship in vascular smooth muscle. Time- and stimulus-dependent dissociation. Am J Hypertens 3:253S–256S
5. Karaki H (1991) $Ca^{2+}$ regulation of vascular smooth muscle and release of endothelium-derived relaxing factor. In: Sperelakis N, Kuriyama H (eds) Ion channels of vascular smooth muscle cells and endothelial cells. Elsevier, New York, pp 297–315
6. Hori M, Sato K, Sakata K, Ozaki H, Takano-Ohmuro H, Tsuchiya T, Sugi H, Kato I, Karaki H (1992) Receptor agonists induce myosin phosphorylation-dependent and phosphorylation-independent contractions in vascular smooth muscle. J Pharmacol Exp Ther 261:506–512
7. Sato K, Hori M, Ozaki H, Takano-Ohmuro H, Tsuchiya T, Sugi H, Karaki H (1992) Myosin phosphorylation-independent contraction induced by phorbol ester in vascular smooth muscle. J Pharmacol Exp Ther 261:497–505
8. Sato K, Ozaki H, Karaki H (1988) Changes in cytosolic calcium level in vascular smooth muscle strips measured simultaneously with contraction using fluorescent calcium indicator fura 2. J Pharmacol Exp Ther 246:294–300
9. Kitazawa T, Kobayashi S, Horiuchi K, Somlyo AV, Somlyo AP (1989) Receptor-coupled, permeabilized smooth muscle. J Biol Chem 264:5339–5342
10. Nishimura J, Kolber M, Van Breemen C (1988) Norepinephrine and GTPγS increase myofilament $Ca^{2+}$ sensitivity in α-toxin permeabilized arterial smooth muscle. Biochem Biophys Res Commun 157:677–683
11. Moreland S, Nishimura J, Van Breemen C, Ahn HY, Moreland RS (1992) Transient myosin phosphorylation at constant Ca during agonist activation of permeabilized arteries. Am J Physiol 263:C540–C544
12. Hirata K, Kikuchi A, Sasaki T, Kuroda S, Kaibuchi K, Matsuura Y, Seki H, Saida K, Takai K (1992) Involvement of rho p21 in the GTP-enhanced calcium ion sensitivity of smooth muscle. J Biol Chem 267:8719–8722
13. Kitazawa T, Gaylinn BD, Denny GH, Somlyo AP (1991) G-protein-mediated Ca sensitization of smooth muscle contraction through myosin light chain phosphorylation. J Biol Chem 266: 1708–1715
14. Gong MC, Fuglsang A, Alessi D, Kobayashi S, Cohen P, Somlyo AV, Somlyo AP (1992) Arachidonic acid inhibits myosin light chain phosphatase and sensitizes smooth muscle to calcium. J Biol Chem 267:21492–21498

15. Tansey MG, Word A, Hidaka H, Singer HA, Schworer, CM, Kamm KE, Stull JT (1992) Phosphorylation of myosin light chain kinase by the multifunctional calmodulin-dependent protein kinase II in smooth muscle cells. J Biol Chem 267:12511–12516
16. Stull JT, Hsu L-C, Tansey MG, Kamm KE (1990) Myosin light chain kinase phosphorylation in tracheal smooth muscle. J Biol Chem 265:16683–16690
17. Gilbert EK, Weaver BA, Rembold CM (1991) Depolarization decreases $[Ca]_i$ sensitivity of myosin light chain kinase in arterial smooth muscle; comparison of aequorin and fura 2 [Ca] estimates. FASEB J 5:2593–2599
18. Nakanishi S, Kakita S, Takahashi I, Kawahara K, Tsukida E, Sano T, Yamada K, Yoshida M, Kase H, Matsuda Y, Hashimoto Y, Nonomura Y (1992) Wortmannin, a microbial product inhibitor of myosin light chain kinase. J Biol Chem 267:2157–2163
19. Takayama M, Takizawa S, Ozaki H, Karaki H (1991) Effects of wortmannin on cytosolic $Ca^{2+}$ level and force in vascular smooth muscle. Jpn J Pharmacol 64(suppl 1):216P
20. Ozaki H, Ishihara H, Kohama K, Nonomura Y, Shibata S, Karaki H (1987) Calcium-independent phosphorylation of smooth muscle myosin light chain by okadaic acid isolated from black sponge (*Halicondria okadai*). J Pharmacol Exp Ther 243:1167–1173
21. Ishihara H, Ozaki H, Sato K, Hori M, Karaki H, Watabe S, Kato Y, Fusetani N, Hasgimoto K, Uemura D, Hartshorne D (1989) Calcium-independent activation of contractile apparatus in smooth muscle by calyculin-A. J Pharmacol Exp Ther 250:388–396
22. Hori M, Sato K, Miyamoto S, Ozaki H, Karaki H (1993) Different pathways of calcium sensitization activated by receptor agonists and phorbol esters in vascular smooth muscle. Br J Pharmacol 110:1527–1531

# Cytosolic Calcium Concentration–Force Relation in Vascular Smooth Muscle

Hideo Kanaide

*Summary.* Front-surface fluorometry and fura-2-loaded medial strips of various arteries were used to determine the cytosolic calcium concentration ($[Ca^{2+}]_i$)–force relation during contraction and relaxation. During both high external $K^+$- and stable thromboxane $A_2$ analog (U-46619)-induced contractions the actively developed force was much greater at a high resting load, although the $[Ca^{2+}]_i$ transients were not affected by changes in the resting load. The basic $[Ca^{2+}]_i$–force relation was obtained by the cumulative application of extracellular $Ca^{2+}$ during 118 mM $K^+$ depolarization ($Ca^{2+}$ contraction). During the initial $[Ca^{2+}]_i$-rising phase of contractions, mainly due to the release of $Ca^{2+}$ from the store (induced by serotonin or norepinephrine), the $[Ca^{2+}]_i$–force relation was similar to that of the $Ca^{2+}$ contraction; however, during the steady-state phases of contractions ($Ca^{2+}$ influx) there was enhancement of the force for a given level of $[Ca^{2+}]_i$, the extent of which depended on the species, type of artery, and stimuli examined. Thus there is time-dependent and the species-, organ-, and stimuli-specific modulation of $Ca^{2+}$ sensitivity of the contractile apparatus in arterial smooth muscle. Nitroglycerin, isoproterenol, and diltiazem reduced both $[Ca^{2+}]_i$ and force. Nitroglycerin and isoproterenol, but not diltiazem, depleted the stored $Ca^{2+}$ and inhibited the release of $Ca^{2+}$ from the store. Nitroglycerin and isoproterenol, but not diltiazem, shifted the $[Ca^{2+}]_i$–force relation of the steady-state phase of $Ca^{2+}$ contractions to the right from that of the $Ca^{2+}$ contraction. Thus the $[Ca^{2+}]_i$–force relation ($Ca^{2+}$ sensitivity of the contractile apparatus) varies during relaxation, depending on the type of relaxants and the length of elapsed time after the application.

*Key Words.* Vascular smooth muscle—Fura-2—Calcium sensitivity—Cytosolic calcium–Force relation

## Introduction

The contraction of vascular smooth muscle is regulated mainly by changes in the cytosolic calcium concentration ($[Ca^{2+}]_i$), and so the development of force has been regarded as a good indicator of an increase in $[Ca^{2+}]_i$ [1]. The contraction is initiated by an increase in $[Ca^{2+}]_i$ and subsequent phosphorylation of the myosin light chain by

Division of Molecular Cardiology, Research Institute of Angiocardiology, Faculty of Medicine, Kyushu University, 3-1-1 Maidashi, Fukuoka, 812 Japan

the complex of $Ca^{2+}$/calmodulin and myosin light chain kinase (MLCK) [2]. It has been demonstrated that, for a given elevation of $[Ca^{2+}]_i$, the receptor-mediated stimulation by various agonists caused a proportionally greater force development than did depolarization with high-$K^+$ solutions [3–8]. It is generally accepted that the force development mainly depends on $Ca^{2+}$-mediated myosin light chain phosphorylation, although it has been demonstrated that force development could be maintained under lower levels of $[Ca^{2+}]_i$ and myosin light chain phosphorylation through other $Ca^{2+}$-insensitive mechanisms, such as latch bridges [9] or cooperating cycling of cross-bridges [10]. Using receptor-coupled permeabilized vascular smooth muscle with staphylococcal α-toxin [11] and β-escin [12], it was demonstrated that the $Ca^{2+}$ sensitivity of the contractile apparatus is modulated by the receptor–G protein-coupled mechanism. A protein kinase C-mediated mechanism was proposed to explain the agonist-modulated $Ca^{2+}$ sensitivity of vascular smooth muscle contractility [6–8]. It was reported that second messengers, cyclic adenosine monophosphate (cAMP) and cyclic guanosine-3′,5′-monophosphate (cGMP), relaxed the permeabilized rat mesenteric artery at a $[Ca^{2+}]_i$ held constant with a $Ca^{2+}$-ethyleneglycol-bis (β-aminoethy(ether)-*N,N,N′,N′*-tetraacetic acid (EGTA) buffer and a $Ca^{2+}$ ionophore, ionomycin, which indicated that second messengers exert a direct modulation of smooth muscle $Ca^{2+}$ sensitivity of the contractile apparatus [13].

This study focuses on the $[Ca^{2+}]_i$–force relation during the contraction and relaxation of vascular smooth muscle. Particular emphasis is placed on the factors that modulate the $[Ca^{2+}]_i$–force relation during the contraction and on the vasorelaxants that modulate the $[Ca^{2+}]_i$–force relation. We utilized front-surface fluorometry and fura-2 [6–8, 14–16] to determine simultaneously $[Ca^{2+}]_i$ and the force of the vascular strips.

## Methods and Materials

### Tissue Preparation

Bovine middle cerebral arteries and ophthalmic arteries and porcine coronary arteries were obtained immediately after the animals were slaughtered at a local abattoir. The tissues were then transported to the laboratory in an ice-cold physiologic salt solution (PSS). Japanese white rabbits (male, 16–20 weeks, body weight 2.5–3.0 kg) were killed with pentobarbital (100 mg/kg IV), and the femoral arteries were immediately excised. (This protocol was approved by the Animal Care and Committee of Research Institute of Angiocardiology, Faculty of Medicine, Kyushu University.)

The fat and adventitia were removed by dissection under a binocular microscope. The endothelial cells were removed by rubbing the intraluminal surface with a cotton swab. Medial preparations were cut into approximately 1 × 5 mm circular strips 0.2 mm thick.

### Fura-2 Loading

The vascular strips thus obtained were loaded with a $[Ca^{2+}]_i$ indicator dye, fura-2, by incubation in medium containing 50 μM fura-2/AM (an acetoxymethyl ester form of fura-2) and 2.5% fetal bovine serum for 4–5 h at 37°C. Subsequently, the strips were washed with PSS containing 1.25 mM $Ca^{2+}$ at 37°C to remove the dye from the extracellular space and then were equilibrated in normal PSS for at least 1 h before starting the measurements. The loading of the vascular strips with fura-2 under these conditions did not alter the time course or the maximum levels of the force development during 118 mM $K^+$ depolarization, suggesting that no tissue damage occurred via possible acidification of the cells due to formaldehyde release during AM-ester hydrolysis [17, 18].

## Measurement of Force

The strips were mounted vertically in a quartz organ bath, and a force transducer (strain gauge TB-612, Nihon Koden, Tokyo, Japan) was used. During a 1 h equilibration period, the strips were stimulated with 118 mM $K^+$ depolarization every 15 min; then the resting tension was increased in a stepwise manner. After equilibration, the resting tension was adjusted to about 300 mg. The force development was expressed as a percentage, assuming the values in normal (5.9 mM $K^+$) and 118 mM $K^+$ PSS to be 0% and 100%, respectively.

## Measurement of $[Ca^{2+}]_i$

Changes in the fluorescence intensity of the fura-2–$Ca^{2+}$ complex were monitored simultaneously with measurement of the force, using front-surface fluorometry [6–8, 14–16]. The fluorescence intensity was measured by a fluorometer which was specially designed for fura-2 fluorometry (CAM-OF series 1 and 2) in collaboration with the Japan Spectroscopic, (Tokyo, Japan). In brief, the strips were illuminated by guiding the alternating (400 Hz) 340 and 380 nm excitation light from a xenon light source through quartz optic fibers arranged in a concentric inner circle (diameter 3 mm). Surface fluorescence of the strips was collected by glass optic fibers arranged in an outer circle (diameter 7 mm) and introduced through a 500 nm band-pass filter (full width at half maximum transmission was 10 nm) into a photon-counting photomultiplier. The ratio of the 500 nm fluorescence at 340 nm excitation to that at 380 nm excitation was expressed as a percentage, assuming the values in normal (5.9 mM $K^+$) and 118 mM $K^+$ PSS to be 0% and 100%, respectively. The absolute values of $[Ca^{2+}]_i$ were calculated as described by Grynkiewicz et al. [19] with minor modifications. The mean values obtained during measurements of $[Ca^{2+}]_i$ at rest (0%) and with 118 mM $K^+$ depolarization (100%) were about 100 and 800 nM, respectively.

## "Basic" $[Ca^{2+}]_i$–Force Relation of the $Ca^{2+}$ Contraction

The $[Ca^{2+}]_i$–force relation during contraction and relaxation was evaluated by comparing it with the "basic" $[Ca^{2+}]_i$–force relation. The basic $[Ca^{2+}]_i$–force relation was determined as follows: Vascular strips were incubated in $Ca^{2+}$-free PSS containing 2 mM EGTA for 10 min, followed by 5 min exposure to $Ca^{2+}$-free PSS without EGTA; the solution was then changed to a $Ca^{2+}$-free 118 mM $K^+$ solution, and cumulative applications of extracellular $Ca^{2+}$ (0–10 mM) were initiated during 118 mM $K^+$ depolarization. Both the $[Ca^{2+}]_i$ and force increased stepwise with elevations of extracellular $Ca^{2+}$ concentration ($[Ca^{2+}]_o$). The $[Ca^{2+}]_i$ increased from −25% (at 0 mM $[Ca^{2+}]_o$) to 119% (at 10 mM $[Ca^{2+}]_o$). The force increased from 0% to 122%. Making

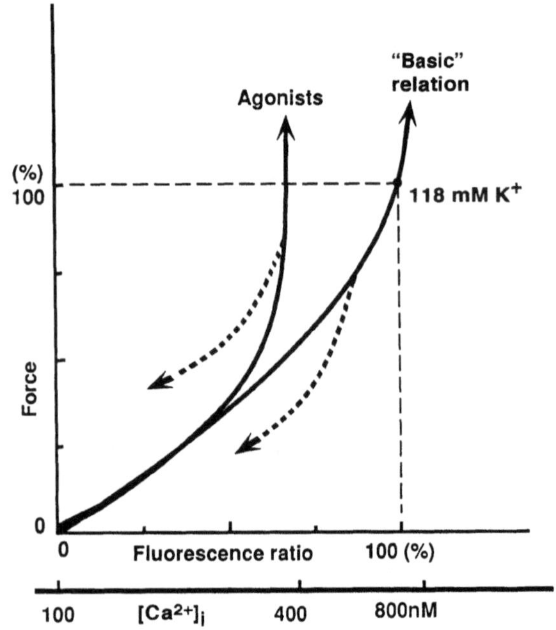


**Fig. 1.** Cytosolic calcium concentration ($[Ca^{2+}]_i$)-force relation. Right and left *solid lines* indicate the $[Ca^{2+}]_i$-force relations of $Ca^{2+}$ contraction (*"basic" relation*) and agonists-induced contraction, respectively. *Broken lines* at *left* and *right* of the solid lines indicate the left and the right shifts of the relation curves observed during the diltiazem-induced relaxation in histamine-induced contraction and nitroglycerin-induced relaxation in $K^+$ depolarization-induced contraction, respectively

use of these data, the basic $[Ca^{2+}]_i$-force relation of the $Ca^{2+}$-induced contractions ($Ca^{2+}$ contraction) was obtained as shown in Figure 1. The basic $[Ca^{2+}]_i$-force relation of the $Ca^{2+}$ contraction was similar to the $[Ca^{2+}]_i$-force relation noted with saponin-skinned fibers [8].

When the vascular strips were exposed to various concentration (30, 40, 60, and 118 mM) of high-$K^+$ solutions, the $[Ca^{2+}]_i$ rose abruptly and then decreased until it reached a lower steady level. The force also rose rapidly and then gradually increased to a maximum, steady level. The steady levels of $[Ca^{2+}]_i$ and force were dependent on the $K^+$ concentration. $[Ca^{2+}]_i$ and force remained at their steady levels for as long as the stimulus was maintained (at least 30 min). The $[Ca^{2+}]_i$-force relation, at the steady levels of contractions induced by the single-dose applications of any high-$K^+$ solutions, overlapped with the line of basic $[Ca^{2+}]_i$-force relation of the $Ca^{2+}$ contraction.

Thus the basic $[Ca^{2+}]_i$-force relation of the $Ca^{2+}$ contraction is defined as follows: (1) It was obtained by the cumulative applications of extracellular $Ca^{2+}$ during 118 mM $K^+$ depolarization. (2) The $[Ca^{2+}]_i$-force relation of the steady level of contractions induced by single-dose applications of high-$K^+$ solutions overlapped with that of the $Ca^{2+}$ contraction. (3) The basic $[Ca^{2+}]_i$-force relation of the $Ca^{2+}$ contraction was similar to the one noted with the skinned fibers.

## Drugs and Solutions

The physiologic salt solution used was of the following composition (in millimoles): NaCl 123, KCl 4.7, $NaHCO_3$ 15.5, $KH_2PO_4$ 1.2, $MgCl_2$ 1.2, $CaCl_2$ 1.25, and D-glucose 11.5. High-$K^+$ PSS was identical with normal PSS, except for an equimolar substitution of KCl for NaCl. $Ca^{2+}$-free solution was identical with normal PSS, except for a substitution of 2 mM ethyleneglycol-bis(β-aminoethylether)-*N*,*N*,*N*′,*N*′-tetraacetic acid (EGTA) for 1.25 mM $CaCl_2$. All solutions were gassed with a mixture of 95% $O_2$ and 5% $CO_2$ (pH 7.4 at 37°C). Fura-2/AM was purchased from the Dojindo Laboratories (Kumamoto, Japan).

## Data Analysis

The values are expressed as the mean ± standard errors (SE). Student's *t*-test was used to determine the statistical significance. The statistical analysis of the shift of $[Ca^{2+}]_i$–force curves was carried out by an analysis of covariance. $P$ values of less than 0.05 were considered to be significant.

# Results and Discussion

## $[Ca^{2+}]_i$–Force Relation During Contraction

### Effects of Preload and Resting Load on the Force and $[Ca^{2+}]_i$ in Medial Strips of Bovine Middle Cerebral Artery

Assuming the length of the medial strips to be 1 at the resting load of 10 mg, the relative lengths at the resting load of 50 mg and the optimal load of 200 mg were 1.14 and 1.36, respectively. Thus there was a close correlation between the resting load and the resting length, and a simple exponential relation was maintained between these two factors up to the resting load of 400 mg. During the contraction induced by the 118 mM $K^+$ depolarization, the actively developed force (total tension minus resting tension) was much larger at a higher resting load, whereas $[Ca^{2+}]_i$ transients were not affected by any changes in the resting load. The stepwise elevations of $[Ca^{2+}]_i$ induced by the cumulative applications of extracellular $Ca^{2+}$ during 118 mM $K^+$ depolarization ($Ca^{2+}$ contraction) were not affected by the change in the resting load, but the concomitant development of force was larger with a resting load of 200 mg than with 50 mg. The extent of the augmentation of 118 mM $K^+$-induced force development by increasing the resting load from 50 mg to 200 mg was not affected by H-7, which is a relatively specific protein kinase C inhibitor.

At a given $[Ca^{2+}]_i$ elevation, the developed force induced by U-46619, a stable analog of thromboxane $A_2$, was greater than that induced by high-$K^+$ depolarization. An increase in the resting load from 50 mg to 200 mg enhanced not only the 118 mM $K^+$-induced contraction but also that induced by $10^{-7}$ M U-46619 (50 mg 160%; 200 mg 301%) with no significant change in the $[Ca^{2+}]_i$ elevation (50 mg 98%; 200 mg 95%). In $Ca^{2+}$-free PSS, U-46619 induced a rapid and transient increase in $[Ca^{2+}]_i$ owing to the release of $Ca^{2+}$ from the intracellular store. There was no significant difference in the peak of $[Ca^{2+}]_i$ elevation induced by U-46619 between loads of 50 or 200 mg. However, the concomitant force development induced by U-46619 in $Ca^{2+}$-free PSS was enhanced by an increase in the resting load from 50 mg to 200 mg.

It was concluded that the $[Ca^{2+}]_i$–force relation observed during the contraction induced by U-46619 was shifted to the left from that of the $Ca^{2+}$ contraction. The preload, or resting load, regulates the contractile responsiveness of cerebrovascular smooth muscle to various stimulations, mainly by H-7-independent modulation of the $Ca^{2+}$ sensitivity of the contractile apparatus, without affecting the extent of $[Ca^{2+}]_i$ elevation.

## $[Ca^{2+}]_i$–Force Relation During Contractions in the Rabbit Femoral Artery and Bovine Ophthalmic Artery

When rabbit femoral artery strips were exposed to $10^{-5}$ M norepinephrine (NE) and to $10^{-4}$ M serotonin (5-HT), the $[Ca^{2+}]_i$ rose abruptly, reaching peaks of 114% and 102% after 1 and 2 min, respectively (the $[Ca^{2+}]_i$-rising phase) and then declining with a similar time course ($[Ca^{2+}]_i$-declining phase) to reach a low, steady level (steady-state phase). The force induced by NE and 5-HT reached peaks at 4 min (134%) and 2 min (118%) and then gradually declined. In contrast to the similarity in $[Ca^{2+}]_i$ transient between NE and 5-HT, the force induced by NE declined more slowly and then reached higher steady levels than that seen with 5-HT; the levels of force 20 min after the application of NE and 5-HT were 117% and 75%, respectively. Over the entire time course of the 5-HT-induced contraction ($[Ca^{2+}]_i$-rising, $[Ca^{2+}]_i$-declining, and steady-state phases), the $[Ca^{2+}]_i$–force relation was almost the same as that of the $Ca^{2+}$ contraction. However, during the $[Ca^{2+}]_i$-declining and steady-state phases, NE produced a greater force than that expected from a given change in $[Ca^{2+}]_i$ of the $Ca^{2+}$ contraction, resulting in a left shift of the $[Ca^{2+}]_i$–force relation. The extent of the left shift depended on the concentration of NE.

When bovine ophthalmic artery strips were exposed to NE and 5-HT, changes in $[Ca^{2+}]_i$ and force were almost the same as those seen in rabbit femoral artery strips. In contrast to the rabbit femoral artery strips, the left shift of the $[Ca^{2+}]_i$–force relation of the $[Ca^{2+}]_i$-declining and steady-state phases of 5-HT-induced contraction was much greater than that of the NE-induced contraction. It is therefore suggested that there is a time—dependent and species-, organ-, or stimulus-specific modulation of $Ca^{2+}$ sensitivity of the contractile apparatus in arterial smooth muscle.

## Role of Protein Kinase C in the Modulation of the $[Ca^{2+}]_i$–Force Relation of High External $K^+$ Depolarization- or Histamine-Induced Contractions in Porcine Coronary Artery Strips

In normal PSS, histamine induced abrupt (first component) and then sustained (second component) elevations of $[Ca^{2+}]_i$, in a concentration dependent manner, in porcine coronary artery strips. In $Ca^{2+}$-free PSS, histamine induced transient elevations of $[Ca^{2+}]_i$, and the time course was similar to that of the first component observed in normal PSS. Thus the first and second components were related to $Ca^{2+}$ release from the store and $Ca^{2+}$ influx through the sarcolemma, respectively. In the second component, histamine caused a greater force for a given change in $[Ca^{2+}]_i$ than did the $Ca^{2+}$ contraction at $[Ca^{2+}]_i$ higher than 300 nM [8].

Prior to the depolarization or histamine stimulation in normal PSS, the vascular strips were treated with calphostin C, a specific inhibitor of protein kinase C, for 60 min. Calphostin C affected either the $[Ca^{2+}]_i$ transient or the force induced by 40 mM $K^+$ depolarization. In the case of $10^{-5}$ M histamine-induced contraction, $10^{-6}$ M calphostin C slightly inhibited the first component of $[Ca^{2+}]_i$ elevation ($Ca^{2+}$

release component) and markedly but partially (about 50%) inhibited the second component of $[Ca^{2+}]_i$ elevation ($Ca^{2+}$ influx component). The first component of force was not affected. The second component of force was almost fully inhibited, which resulted in a right shift of the $[Ca^{2+}]_i$–force relation at a sustained contraction. Thus protein kinase C may be related to the second component of the contraction induced by histamine and to increases in the $Ca^{2+}$ sensitivity of the contractile apparatus.

## $[Ca^{2+}]_i$–Force Relation During Relaxation

### Effects of Diltiazem on $[Ca^{2+}]_i$ and Force in Porcine Coronary Artery Strips

Diltiazem $10^{-6}$ M, a $Ca^{2+}$ entry blocker, inhibited the increases in $[Ca^{2+}]_i$ and force induced by the cumulative applications of extracellular $Ca^{2+}$ during high $K^+$ depolarization ($Ca^{2+}$ contraction) [8]. The $[Ca^{2+}]_i$–force relation curve obtained during a $Ca^{2+}$ contraction of diltiazem-treated strips overlapped with that obtained in untreated strips, suggesting that diltiazem has no direct effect on the contractile elements.

Diltiazem $10^{-8}$ M to $10^{-5}$ M, concentration-dependently inhibited the second component ($Ca^{2+}$ influx) of the $[Ca^{2+}]_i$ elevation and force induced by $10^{-5}$ M histamine. Only at higher concentrations ($>10^{-5}$ M) did diltiazem inhibit the first component of the increases in $[C^{2+}]_i$ and force induced by $10^{-5}$ M histamine in, both normal and $Ca^{2+}$-free PSS. It was reported elsewhere that diltiazem blocks the binding of histamine to specific receptors at high concentrations [20]. Thus the inhibition of histamine-receptor-mediated release of $Ca^{2+}$ from the intracellular store site that is seen with a high concentration of diltiazem may be due to competitive binding to receptor sites. The histamine-induced $[Ca^{2+}]_i$–force relation curve at the second component of the contraction was shifted in parallel to the left by diltiazem. This observation suggests that diltiazem caused a reduction in the force that was proportional to the reduction of $[Ca^{2+}]_i$, without directly affecting either the contractile apparatus or mechanisms by which histamine produces a greater force for the given changes in $[Ca^{2+}]_i$.

It is concluded that diltiazem, at therapeutic concentrations, specifically inhibits extracellular $Ca^{2+}$-dependent ($Ca^{2+}$ influx) increases in $[Ca^{2+}]_i$, with no effect on the release of $Ca^{2+}$ from the intracellular store site or on $Ca^{2+}$ sensitivity of the contractile apparatus. At high concentrations, diltiazem inhibited $Ca^{2+}$ release from the intracellular store sites, possibly by inhibiting the binding of the agonist at receptor sites.

### Effects of Nitroglycerin on $[Ca^{2+}]_i$ and Force in Porcine Coronary Artery Strips

Irrespective of whether the vascular strips were in a resting state or were exposed to high external $K^+$ depolarization or to histamine stimulation, nitroglycerin reduced $[Ca^{2+}]_i$ and force in a concentration-dependent manner [7]. Nitroglycerin did not enhance the decrease in $[Ca^{2+}]_i$ observed when vascular strips were exposed to $Ca^{2+}$-free PSS and had no effect on the $[Ca^{2+}]_i$ transient induced by the repetitive applications of caffeine in $Ca^{2+}$-free PSS. These findings indicate that nitroglycerin does not affect $Ca^{2+}$ release by caffeine, nor does it deplete the caffeine-sensitive stored $Ca^{2+}$ of vascular strips. The force induced by the first application of caffeine, however, was significantly inhibited by nitroglycerin, suggesting that nitroglycerin inhibited the caffeine-induced force without affecting $[Ca^{2+}]_i$. Nitroglycerin strongly inhibited the transient elevations of $[Ca^{2+}]_i$ and force induced by histamine in $Ca^{2+}$-free PSS. However, when nitroglycerin was washed out in $Ca^{2+}$-free PSS, the second application of histamine induced transient elevations of both $[Ca^{2+}]_i$ and force; the extent of these elevations was greater than those observed during the first application of histamine in the presence of nitroglycerin. In addition, the second application of histamine induced a marked force despite evidence that the peak $[Ca^{2+}]_i$ was significantly lower than that at the resting level. It is therefore suggested that nitroglycerin attenuated the transient elevation of $[Ca^{2+}]_i$ induced by histamine in $Ca^{2+}$-free PSS not only by actively depleting the histamine-sensitive stored $Ca^{2+}$ but also by inhibiting the release of $Ca^{2+}$.

In addition to the findings that $[Ca^{2+}]_i$ of vascular strips decreased during relaxation induced by nitroglycerin, nitroglycerin relaxed the vascular strips to a greater extent than that expected from the reduction in $[Ca^{2+}]_i$ during the $Ca^{2+}$ contraction, whether the vascular strips were at rest or were being exposed to high-$K^+$ depolarization or histamine stimulation. The $[Ca^{2+}]_i$–force relation curve of the $Ca^{2+}$ contraction in the presence of nitroglycerin was shifted to the right of that obtained without nitroglycerin. The $[Ca^{2+}]_i$–force relation curve of the contraction obtained by the cumulative application of histamine in the presence of nitroglycerin shifted slightly to the right of that without nitroglycerin. Thus nitroglycerin relaxes the porcine coronary artery strips by reducing $[Ca^{2+}]_i$ and by directly controlling the $Ca^{2+}$ sensitivity of the contractile apparatus mediated through second messengers not related to changes in $[Ca^{2+}]_i$.

## Effect of Isoproterenol on $[Ca^{2+}]_i$ and Force in Porcine Coronary Artery Strips

Isoproterenol actively decreased the $[Ca^{2+}]_i$ of the strips at rest in normal and $Ca^{2+}$-free PSS [15]. When isoproterenol was applied at the time of steady state of high-$K^+$-induced $[Ca^{2+}]_i$ elevation, there was an initial transient reduction (first component) followed by a subsequent sustained reduction (second component) of $[Ca^{2+}]_i$. For a given $[Ca^{2+}]_i$ level during high-$K^+$ depolarization, the force developed in the presence of isoproterenol was smaller than that in its absence. Thus the $[Ca^{2+}]_i$–force relation during the steady state of high-$K^+$-induced contraction was shifted to the right by isoproterenol. Pretreatment with ryanodine, a compound that depletes $Ca^{2+}$ stored in the sarcoplasmic reticulum, abolished the first component but not the second sustained decrease in $[Ca^{2+}]_i$ by isoproterenol. Isoproterenol inhibited, in a concentration-dependent manner, the first and second components of histamine-induced $[Ca^{2+}]_i$ elevation and force. The $[Ca^{2+}]_i$–force relation during the histamine-induced contraction was shifted to the right by isoproterenol. During contractions induced by a higher concentration of histamine, despite the negligible decrease in $[Ca^{2+}]_i$, isoproterenol could relax the vascular strips in a concentration-dependent manner. In $Ca^{2+}$-free PSS, isoproterenol inhibited the transient elevation of $[Ca^{2+}]_i$ and force induced by histamine in a concentration-dependent manner but could not inhibit the transient elevations of $[Ca^{2+}]_i$ and force induced by caffeine. In the case of isoproterenol-induced relaxation of the porcine coronary artery strips, the following mechanisms are operative in relation to $[Ca^{2+}]_i$ homeostasis: (1) As for the case of nitroglycerin there was a reduction in the histamine-induced release of stored $Ca^{2+}$, presumably by interference with the histamine-stimulated signal-transduction pathway and by a decrease in the stored $Ca^{2+}$ in the histamine-sensitive store. (2) There

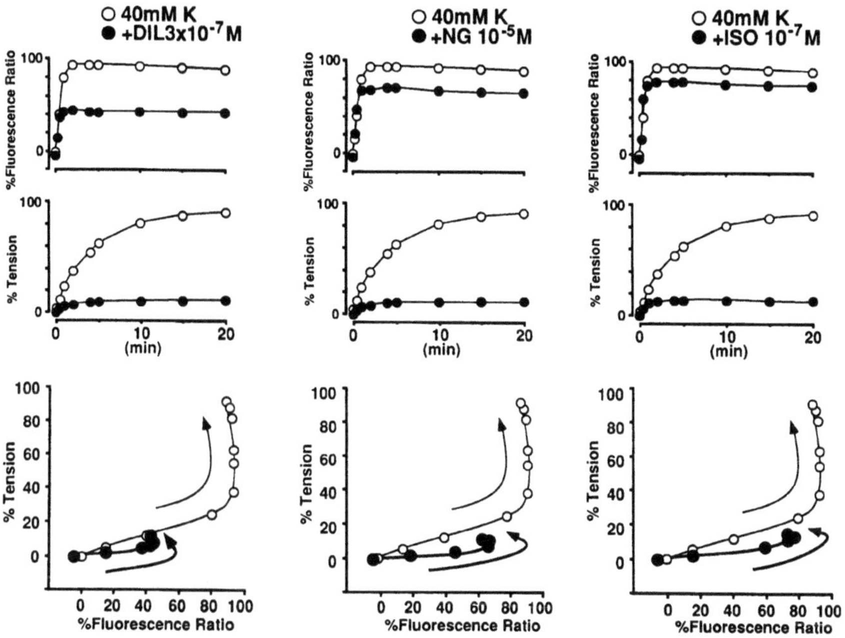


**Fig. 2.** Alterations by $3 \times 10^{-7}$ M diltiazem (*DIL*) (**left panels**), $10^{-5}$ M nitroglycerin (*NG*) (**center panels**), and $10^{-7}$ M isoproterenol (*ISO*) (**right panels**) of time courses of $[Ca^{2+}]_i$ (*top*) and force (*middle*) and their relations (*bottom*) of the 40 mM $K^+$ depolarization-induced contraction. *Arrows* show the direction of time

were active decreases in the $[Ca^{2+}]_i$ of the strips, presumably via stimulation of $Ca^{2+}$ extrusion from cells and inhibition of $Ca^{2+}$ influx stimulated by high-$K^+$ depolarization and histamine. (3) There was a transient increase in $Ca^{2+}$ uptake into the intracellular ryanodine-sensitive storage site(s). In addition, in the case of relaxation induced by isoproterenol, the $[Ca^{2+}]_i$–force relation always shifted to the right, indicating that there is decreased $Ca^{2+}$ sensitivity of certain contractile apparatuses in the smooth muscle cells.

## Effects of Pretreatment with Diltiazem, Nitroglycerin, and Isoproterenol on the Elevation of $[Ca^{2+}]_i$ and Force Induced by High-$K^+$ Depolarization and Histamine

Figure 2 demonstrates the effects of pretreatment with $3 \times 10^{-7}$ M diltiazem, $10^{-5}$ M nitroglycerin, and $10^{-7}$ M isoproterenol on the time courses of the elevations of $[Ca^{2+}]_i$ and force induced by 40 mM $K^+$ depolarization. During the steady state of the contraction, although the extent of the reduction of force by these three treatments were marked and similar, diltiazem induced the greatest and isoproterenol the smallest reduction of $[Ca^{2+}]_i$. The time courses of the $[Ca^{2+}]_i$–force relation during the contraction induced by 40 mM $K^+$ depolarization in the presence of nitroglycerin or isoproterenol were shifted to the right compared to those without nitroglycerin or isoproterenol. These findings were consistent with the characteristic features of each vasorelaxant as described above.

Figure 3 demontrates the effect of pretreatment with $3 \times 10^{-7}$ M diltiazem, $10^{-5}$ M nitroglycerin, and $10^{-7}$ M isoproterenol on the time course of elevations of $[Ca^{2+}]_i$ and

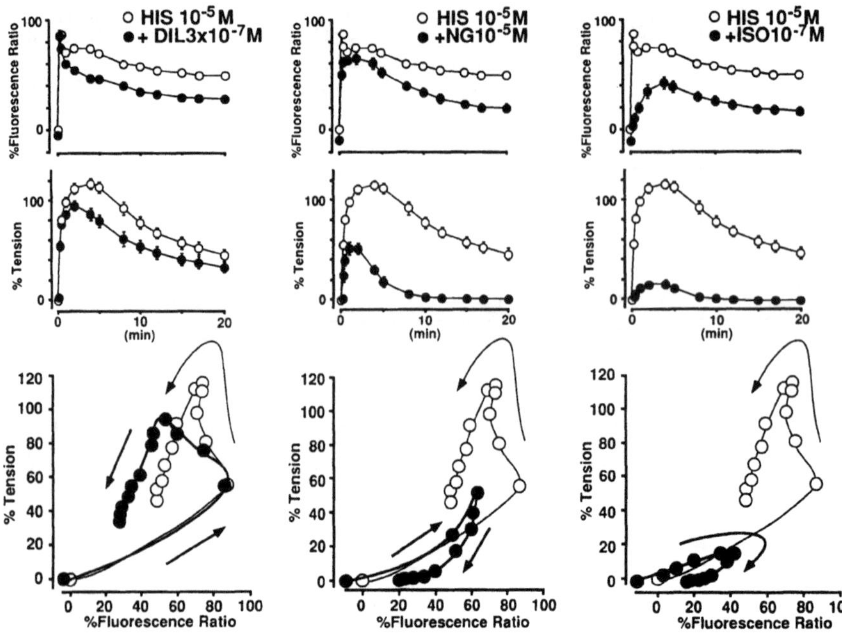


**Fig. 3.** Alterations by $3 \times 10^{-7}$ M diltiazem (**left panels**), $10^{-5}$ nitroglycerin (**center panels**), and $10^{-7}$ M isoproterenol (**right panels**) of time courses of $[Ca^{2+}]_i$ (*top*) and force (*middle*), and their relations (*bottom*) of the $10^{-5}$ M histamine (*HIS*)-induced contraction. *Arrows* show the direction of time

force induced by $10^{-5}$ M histamine. At the concentrations examined, these three vasodilators induced almost the same extent of reduction of $[Ca^{2+}]_i$ during the second component of the contraction induced by histamine. It was obvious, however, that the reduction of force by diltiazem was much smaller than that by nitroglycerin or isoproterenol. Nitroglycerin and isoproterenol, but not diltiazem, inhibited the first component of $[Ca^{2+}]_i$ elevation at the concentrations examined. When the strips were pretreated with diltiazem, the $[Ca^{2+}]_i$–force relation of the first component of the histamine-induced contraction was not affected, but that of the second component did shift to the left from that without diltiazem treatment and thus showed a counterclockwise rotation of the time course of the $[Ca^{2+}]_i$–force relation curve. In the case of nitroglycerin or isoproterenol treatment, however, the $[Ca^{2+}]_i$–force relation of the second component of the histamine-induced contraction shifted to the right from that without treatment and thus showed a clockwise rotation of the time course for the $[Ca^{2+}]_i$–force relation curve. These findings were also consistent with the characteristic features of each vasorelaxant as described above.

During the contraction induced by high-$K^+$ depolarization, diltiazem ($Ca^{2+}$ channel blocker) reduced both $[Ca^{2+}]_i$ and force. Nitroglycerin and isoproterenol (increased cGMP and cAMP, respectively) reduced the force in preference to $[Ca^{2+}]_i$. During the contraction induced by histamine, the release of $Ca^{2+}$ from the store (first component) was inhibited by nitroglycerin and isoproterenol. In contrast, diltiazem hardly inhibited $Ca^{2+}$ release from the store by histamine. During the second sustained component of histamine-induced contraction, nitroglycerin and isoproterenol shifted the $[Ca^{2+}]_i$–force relation curve to the right.

In conclusion, for contraction and relaxation of vascular smooth muscle, there is a time-dependent, species-, organ-, and stimuli-specific modulation of the $[Ca^{2+}]_i$–force relation ($Ca^{2+}$ sensitivity of the contractile apparatus), highlighting the marked complexity of the underlying regulatory mechanism of the contractile cycle.

*Acknowledgments.* I thank Brian T. Quinn for comments. This work was supported in part by grants-in-aid for Scientific Research on a Priority Area (06274221), Developmental Scientific Research (06557045), and General Scientific Research (04454268).

## References

1. Brading AF (1979) Maintenance of ionic composition. Br Med Bull 35:227–234
2. Kamm KE, Stull JT (1985) The function of myosin and light chain kinase phosphorylation in smooth muscle. Annu Rev Pharmacol Toxicol 25:593–620
3. Morgan JP, Morgan KG (1984) Stimulus-specific patterns of intracellular calcium levels in smooth muscle of ferret portal vein. J Physiol (Lond) 351:155–167
4. Himpens B, Somlyo AP (1988) Free-calcium and force transients during depolarization and pharmacological coupling in guinea-pig smooth muscle. J Physiol (Lond) 395:507–530
5. Rembold CW, Murphy RA (1988) Myoplasmic $[Ca^{2+}]$ determines myosin phosphorylation in agonist-stimulated swine arterial smooth muscle. Circ Res 63:593–603
6. Kodama M, Kanaide H, Abe S, Hirano K, Kai H, Nakamura M (1989) Endothelin-induced Ca-independent contraction of the porcine coronary artery. Biochem Biophys Res Commun 160:1302–1308
7. Abe S, Kanaide H, Nakamura M (1990) Front-surface fluorometry with fura-2 and effects of nitroglycerin on cytosolic calcium concentrations and on tension in the coronary artery of the pig. Br J Pharmacol 101:545–552
8. Hirano K, Kanaide H, Abe S, Nakamura M (1990) Effects of diltiazem on calcium concentrations in the cytosol and on force of contractions in porcine coronary arterial strips. Br J Pharmacol 101:273–280
9. Chatterjee M, Murphy RA (1983) Calcium-dependent stress maintenance without myosin phosphorylation in skinned smooth muscle. Science 221:464–466
10. Somlyo AV, Goldman YE, Fujimori T, Bond M, Trentham DR, Somlyo AP (1988) Cross-bridge kinetics, cooperativity, and negatively strained cross-bridges in vertebrate smooth muscle: a laser-flash photolysis study. J Gen Physiol 91:165–192
11. Nishimura J, Kolber M, van Breemen C (1988) Norepinephrine and GTP-$\gamma$-S increase myofilament calcium sensitivity in $\alpha$-toxin permeabilized arterial smooth muscle. Biochem Biophys Res Commun 157:677–683
12. Kobayashi S, Kitazawa T, Somlyo AP, Somlyo AY (1989) Cytosolic heparin inhibits muscarinic and $\alpha$-adrenergic $Ca^{2+}$ release in smooth muscle: physiological role of inositol 1,4,5-trisphosphate in pharmacomechanical coupling. J Biol Chem 264:17997–18004
13. Nishimura J, van Breemen C (1989) Direct regulation of smooth muscle contractile elements by second messengers. Biochem Biophys Res Commun 163:929–935
14. Watanabe C, Yamamoto H, Hirano K, Kobayashi S, Kanaide H (1992) Mechanisms of caffeine-induced contraction and relaxation of rat aortic smooth muscle. J Physiol (Lond) 456:193–213
15. Ushio-Fukai M, Abe S, Kobayashi S, Nishimura J, Kanaide H (1993) Effects of isoprenaline on cytosolic calcium concentrations and on tension in the porcine coronary artery. J Physiol (Lond) 462:679–696
16. Kuroiwa M, Aoki H, Kobayashi S, Nishimura J, Kanaide H (1993) Role of GTP-protein and endothelium in contraction induced by ethanol in pig coronary artery. J Physiol (Lond) 470:521–537

17. Tsien RY, Pozzan T, Rink TJ (1982) Calcium homeostasis in intact lymphocytes: cytosolic free calcium monitored with a new, intracellular trapped fluorescent indicator. J Cell Biol 94:325–334
18. Rink TJ, Pozzan T (1985) Using quin2 in cell suspensions. Cell Calcium 6:133–145
19. Grynkiewicz G, Poenie M, Tsien RY (1985) A new generation $Ca^{2+}$ indicators with greatly improved fluorescence properties. J Biol Chem 260:3440–3450
20. Matsumoto T, Kanaide H, Nishimura J, Kuga T, Kobayashi S, Nakamura M (1989) Histamine-induced calcium transients in vascular smooth muscle cells: effects of verapamil and diltiazem. Am J Physiol 257:H563–H570

# Cross-Bridge Cycle in Phasic and Tonic Smooth Muscle

A.V. Somlyo[1], E. Nishiye[2], A. Khromov[1], and A.P. Somlyo[1]

*Summary.* Comparison of the cross-bridge cycle and its regulation in phasic and tonic smooth muscles reveals the following: (1) ATP binding and the detachment of cross-bridges by ATP is fast and not rate-limiting to the overall catalytic cycle. (2) The rate of force development is faster in phasic than in tonic smooth muscles under two physiologic conditions: (a) when myosin light chain ($MLC_{20}$) phosphorylation is rate-limiting; and (b) in thiophosphorylated smooth muscle in which the actomyosin reactions are rate-limiting. (3) $MLC_{20}$ kinase and phosphatase activities are higher in phasic than in tonic smooth muscle. (4) MgADP, at concentrations thought to be within the physiologic range, reduces the amplitude of the rapid component and prolongs the slow component of relaxation from rigor, whereas inorganic phosphate ($P_i$) accelerates relaxation. An interpretation of these effects based on a scheme of the cross-bridge cycle is presented. (5) The affinity of cross-bridges for MgADP is much higher in tonic than in phasic smooth muscles. (6) Creatine phosphate (CP) accelerates relaxation induced by flash photolysis of Diazo-2 in isometrically contracted, permeabilized smooth muscle. We suggest that this effect is due to the removal by CP and creatine kinase of endogenous MgADP, which in the absence of CP slows the rate of relaxation. (7) It is suggested that cooperative reattachment of nonphosphorylated cross-bridges, through the action of attached bridges and the maintenance of force-bearing AM′.ADP states by MgADP contribute to the high force, low phosphorylation, and shortening velocity state referred to as "latch." (8) It is further suggested that the high affinity of myosin for MgADP in tonic smooth muscle is related to the presence of isoforms containing largely the more basic $LC_{17b}$ light chain and lacking the 7-amino-acid insert in the myosin head region.

*Key Words.* Myosin—Light chain kinase—Adenosine diphosphate—Myosin isozymes

## Introduction

Smooth muscles have been classified as fast-contracting phasic and slow-contracting tonic subtypes [1]. These two extremes, obviously subject to gradations, differ in their

[1] Department of Molecular Physiology and Biological Physics, University of Virginia Health Sciences Center, Box 449, Jordan Hall, Charlottesville, VA 22908, USA

[2] Department of Neurosurgery, Kyushu University 60, 3-1-1 Maidashi, Higashi-Ku, Fukuoka, 812 Japan

activation and contractile properties. Phasic smooth muscles generate action potentials and respond to a high potassium concentration with a rapid transient contraction, whereas tonic muscles do not normally generate action potentials, undergo graded depolarization by agonists, and respond to depolarization with slow, sustained contractions. Structural differences, including the extent and distribution of the sarcoplasmic reticulum and the ratio of thin to thick filaments, also contribute to these two phenotypes. In this review we concentrate on differences in the cross-bridge cycle and its regulation, with the goal of understanding the underlying mechanisms and the rate-limiting steps.

Femoral artery and trachealis muscles were chosen as representative of tonic smooth muscle and bladder and portal vein or longitudinal muscle of the ileum as phasic smooth muscles. To have access and to control the solutions bathing the myofilaments and their regulatory proteins, muscles were permeabilized with staphylococcal α-toxin [2] or β-escin [3]. α-Toxin protein inserts into the plasma membrane, forming hexagonal pores that permit the passage of molecules ≤ 1000 daltons and so retain the physiologic concentrations of endogenous calmodulin, myosin light chain kinase, and phosphatase. Permeabilizaton with β-escin allows the passage of larger molecules with a cutoff of approximately 150 kDa (Iizuka, Somlyo, and Somlyo, in preparation). Ultrastructural studies reveal no abnormality in the number and distribution of myosin and actin filaments after these permeabilization protocols. Laser flash photolysis of caged ATP([$P_3$-1]2-nitrophenyl[ethyladenosine 5′-triphosphate]) was used in permeabilized phasic and tonic smooth muscles to eliminate diffusional delays that would affect the kinetics of contraction and relaxation [4–6]. The experimental protocols, solutions, and flash photolysis apparatus were described by Somlyo and Somlyo [6], Nishiye et al. [7], and Fuglsang et al. [8].

## Contraction Kinetics in Smooth Muscle

It is generally accepted that in response to a physiologic excitatory stimulus and the resultant rise in intracellular [$Ca^{2+}$], phosphorylation of the regulatory light chains ($MLC_{20}$) is the necessary switch that turns on the myosin motor fueled by hydrolysis of ATP, leading to force development [review: 9]. The extent of $MLC_{20}$ phosphorylation reflects the sum of the forward myosin light chain kinase (MLCK) and backward myosin light chain phosphatase reactions. We sought to test the hypothesis that the significantly different rates of force development in, respectively, phasic and tonic smooth muscles [10] reflect differences in both the respective rates of myosin light chain phosphorylation and differences in the intrinsic properties of the actomyosin per se.

To distinguish these two processes, the rate of force development in control muscles was compared to that in muscles in which $MLC_{20}$ was thiophosphorylated with ATPγS and hence not dephosphorylated by myosin phosphatase. The muscles were depleted of ATP and subsequently exposed to $Ca^{2+}$ and caged ATP; photolysis of the latter rapidly (about $100\,s^{-1}$) released ATP (about 1.5 mM), causing contraction (Fig. 1).

Thiophosphorylation prior to photoloysis of caged ATP markedly increased the rate of contraction by the tonic trachealis muscle, shortening the half-time ($t_{1/2}$) by approximately sevenfold (Fig. 1, Table 1). Inclusion of the phosphatase inhibitor okadaic acid with the caged ATP did not significantly affect the rate of force development in this muscle. Gel electrophoresis of the thiophosphorylated muscles

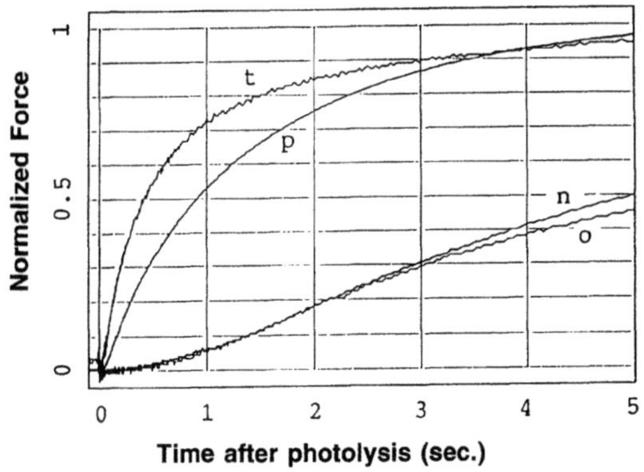


**Fig. 1.** Force responses of trachealis muscles after photolysis of caged ATP. *t*, muscle thiophosphorylated before photolysis; *p*, muscle prephosphorylated using ATP and okadaic acid; *n*, muscle not phosphorylated before photolysis; *o*, another not prephosphorylated muscle, but okadaic acid (10 μM) was added to the photolysis solution. *t* and *n* are records obtained from the same strip; *o* was obtained from another strip from the same animal; and *p* was obtained from a different animal. Each force record was normalized to its amplitude of response. A high performance liquid chromatography (HPLC) analysis revealed that the amount of liberated ATP was 1.4 mM in *t*, 1.0 mM in *p*, 1.5 mM in *n*, and 1.4 mM in *o*. (From [10], with permission)

showed almost complete thiophosphorylation of $MLC_{20}$. The initial slow phase (delay) of force development ($t_d$) was also shortened by approximately an order of magnitude in the thiophosphorylated trachealis muscles. The $t_{1/2}$ of force development in the phasic, not prephosphorylated, ileum smooth muscle was 2.4–4.4 times faster than comparable values in the tonic trachealis smooth muscle (Table 1). The light chains of thiophosphorylated ileum specimens also showed almost complete thiophosphorylation of $MLC_{20}$, and the $t_{1/2}$ for force development was speeded up by a factor of approximately 8 (Table 1). Thus the rates of force development by phasic

**Table 1.** Half-time and delay of force development in tonic and phasic smooth muscles.

| Muscle | Half-time (s) | Delay (s) |
|---|---|---|
| Trachealis muscle | | |
| Not prephosphorylated | | |
| Without okadaic | 5.4 ± 0.80 (16) | 0.30 ± 0.05 (15) |
| With okadaic | 5.3 ± 0.70 (8) | 0.51 ± 0.08[a] (8) |
| Thiophosphorylated[b] | 0.8 ± 0.17 (9) | 0.05 ± 0.01 (9) |
| Ileum longitudinal muscle | | |
| Not prephosphorylated | | |
| Without okadaic | 2.1 ± 0.40 (7) | 0.26 ± 0.07 (8) |
| With okadaic | 1.2 ± 0.20 (9) | 0.24 ± 0.06 (8) |
| Thiophosphorylated[c] | 0.3 ± 0.02 (8) | 0.03 ± 0.00 (8) |

Values are given as mean ± SEM (number of observations).
[a] Photolysis was carried out in the presence of $Ca^{2+}$ in four runs and in its absence in five runs. Results were not significantly different.
[b] Six runs in the presence of $Ca^{2+}$, two runs in its absence.
[c] Of nine observed delays, one outlier was rejected ($P < 0.025$).
From [10], with permission.

ileum and portal vein muscle are markedly faster in both the nonprephosphorylated and the thiophosphorylated states compared to the tonic trachealis muscle, supporting the hypothesis that both the contractile machinery and the regulatory system are slower in the tonic than in the phasic smooth muscles. The markedly enhanced rate of force development in thiophosphorylated muscle supports the concept [10–12] that under physiologic (not prephosphorylated) conditions slow light chain phosphorylation and prephosphorylation reactions, rather than the intrinsic kinetics of (phosphorylated) myosin activated by actin, limit the rate of force development by smooth muscle.

## Relaxation Kinetics in Smooth Muscle

In intact smooth muscle, the rates of relaxation, like the rates of force development, are also two to three times faster in phasic than in tonic smooth muscle. This could reflect differences in $Ca^{2+}$ sequestration and buffering, including dissociation of the $Ca^{2+}$/calmodulin complex from MLCK, the dephosphorylation rate of $MLC_{20}$, or different cross-bridge kinetics. We initially examined the kinetics of cross-bridge detachment from rigor and the effects of MgADP on these kinetics to determine if they also differ in the two types of smooth muscle. Apart from differences in regulation through a different $Ca^{2+}$-sensitive mechanism, the qualitative cross-bridge cycle scheme of smooth muscle is considered to be well fitted to the skeletal muscle model (Fig. 2) [13–15], albeit with different rate constants [10, 12, 16–18]. According to this cross-bridge cycling model, ATP is bound to actomyosin (AM), causing the cross-bridges to detach. ATP is then hydrolyzed and the heads reattach (steps 1–4). A conversion from weakly to tightly bound heads occurs at the time of inorganic phosphate ($P_i$) release, followed by ADP release and repetition of the cycle (steps 5–7). The overall rate of ATP hydrolysis by smooth muscle myosin in solution is $1\text{–}2\,s^{-1}$ per myosin head [9, 17] and $0.13\,s^{-1}$ in isometrically contracting muscle [19]. The rate-limiting step(s) of the cross-bridge cycle in smooth muscle has not been identified. The time course of relaxation initiated by photolysis of caged ATP in the phasic portal vein smooth muscle from a high rigor state could be well fitted to two exponentials during the initial 200 ms with time constant of $\tau_1 = 34 \pm 1.2$ ms and $\tau_2 = 1.2 \pm 0.06$ s, respectively, and relative amplitudes of 0.14 and 0.86, respectively. The estimated rates of detachment at in vivo ATP concentrations (3–5 mM), based on the measured

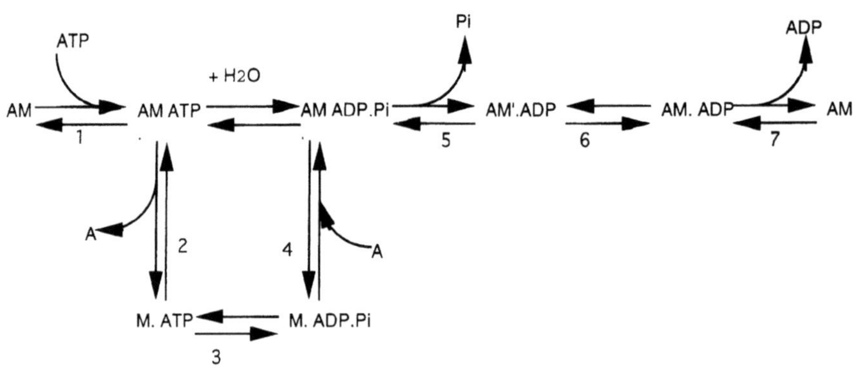


**Fig. 2.** Cross-bridge cycle scheme of smooth muscle

second-order rate constant, range between 12 and $115\,s^{-1}$. Thus the initial rate of detachment measured in fibers (step 2, Fig. 2) is much faster than the overall rate of ATP hydrolysis and supports the conclusion that ATP-induced detachment of nucleotide-free cross-bridges does not limit the rate of the cross-bridge cycle or the catalytic ATPase activity in smooth muscle [12, 16].

The approximate rate constant of MgADP release ($k_{-D}$), $0.7\,s^{-1}$ with 95% confidence limits of $0.35–7.70\,s^{-1}$, was estimated from the relation between MgADP concentration and the time constant of the slow component of the initial relaxation during the initial 200 ms after photolysis. The rate constant of ADP dissociation obtained in isolated gizzard actomyosin is faster ($15\,s^{-1}$) [16]. The slower rate in fibers could reflect differences due to the assembly of the proteins into filaments and strain on the myosin heads or contributions of other regulatory proteins in the fibers. Interestingly, the rates of product release ($3–4\,s^{-1}$) are slower even in solution, when actin is cross-linked to the S1 subfragment of myosin [20].

The shape and amplitude (Fig. 3) of the detachment transient that follows photolysis of caged ATP from rigor is markedly altered by MgADP in both phasic and tonic smooth muscles and, in the opposite direction, by pretreatment of the muscles, while in rigor, with apyrase to deplete endogenous MgADP [7, 8]. The amplitude of the rapid, initial first component is significantly decreased by MgADP concentrations as low as 5 μM (Fig. 3), and the rate constant of the slow component is dramatically slowed by [MgADP] ranging between 5 to 200 μM. In general, the amplitude of the first component ($A_f$) is greater in phasic (bladder and portal vein) smooth muscle than in tonic (femoral artery) smooth muscle (Fig. 4). The half-times of relaxation from rigor in α-toxin-permeabilized muscles in the absence of added MgADP are three to four times longer in the tonic muscles than in the phasic muscles. Pretreatment with apyrase accentuates this difference, with the $t_{1/2}$ of the phasic portal vein being approximately 10-fold less than that of the tonic femoral artey. The rabbit femoral artery is much more sensitive to ADP than rabbit bladder muscle: 20 μM ADP causes a much greater slowing of relaxation from rigor in smooth muscle of the femoral artery than that of the bladder. Even at MgADP concentrations as high as 200 μM, its effect on bladder muscle is significantly lower than on femoral artery muscle (Fig. 4). Fitting the relation between the relative amplitudes of the initial fast component of relaxation as a function of [MgADP] yields apparent dissociation constants of $4.9 \pm 1.0\,\mu M$ in bladder and $1.1 \pm 0.3\,\mu M$ in femoral artery. While the initial amplitude of the first component is sensitive to MgADP in both types of muscle, in the bladder the overall relaxation rate (reflected in the $t_{1/2}$ values) is insensitive to concentrations of MgADP up to 200 μM; this finding contrasts with the marked sensitivity of the femoral artery muscle, in which the $t_{1/2}$ value increased from 18 s to 50 s. Thus there is a dramatic difference between the effects of MgADP on the overall relaxation rate from rigor in the tonic, compared to the phasic, smooth muscle.

Examination of the cross-bridge cycle (Fig. 2) suggests two possible explanations. One possibility is that there are two populations of cross-bridges with different affinities for MgADP, and the photo-released caged ATP competes more successfully with the ADP bound to the heads of one population than the other, resulting in different rates of detachment. The femoral artery would then be expected to have a larger proportion of bridges with the slower ADP off-rate. The other possible mechanism, and in our view the more likely, through which ADP could slow the overall rate of relaxation is through inhibition of ADP release from cycling cross-bridges (reversal of step 7), thereby prolonging the lifetime of the strongly bound AM′.ADP force gener-

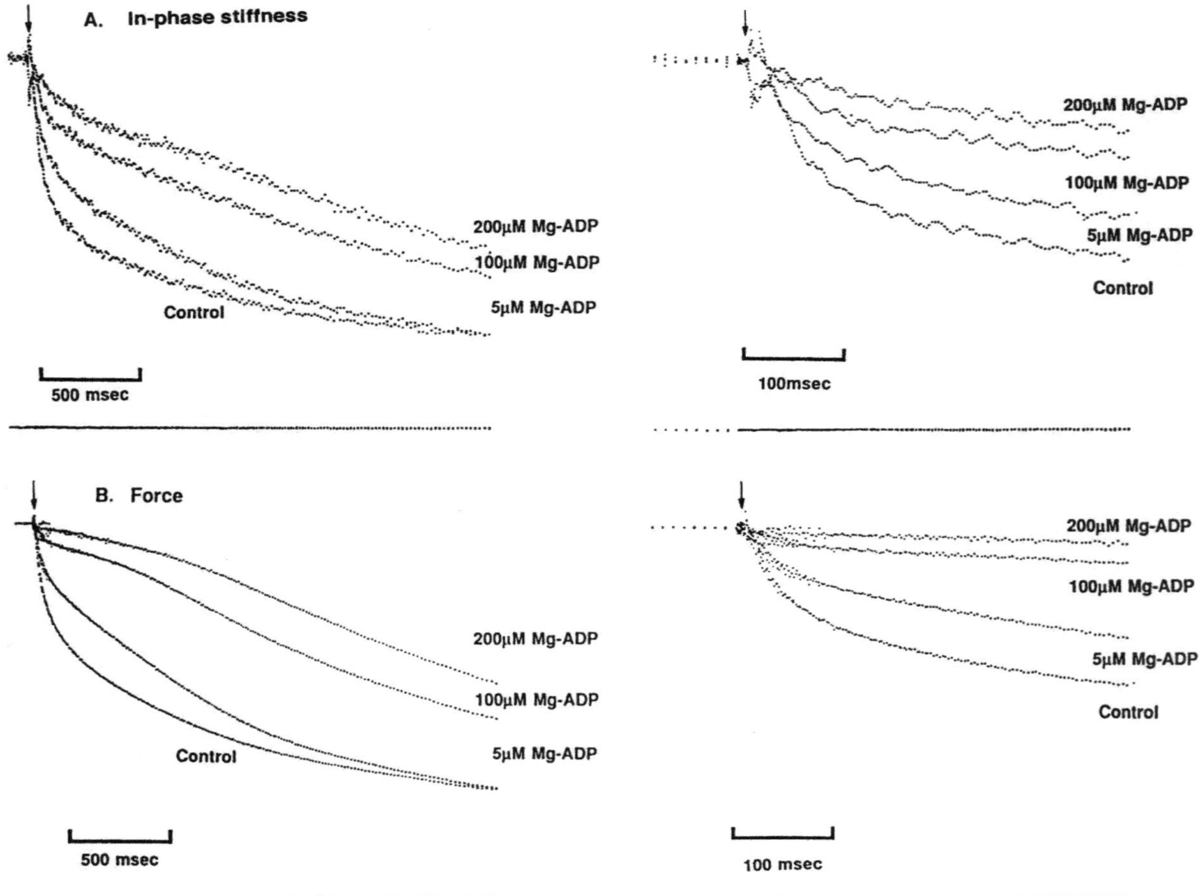


**Fig. 3.** Effect of MgADP on in-phase stiffness (**A**) and force (**B**) recordings after photolysis of caged ATP. After incubation in caged ATP-containing solution with various concentrations of MgADP, the laser was flashed at the point indicated by the *arrows*. Bottom lines show relaxed baseline obtained from steady level of tension after photolysis. (From [7], with permission)

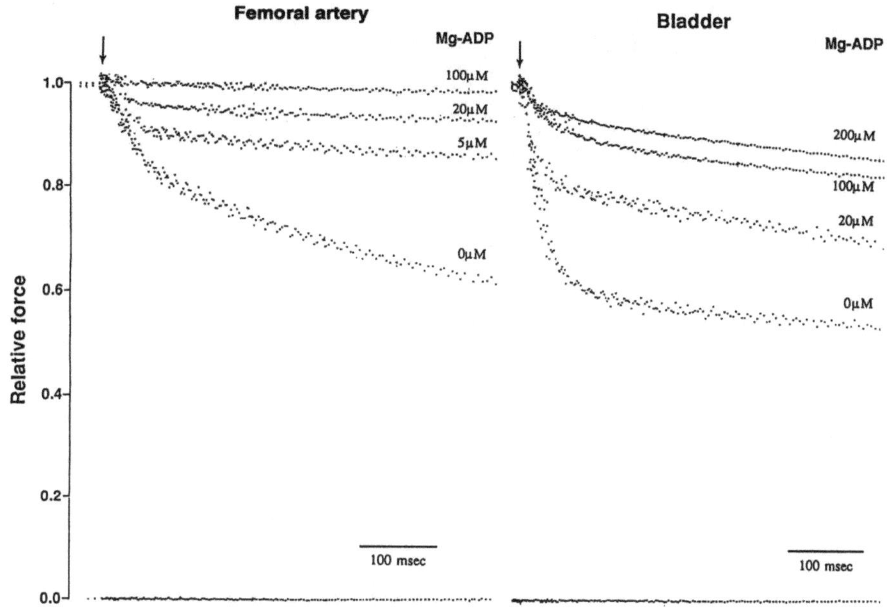


**Fig. 4.** Effect of MgADP on force recordings after photolysis of caged ATP in phasic rabbit bladder and tonic femoral artery smooth muscle. After incubation in caged ATP-containing solution with various concentrations of MgADP, the laser was flashed; relaxation from rigor was observed (representative of two to six experiments; individual traces are from different experiments). Note that although MgADP slowed relaxation in both types of smooth muscle, it affected both the amplitude and the slope of the slow phase of relaxation in the femoral artery, but only the amplitude in bladder. (From [8], with permission)

ating states. The affinity of the latter state may not be as high as that of the nucleotide-free rigor cross-bridge (1–5 μM in both phasic and tonic smooth muscle) but is clearly well below 100 μM in tonic, but not phasic, smooth muscle. The higher affinity of cycling (nonrigor) cross-bridges in tonic smooth muscle for MgADP would slow product release from a force-bearing AM.ADP state and so result in a larger population of strongly bound heads. Similar interpretations of the effects of MgADP on relaxation from rigor have been suggested for skeletal muscle [21] as well as for smooth muscle [7, 12, 22]. One group [23] concluded that in skeletal muscle the predominant effect of MgADP is through competition with ATP for binding to the AM state. The predominant effect of MgADP on the amplitude of the initial fast phase in the bladder may be interpreted in a similar fashion.

Inorganic phosphate ($P_i$) markedly accelerates the relaxation rate in both femoral artery and bladder smooth muscle (Fig. 5), presumably through inhibition of step 5 ($P_i$ release) and reversal of step 4 (attachment) in Figure 2, thereby increasing the population of weakly bound states. In the presence of 30 mM $P_i$ and treatment with apyrase to deplete endogenous ADP, the rate constants of the slow components were similar in both types of smooth muscle (phasic and tonic) (Fig. 5), and $P_i$ was shown to antagonize the effects of MgADP. Therefore the difference in the rate constants of the slow component of the two types of muscle is probably largely due to the ability of MgADP to increase the population of bridges in the strongly bound state in the tonic muscles and the enhancement of cooperative reattachment [12; review: 24) of cross-bridges. Based on this argument, the slow phase of relaxation in latch, which is also

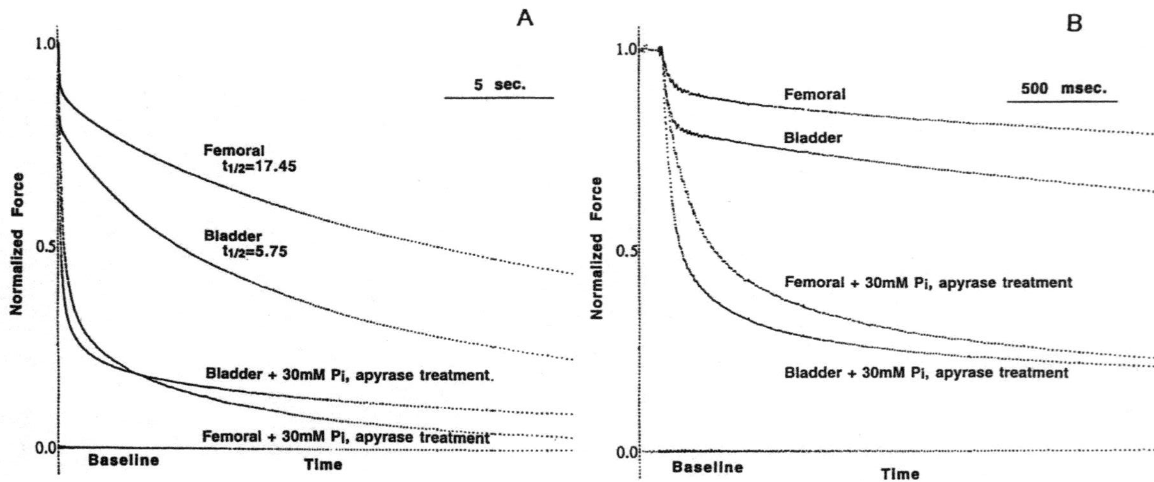


**Fig. 5.** Effect of 40 min of treatment with apyrase and 30 mM Pi + 20 mM CP on force after photolysis of caged ATP in rabbit femoral artery and rabbit bladder. Note that the **A** and **B** panels show identical force traces at different time scales. Bottom lines show the relaxed baseline obtained from a steady level of tension after photolysis. (From [8], with permission)

accelerated by $P_i$, is at least partially due to continued, perhaps cooperative, reattachment of cross-bridges into force-bearing states [7, 8, 12]. High concentrations of, respectively, MgADP and MgADPβS can cause force development in the absence of $MLC_{20}$ phosphorylation [25] and a left shift of the force–$Ca^{2+}$ relation [26] in permeabilized muscle. We have suggested that the combination of cooperative attachment of nonphosphorylated cross-bridges and the increases in the lifetime of strongly bound AM.ADP cross-bridges by MgADP play a significant role in the maintenance of the "latch" state [7, 8, 12]. The longer lifetime of the strongly bound attached state is compatible with the high force, low energy cost and low shortening velocity characteristic of latch. Experiments in progress, designed to examine the role of MgADP on cycling cross-bridges in α-toxin-permeabilized muscles, indicate that MgADP also slows the rate of relaxation under physiologic (nonrigor) conditions [27].

## Regulation in Phasic and Tonic Smooth Muscle

The overall rate-limiting step in force development after a rise in cytosolic $Ca^{2+}$ is the rate of $MLC_{20}$ phosphorylation [11, 12, 28]. The rate and magnitude of $MLC_{20}$ phosphorylation obviously depend on the sum of the MLCK and the myosin light chain phosphatase activities. These activities are higher in fast phasic muscle than in tonic smooth muscle, and the sensitivity to each of three phosphatase inhibitors is significantly higher in the femoral artery smooth muscle than in ileum smooth muscle. Each of the phosphatase inhibitors (microcystin-LR > tautomycin > okadaic acid) cause contraction and increase 20-kDa $MLC_{20}$ phosphorylation in $Ca^{2+}$-free solutions in both phasic and tonic smooth muscle permeabilized with β-escin, and they inhibit the heavy meromyosin (HMM) phosphatase activity of smooth muscle homogenates [29]. When assayed at high enzyme dilution, the potencies of inhibitors are identical for homogenates of ileum and femoral artery, suggesting that the respective phosphatases are not qualitatively, only quantitatively, different, and that the lower $MLC_{20}$ phosphatase actively reflects the lower enzyme content of tonic smooth muscles. The lower phosphatase activity of the tonic smooth muscle was also verified by the finding that the half-times of relaxation and dephosphorylation induced by exposing contracted muscles to a relaxing solution containing the protein kinase inhibitor ML-9 and 10 mM EGTA are four to six times longer in the femoral artery than in the phasic ileum or portal vein smooth muscle (Fig. 6). These differences were found to be present in β-escin-permeabilized as well as in α-toxin-permeabilized smooth muscles in which endogenous calmodulin and the intrinsic kinase and phosphatase were retained in the cells. Indeed, the HMM phosphatase and $MLC_{20}$ kinase activity/smooth muscle cell weight were approximately twofold lower in the tonic femoral artery than in the phasic ileum or portal vein smooth muscle. This higher activity of both $MLC_{20}$ kinase and $MLC_{20}$ phosphatase in the phasic smooth muscle is consistent with its more rapid force development (Table 1) [10]. The higher kinase activity results in higher initial phosphorylation rates and activation of actomyosin ATPase, whereas the high phosphatase activity results in the lower steady-state values of $MLC_{20}$ phosphorylation and force. On the other hand, when removal of [$Ca^{2+}$] turns off MLCK, the decline in $MLC_{20}$ phosphorylation is determined by the phosphatase activity, which is higher in the ileum, accounting for its faster rates of dephosphorylation. Relaxation is generally preceded by dephosphorylation; and the addition of exogenous phosphatase to contracted

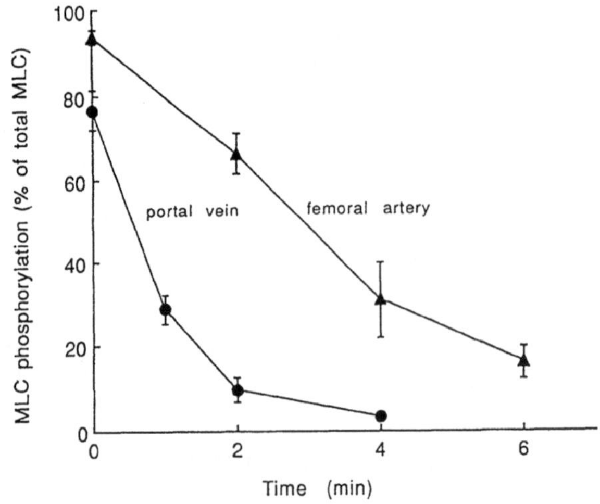


**Fig. 6.** Time course of myosin light chain (MLC) dephosphorylation in α-toxin-permeabilized rabbit femoral artery and portal vein smooth muscle at 15°C. pCa5-activated, α-toxin-permeabilized rabbit femoral artery and portal vein strips were frozen after incubation in pCa > 8, ATP-free solution containing 100 μM ML-9 and 10 mM EGTA for 1, 2, 4, or 6 min. Phosphorylation levels were expressed as percent of total $MLC_{20}$. (From [29], with permission)

permeabilized muscles can cause relaxation [9, 30–32]. The delay between $MLC_{20}$ dephosphorylation and relaxation reflects other processes, such as cooperative reattachment of nonphosphorylated bridges, prolongation of the lifetimes of strongly bound dephosphorylated AM.ADP bridges by high-affinity MgADP binding, and, possibly, contributions from thin filament-associated proteins.

## Myosin Light Chain and Heavy Chain Isoforms

The molecular basis of the differences in rates of contraction and relaxation in phasic and tonic smooth muscles may reside in differences in the light chain and heavy chain isoforms and their combinations [review: 33]. The concentration of the more basic isoform, $LC_{17b}$, was found to be 58% in the tonic femoral artery, which also had the highest affinity for MgADP, whereas the phasic bladder smooth muscle, which has a much lower affinity for MgADP, contained predominantly the acidic $LC_{17a}$ isoform (Fig. 7). Muscles with the $LC_{17a}$ isoform have higher shortening velocities [34], rates of force development [10], and ATPase activity [35]. The trachealis smooth muscle is the one outlier in this correlation [34]. Thus $LC_{17}$ isoforms may not be the sole determinant of the differences in phasic and tonic muscles, and recent reports suggest that the presence of a 7-amino-acid insert near the ATP binding site in the smooth muscle myosin heavy chain is correlated with high myosin ATPase activity [36–38]. It is likely, as suggested elsewhere [33], that the fastest smooth muscles are equipped with myosins containing both the heavy chain insert and the $LC_{17a}$ light chain, each of which may exert its effects through accelerating the ADP off-rate (decreasing the affinity for MgADP) from cycling cross-bridges.

*Acknowledgments.* We gratefully acknowledge the expertise Barbara Nordin lent to the preparation of the manuscript. We thank Mary Alice Spina for technical assistance

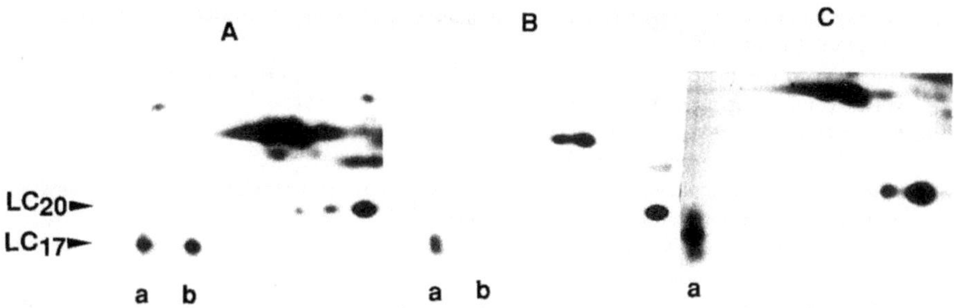


**Fig. 7.** Isoelectric focusing/sodium dodecyl sulfate polyacrylamide gel electrophoresis (SDS-PAGE) of whole muscle extracts of rabbit femoral artery (**A**), guinea pig portal vein (**B**), and rabbit bladder (**C**). $LC_{20}$, 20-kDa myosin light chain; $LC_{17a}$, $LC_{17b}$, two isoforms of the 17-kDa myosin light chain. (From [8], with permission)

and preparation of the illustrations. This work was supported by National Institutes of Health grant P01-HL19242-16.

## References


1. Somlyo AP, Somlyo AV (1968) Vascular smooth muscle. I. Normal structure, pathology, biochemistry and biophysics. Pharmacol Rev 20:197–272
2. Kitazawa T, Kobayashi S, Horiuti K, Somlyo AV, Somlyo AP (1989) Receptor-coupled, permeabilized smooth muscle: role of the phosphatidylinositol cascade, G-proteins and modulation of the contractile response. J Biol Chem 264:5339–5342
3. Kobayashi S, Kitazawa T, Somlyo AV, Somlyo AP (1989) Cytosolic heparin inhibits muscarinic and adrenergic $Ca^{2+}$-release in smooth muscle. J Biol Chem 264:17997–18004
4. Kaplan JH, Forbush III B, Hoffman JF (1978) Rapid photolytic release of adenosine 5′-triphosphate from a protected analog: utilization by the Na:K pump of human red blood cell ghosts. Biochemistry 17:1929–1935
5. Goldman YE, Hibbered MG, Trentham DR (1984) Relaxation of rabbit psoas muscle fibres from rigor by photochemical generation of adenosine-5′-triphosphate. J Physiol (Lond) 354: 577–604
6. Somlyo AP, Somlyo AV (1990) Flash photolysis studies of excitation-contraction coupling, regulation, and contraction in smooth muscle. Annu Rev Physiol 52:857–874
7. Nishiye E, Somlyo AV, Török K, Somlyo AP (1993) The effects of MgADP on cross-bridge kinetics: a laser flash photolysis study of guinea-pigs smooth muscle. J Physiol (Lond) 460:24–271
8. Fuglsang A, Khromov A, Török K, Somlyo AV, Somlyo AP (1993) Flash photolysis studies of relaxation and cross-bridge detachment: higher sensitivity of tonic than phasic smooth muscle to MgADP. J Muscle Res Cell Motil 14:666–673
9. Hartshorne DJ (1987) Biochemistry of the contractile process in smooth muscle. In: Johnson LR (ed) Physiology of the gastrointestinal tract, 2nd edn. Raven, New York, pp 423–482
10. Horiuti K, Somlyo AV, Goldman YE, Somlyo AP (1989) Kinetics of contraction initiated by flash photolysis of caged adenosine triphosphate in tonic and phasic smooth muscle. J Gen Physiol 94:769–781
11. Kamm KE, Stull JT (1986) Activation of smooth muscle contraction: relation between myosin phosphorylation and stiffness. Science 232:80–82
12. Somlyo AV, Goldman YE, Fujimori T, Bond M, Trenham DR, Somlyo AP (1988) Cross-bridge kinetics: cooperativity and negatively strained cross-bridges in vertebrate smooth muscle. J Gen Physol 91:165–192
13. Lymn RW, Taylor EW (1971) Mechanisms of adenosine triphosphate hydrolysis by actomyosin. Biochemistry 10:4617–4624

14. Eisenberg E, Green LE (1980) The relation of muscle biochemistry to muscle physiology. Annu Rev Physiol 42:293–309
15. Hibberd MG, Trentham DR (1986) Relationships between chemical and mechanical events during muscle contraction. Annu Rev Biophys Biophys Chem 15:119–161
16. Marston SB, Taylor EW (1980) Comparison of the myosin and actomyosin ATPase mechanisms of the four types of vertebrate muscles. J Mol Biol 139:573–600
17. Sellers JR (1985) Mechanism of the phosphorylation-dependent regulation of smooth muscle heavy meromyosin. J Biol Chem 260:15815–15819
18. Butler TM, Siegman MJ, Mooers SU, Narayan SR (1990) Myosin-product complex in the resting state and during relaxation of smooth muscle. Am J Physiol 258:C1092–C1099
19. Siegman MJ, Butler TM, Mooers SU, Davies RE (1980) Chemical energetics of force development, force maintenance, and relaxation in mammalian smooth muscle. J Gen Physiol 76:609–629
20. Rosenfeld SS, Taylor EW (1984) The ATPase mechanism of skeletal and smooth muscle acto-subfragment 1. J Biol Chem 259:11901–11919
21. Dantzig JA, Hibberd MG, Trentham DR, Goldman YE (1991) Cross-bridge kinetics in the presence of MgADP investigated by photolysis of caged ATP in rabbit psoas muscle fibers. J Physiol (Lond) 432:639–680
22. Arner A, Hellstrand P, Rüegg JC (1987) Influence of ATP, ADP and AMPPNP on the energetics of contraction in skinned smooth muscle. Prog Clin Biochem Res 245:43–57
23. Lu Z, Moss RL, Walker JW (1993) Tension transients initiated by photogeneration of MgADP in skinned skeletal muscle fibers. J Gen Physiol 101:867–888
24. Somlyo AP, Kitazawa T, Kobayashi S, Gong MC, Somlyo AV (1991) Pharmacomechanical coupling: the membranes talk to the cross-bridges. In: Moreland RS (ed) Regulation of smooth muscle contraction, vol 304. Plenum, New York, pp 185–208
25. Kerrick WGL, Hoar PE (1987) Non-$Ca^{2+}$-activated contraction in smooth muscle. Prog Clin Biol Res 245:437–448
26. Nishimura J, van Breemen C (1989) Possible involvement of actomyosin ADP complex in regulation of $Ca^{2+}$-sensitivity in $\alpha$-toxin-permeabilized smooth muscle. Biochem Biophys Res Commun 165:408–415
27. Khromov A, Somlyo AV, Somlyo AP (1994) The differential effect of MgADP on relaxation from isometric contraction by photolysis of Diazo 2 in tonic and phasic smooth muscle. Biophys J 66:A411
28. Yagi S, Becker L, Fay FS (1988) Relationship between force and $Ca^{2+}$ concentration in smooth muscle as revealed by measurements on single cells. Proc Natl Acad Sci USA 85:4109–4113
29. Gong MC, Cohen P, Kitazawa T, Ikebe M, Masuo M, Somlyo AP, Somlyo AV (1992) Myosin light chain phosphatase activities and the effects of phosphatase inhibitors in tonic and phasic smooth muscle. J Biol Chem 267:14662–14668
30. Driska SP, Stein PG, Porter R (1989) Myosin dephosphorylation during rapid relaxation of hog carotid artery smooth muscle. Am J Physiol 256:C315–C321
31. Haeberle JR, Hathaway DR, DePaoli-Roach AA (1985) Dephosphorylation of myosin by the catalytic subunit of type-2 phosphatase produces relaxation of chemically skinned uterine smooth muscle. J Biol Chem 260:9965–9968
32. Bialojan C, Rüegg JC, DiSalvo J (1987) A myosin phosphatase modulates contractility in skinned smooth muscle. Pflugers Arch 410:304–312
33. Somlyo AP (1993) Myosin isoforms in smooth muscle: how may they affect function and structure? J Muscle Res Cell Motil 15:557–663
34. Malmqvist U, Arner A (1991) Correlation between myosin isoform composition of the 17 kDa myosin light chain and maximal shortening velocity in smooth muscle. Eur J Physiol 418:523–530
35. Hasegawa Y, Morita F (1992) Role of 17 kDa essential light chain isoforms of aorta smooth muscle myosin. J Biochem 111:804–809
36. Babij P (1993) Tissue-specific and developmentally regulated alternative splicing of a visceral isoform of smooth muscle myosin heavy chain. Nucleic Acids Res 21:1467–1471
37. Kelley CA, Takahashi M, Yu JH, Adelstein RS (1993) An insert of seven amino acids confers enzymatic differences between smooth muscle myosins from the intestines and vasculature. J Biol Chem 268:12848–12854
38. White S, Martin AE, Periasamy M (1993) Identification of a novel smooth muscle myosin heavy chain cDNA: isoform diversity in the S1 head region. Am J Physiol 264:C1252–C1258

# Role of Protein Kinase C in the G Protein-Mediated Increase in Myofilament $Ca^{2+}$ Sensitivity in Skinned Arterial Smooth Muscle of the Rabbit

Takeo Itoh[1], Yoshimasa Watanabe[1], Makoto Yoshida[2], and Akito Suzuki[2]

*Summary.* The role of protein kinase C in the increase mediated by GTP-binding proteins (G proteins) of myofilament $Ca^{2+}$ sensitivity in β-escin-skinned smooth muscle of the rabbit mesenteric artery has been investigated. The effect of a peptide inhibitor of protein kinase C ($PKC_{19-36}$) on $Ca^{2+}$-induced contraction and 20-kDa myosin light chain ($MLC_{20}$) phosphorylation was studied in the presence of both endothelin-1 (ET-1) with GTP and guanosine 5′-O-(3-thiotriphosphate) (GTPγS) alone. For comparison, we observed the effects of $PKC_{19-36}$ on the same two $Ca^{2+}$-induced responses in the presence of phorbol-12,13-dibutylate (PDBu, an activator of PKC). In skinned strips treated with ionomycin, $Ca^{2+}$(0.1–3.0μM) produced contraction in parallel with an increase in $MLC_{20}$ phosphorylation in a concentration-dependent manner. ET-1, GTPγS, and PDBu shifted both the $Ca^{2+}$–force and $Ca^{2+}$–$MLC_{20}$ phosphorylation relations to the left without a significant change in the maximum response induced by $Ca^{2+}$ in either variable. None of the agents modified the relation between force and $MLC_{20}$ phosphorylation, indicating that the sensitivity of $MCL_{20}$ phosphorylation to $Ca^{2+}$ is enhanced by ET-1, GTPγS, and PDBu. $PKC_{19-36}$ itself modified neither the contraction nor the $MLC_{20}$ phosphorylation induced by $Ca^{2+}$, but it did block the PDBu-induced enhancement of these two $Ca^{2+}$-induced responses. By contrast, $PKC_{19-36}$ did not modify, or only slightly attenuated, the enhancement of the two $Ca^{2+}$-induced responses induced by GTPγS or ET-1. Guanosine 5′-O-(2-thiodiphosphate) (GDPβS) greatly attenuated the enhancement of the $Ca^{2+}$-induced contraction induced by ET-1 and GTPγS separately but not that induced by PDBu. These results suggest that PKC does not play a major role on the G protein-mediated increase in myofilament $Ca^{2+}$ sensitivity in β-escin-skinned smooth muscle of the rabbit mesenteric artery.

*Key Words.* G proteins—Protein kinase C—Myosin phosphorylation—Vascular smooth muscle

## Introduction

A $Ca^{2+}$-mobilizing hormone (agonist) binds to its receptor and promotes hydrolysis of phosphatidylinositol 4,5-bisphosphate ($PIP_2$) through activation of heterotrimeric

[1] Department of Pharmacology, Nagoya City University Medical School, 1 Kawasumi, Mizuho-cho, Mizuho-ku, Nagoya, 467 Japan

[2] Department of Pharmacology, Faculty of Medicine, Kyushu University, Fukuoka, 812 Japan

($\alpha\beta\gamma$) G proteins; it achieves this action by catalyzing the replacement of the guanosine 5′-diphosphate (GDP) previously bound to the $\alpha$-subunit by guanosine 5′-triphosphate (GTP). Hydrolysis of $PIP_2$ is followed by the production of inositol 1,4,5-trisphosphate ($IP_3$) and diacylglycerol (DAG). The former releases $Ca^{2+}$ from the cellular storage sites, and the latter activates protein kinase C (PKC).

Endothelin is one of a family of peptides that includes endothelin-1 (ET-1), endothelin-2, endothelin-3, and sarafotoxin; ET-1 is known to be one of the most potent naturally occurring vasoconstrictor agents [1, 2]. The immunoreactive concentration of ET-1 is increased in response to various types of stimuli, such as shear stress, hypoxia, and growth factors. Like other $Ca^{2+}$-mobilizing agonists, ET-1 binds to its receptor and then promotes the hydrolysis of $PIP_2$ through activation of heterotrimeric G proteins, forming $IP_3$ and DAG [3–6]. Guanosine 5′-O-(3-thiotriphosphate) (GTP$\gamma$S) itself also activates heterotrimeric G proteins and thus increases the activity of phospholipase C, causing production of $IP_3$ and DAG in vascular smooth muscle [7].

It has been found that in skinned smooth muscle agonist treatment with GTP or GTP$\gamma$S increases the contraction, the phosphorylation of myosin light chain ($MLC_{20}$), and the unloaded shortening velocity induced by low concentrations of $Ca^{2+}$ through activation of unidentified G proteins [8–17]. Phorbol esters, potent activators of PKC, also increase the contraction, $MLC_{20}$ phosphorylation, and unloaded shortening velocity induced by low concentrations of $Ca^{2+}$ in skinned smooth muscle [18–20]. Because of the similarity of action of GTP$\gamma$S and phorbol esters, it has been suggested that PKC, activated by an application of agonist or GTP$\gamma$S, plays a primary role in the G protein-mediated increase in myofilament $Ca^{2+}$ sensitivity in vascular smooth muscle [6, 8]. However, evidence to the contrary has also been reported [21, 22]. Thus the contribution of PKC to the G protein-mediated increase in the myofilament $Ca^{2+}$ sensitivity in vascular smooth muscle needs further examination.

The peptide inhibitor $PKC_{19-36}$, which corresponds to the autoinhibitory region of PKC, has been recently introduced and shown to selectively inhibit PKC [23, 24]. To investigate the contribution of PKC to the mechanisms by which G proteins mediate enhancement of the contraction induced by $Ca^{2+}$, we have studied the effects of $PKC_{19-36}$ on $Ca^{2+}$-induced contraction and $MLC_{20}$ phosphorylation in the presence of ET-1 with GTP or of GTP$\gamma$S in ionomycin-treated, $\beta$-escin-skinned smooth muscle. For comparison, we also studied the effects of $PKC_{19-36}$ on the action of PDBu (an activator of PKC) on the same $Ca^{2+}$-induced responses.

## Methods

Male albino rabbits weighing 1.9–2.5 kg were anesthetized with pentobarbitone sodium (40 mg/kg IV) and then exsanguinated. A segment of the third branch of the mesenteric artery distributing to the region of the ileum (diameter approximately 70 μm) was excised immediately and cleaned by removal of connective tissue in Krebs solution at room temperature.

### Simultaneous Measurement of $[Ca^{2+}]_i$ and Isometric Force

To enable recording of $[Ca^{2+}]_i$ and isometric force simultaneously, fine circularly cut strips (0.3–0.5 mm length, 0.04–0.05 mm width, 0.02–0.03 mm thickness) were prepared as described elsewhere [25]. Endothelial cells were removed by gently rubbing the internal surface of the vessel using small knives. The strip was mounted horizontally on an invert-microscope (Diaphot TMD with special optics for epifluoresence; Nikon, Tokyo, Japan), and the resting force was adjusted so as to obtain maximal contraction in 128 mM $K^+$.

To load fura-2 into smooth muscle cells of the strip, 1 μM acetoxymethyl ester of fura-2 (fura-2AM) dissolved by dry dimethylsulfoxide (1 mM stock solution) was used. Two alternative excitation wavelengths, 340 nm and 380 nm (each slit 5 nm), were applied by a spectrofluorometer (Spex, Edison, NJ, USA) and the data analyzed using customized software (Spex, DM-3000CM). The ratio of the fura-2 fluorescence intensities excited by 340 and 380 nm was calculated after subtracting the background fluorescence, as reported elsewhere [12]. Cytosolic $Ca^{2+}$ concentrations were calculated using the formula described by Grynkiewicz et al. [26] and using in vitro calibration [12, 27].

## Chemically Skinned Smooth Muscle

Chemically skinned smooth muscle strips were prepared using saponin [25, 28] or β-escin [12]. When the $Ca^{2+}$-force relation was to be determined, the concentration of EGTA in the solution was 4 mM, and 1 μM ionomycin or 3 μM A23187 was applied to avoid spurious effects due to $Ca^{2+}$ release from intracellular storage sites in the skinned muscle. To prevent deterioration of the $Ca^{2+}$-induced contraction, 0.1 μM calmodulin was present throughout the experiments [25, 28]. Various concentrations of $Ca^{2+}$ were cumulatively applied, from low to high concentration. The amplitudes of the contractions induced by these concentrations of $Ca^{2+}$ were normalized with respect to that induced by 10 μM $Ca^{2+}$ in the same strip.

## Measurements of Shortening Velocity

The shortening velocity ($V_{max}$) of the skinned muscle strip was determined using the "slack test" [29, 30]. A contraction was initiated by application of any desired concentration of $Ca^{2+}$. After the tension had reached a plateau, the strip was slackened to enable measurement of the time from length decrease to force redevelopment ($T$). When the strip was then slowly restretched to the initial length ($L_0$), 2 min was required to reach a tension identical to that which had developed before the slackening procedure. The slack test was imposed every 4 min on a single contraction to determine the shortening velocity under unloaded conditions. The relation between $T$ and the relative length change—ratio between the different length steps ($L$) and $L_0$; $L/L_0$—was linear for any tested concentration of $Ca^{2+}$ (0.3–10.0 μM). Therefore $V_{max}$ was calculated from the slope of the relation between $L/L_0$ and $T$; and the immediate elastic recoil was estimated from the $y$-intercept of the line, which was fitted to the data using the least-squares method. Data from experiments in which the coefficient of correlation between $L$ and $T$ was less than 0.9 were excluded from the analysis [14, 20].

## Measurements of Myosin Light Chain Phosphorylation

Muscle strips (0.2–0.3 mm length, 1.2 mm width, 0.02–0.03 mm thickness) were skinned using a relaxing solution containing β-escin 30 μM with ionomycin 1 μM or saponin 25 μg/ml with A23187 3 μM for 20–25 min and washed again with the relaxing solution (containing no detergent). Each skinned muscle strip was then suspended in a solution containing one of several concentrations of $Ca^{2+}$ for 15 min in the presence or absence of ET-1 with GTP, GTPγS alone, or PDBu alone. The strips were quick-frozen with 10% trichloroacetic acid (TCA) in acetone-dry ice containing 10 mM dithiothreitol (DTT) and allowed to reach room temperature. They were then washed three times with acetone to remove residual TCA. Protein, including myosin light chain ($MLC_{20}$), was extracted in a lysing solution containing (1) 1% sodium dodecyl sulfate (SDS), 10% glycerol, and 20 mM DTT for two-dimensional gel electrophoresis [14, 20, 31]; or (2) 8 M urea, 20 mM Tris, 23 mM glycine, 10 mM DTT, 0.004% bromophenol blue, and saturated sucrose (pH 8.6) for immunoblot analysis [32, 33].

In some experiments (e.g., that described in Fig. 6), two-dimensional gel electrophoresis was used to measure the relative amounts of phosphorylation of $MLC_{20}$. In most of the experiments, nonphosphorylated, monophosphorylated, and diphosphorylated forms of $MLC_{20}$ were measured using an immunoblot procedure, as reported elsewhere [32–35]. Antibody against bovine tracheal $MLC_{20}$ was used as a first antibody and biotin-labeled goat anti-rabbit IgG (Histofine; Seikagaku Kogyo, Tokyo, Japan) as a second antibody. Alkaline phosphatase-labeled streptavidine was then used. Relative amounts of nonphosphorylated, monophosphorylated, and diphosphorylated $MLC_{20}$ were quantified densitometrically. The phosphorylation of $MLC_{20}$ was expressed in moles of $PO_4$ per mole of $MLC_{20}$.

## Calculation of Hill Coefficient

The slope of the concentration–response relation for the effect of $Ca^{2+}$ on force and $MLC_{20}$ phosphorylation is shown as the Hill coefficient ($N$) and the midpoint position: $pK = -\log K$, where $K$ is the dissociation constant. These figures were obtained by fitting the data points for each curve to Eq. (1) by a nonlinear least-squares method.

$$F/F_0 = (C/K)^N/[1 + (C/K)^N] \tag{1}$$

where $C$ represents the concentration of $Ca^{2+}$, $F$ is the amplitude of contraction at any given concentration of $Ca^{2+}$, and $F_0$ is the maximum response evoked by 10 μM $Ca^{2+}$ expressed as a relative force of 1.0.

## Solutions

The ionic composition of the Krebs solution was as follows (in millimolars): $Na^+$ 137.5; $K^+$ 5.9; $Mg^{2+}$ 1.2; $Ca^{2+}$ 2.6; $HCO_3^-$ 15.5; $H_2PO_4^-$ 1.2; $Cl^-$ 134.3; glucose 11.5. The concentration of $K^+$ was modified as required by replacing NaCl with KCl, isosmotically. $Ca^{2+}$-free Krebs solution was prepared by substituting an equimolar concentration of $MgCl_2$ for $CaCl_2$ and adding 2 mM EGTA. The solutions were bubbled with 95% $O_2$ and 5% $CO_2$; their pH was maintained at 7.3–7.4.

For experiments on skinned muscle, the composition of the relaxing solution was 87 mM potassium methanesulfonate (KMS), 20 mM piperazine-*N*-*N'*-bis-(2-ethanesulfonic acid) (PIPES), 5.1 mM $Mg(MS)_2$, 5.2 mM ATP, 5 mM phosphocreatine, and 4 mM ethyleneglycol-bis-(β-aminoethyl)-*N,N,N',N'*-tetraacetic acid (EGTA). Various $Ca^{2+}$ concentrations were prepared by adding appropriate amounts of $Ca(MS)_2$ to 4 mM EGTA based on the calculation reported previously [28]. The pH of the solution was adjusted to 7.1 at 25°C with KOH, and the ionic strength was standardized at 0.2 M by changing the amount of KMS added.

## Drugs

Drugs used were ET-1 (Peptide Institute, Osaka, Japan); fura-2, fura-2AM, EGTA, PIPES, and HEPES (Dojin, Kumamoto, Japan); $PKC_{19-36}$ (Peninsula Laboratories, Belmont, CA, USA); calmodulin, NA, and β-escin (Sigma, St. Louis, MO, USA); saponin (ICN, Costa Mesa, CA, USA); GTP, guanosine 5′-O-(2-thiodiphosphate) (GDPβS) and GTPγS (Boehringer Mannheim, Mannheim, Germany); ATP, Na salt (Kohjin, Tokyo, Japan); 12-*o*-tetradecanoylphorbol-13-acetate (PMA) and PDBu (Wako Pure Chemical, Osaka, Japan); propranolol (Nacalai, Kyoto, Japan); phosphatidylserine (PS, beef brain) (Serdary Research Laboratories Cliffs, NJ, USA); and ionomycin (free acid) (Calbiochem, La Jolla, CA, USA).

## Statistics

The values recorded were expressed as means ± SD. The statistical significance was determined using a paired or unpaired Student's *t*-test. Probabilities of less than 5% ($P < 0.05$) were considered significant.

# Results

## Effect of ET-1 on Contraction at Given Concentrations of $[Ca^{2+}]_i$ in Intact Smooth Muscle

Figure 1 shows the effects of 128 mM $K^-$ and 10 nM ET 1 on $[Ca^{2+}]_i$ and force in a thin smooth muscle strip of rabbit mesenteric artery. The resting $[Ca^{2+}]_i$ and force were 118.4 ± 17.1 nM and 0.8 ± 0.3 mg, respectively. High $K^+$ produced a phasic increase followed by a tonic increase in both $[Ca^{2+}]_i$ and force, the maximum increases in $[Ca^{2+}]_i$ and force being 855.1 ± 216.8 nM and 13.1 ± 2.3 mg, respectively ($n = 4$). ET-1 slowly increased both $[Ca^{2+}]_i$ and force, the effects being maintained for more than 30 min. The maximum increases in $[Ca^{2+}]_i$ and force induced by ET-1 were 164.1 ± 18.3 nM and 8.3 ± 2.5 mg, respectively. When 1 μM nicardipine was applied during the ET-1-induced maintained contraction, the ET-1-induced $[Ca^{2+}]_i$ increase was completely blocked, but 10–20% of the ET-1-induced force remained. When the relation between $[Ca^{2+}]_i$ and force was determined during the tonic phase, at 2 min and 10 min after application of 128 mM $K^+$ and 10 nM ET-1, respectively, ET-1 produced a larger contraction with a smaller increase in $[Ca^{2+}]_i$ compared to the situation with high $K^+$ (Fig. 1).

To study further the effect of ET-1 on the $[Ca^{2+}]_i$-force relation, the effect was observed in a solution containing 70 mM $K^+$ in 0.1 μM ionomycin-treated strips, which had lost the function of the intracellular $Ca^{2+}$ storage sites [12, 35]. After application of ionomycin for 30 min in Krebs solution, $Ca^{2+}$-free solution containing 5.9 mM $K^+$ with 2 mM EGTA was then applied for 1 min. Subsequently, $Ca^{2+}$-free solution containing 70 mM $K^+$ with 2 mM EGTA was applied for 2 min. At this point, various concentrations of $Ca^{2+}$ (0.16–2.60 mM) were cumulatively applied from low to high in the presence of 70 mM $K^+$ (Fig. 2).

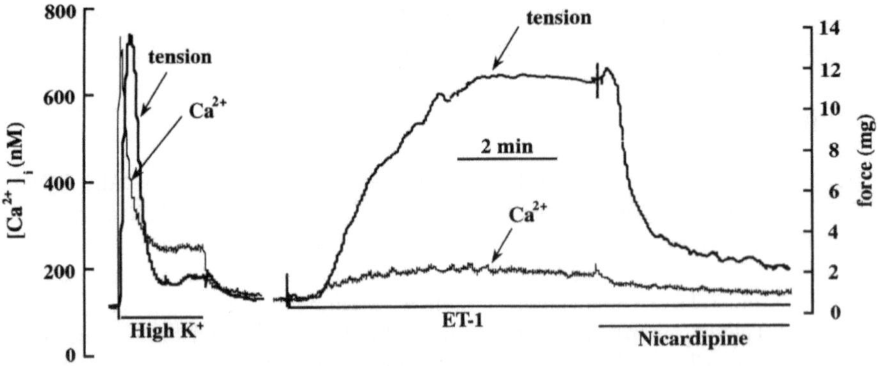


**Fig. 1.** Effects of high $K^+$ (128 mM) and 10 nM ET-1 on $[Ca^{2+}]_i$ and tension in a smooth muscle strip of rabbit mesenteric artery. Each agent was applied as indicated by the bar. *Thin lines*, $[Ca^{2+}]_i$; *thick lines*, tension. Nicardipine 1 μM was applied (as indicated by the *bar*) during the steady-state contraction induced by ET-1

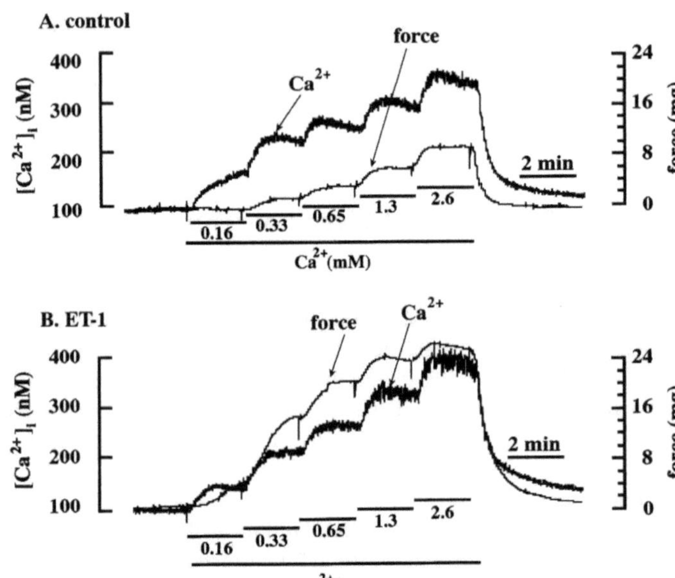


**Fig. 2.** Effects of ET-1 on the $[Ca^{2+}]_i$–force relation in an ionomycin-treated smooth muscle strip. The muscle strip was treated with 0.1 μM ionomycin for 30 min in Krebs solution. *Thick lines*, $[Ca^{2+}]_i$; *thin lines*, tension. $Ca^{2+}$-free solution containing 2 mM EGTA with 5.9 mM $K^+$ was applied for 1 min followed by a 2 min application of $Ca^{2+}$-free solution containing 2 mM EGTA with 70 mM $K^+$ in the presence or absence of 10 nM ET-1. Finally, solutions containing various concentrations of $Ca^{2+}$ (from 0.16 to 2.60 mM) with 70 mM $K^+$ were cumulatively applied, from low to high concentrations, in the presence (B) or absence (A) of 10 nM ET-1

After application of $Ca^{2+}$-free solution with 5.9 mM $K^+$ for 1 min, the resting $[Ca^{2+}]_i$ and force had decreased slightly to 94.9 ± 2.6 nM and 0.4 ± 0.3 mg, respectively. Subsequent application of $Ca^{2+}$-free solution with 70 mM $K^+$ increased neither the $[Ca^{2+}]_i$ (95.1 ± 2.4 nM) nor the force (0.4 ± 0.3 mg). ET-1 slightly reduced the increases in $[Ca^{2+}]_i$ induced by low concentrations of $Ca^{2+}$ (0.16 and 0.33 mM) but slightly enhanced the increase in $[Ca^{2+}]_i$ induced by higher concentrations (1.3 and 2.6 mM $Ca^{2+}$). ET-1 shifted the $[Ca^{2+}]_i$–force relation to the left and increased the maximum amplitude of contraction induced by 2.6 mM $Ca^{2+}$ (Fig. 2). The values of $[Ca^{2+}]_i$ for half-maximum contraction ($ED_{50}$) were 238.8 ± 55.0 nM and 168.7 ± 35.4 nM in the absence and presence of 10 nM ET-1, respectively.

## Effect of ET-1 with GTP and GTPγS Alone on $Ca^{2+}$-Induced Contraction with Mediation of GTP-Binding Proteins in Skinned Smooth Muscle

It has been reported that agonist with GTP and GTPγS alone both enhance the contraction induced by submaximal $Ca^{2+}$ in α-toxin- or β-escin-treated skinned smooth muscle [12, 15, 16, 35, 36]. To study the effects of ET-1 on contractile proteins more directly, its effects on the contraction induced by low concentration of $Ca^{2+}$ were observed in ionomycin-treated β-escin-skinned smooth muscle in which intracellular ion concentrations and pH were well regulated.

In the skinned muscle the minimum concentration of $Ca^{2+}$ needed to produce contraction was 0.1 μM, and the maximum contraction was obtained at 3 μM $Ca^{2+}$. ET-1 10 nM with GTP 100 μM and GTPγS 30 μM alone both enhanced the amplitude of the contraction induced by 0.3 μM $Ca^{2+}$ (Fig. 3). GDPβS 1 mM slightly attenuated the contraction induced by 0.3 μM $Ca^{2+}$ and greatly inhibited the enhancement of the $Ca^{2+}$ contraction induced by ET-1 (Fig. 3A) or GTPγS (Fig. 3C), suggesting that G proteins mediate the enhancement of the $Ca^{2+}$ contraction induced by both ET-1 and GTPγS separately. Neomycin interacts with polyphosphoinositides and inhibits the activation of phospholipase C which causes hydrolysis of phosphatidylinositol 4,5-bisphosphate [16, 37]. Neomycin 0.1 mM did not itself modify the contraction induced by 0.3 μM $Ca^{2+}$; however, it completely blocked the enhancement of the $Ca^{2+}$ contraction induced by ET-1 though not that induced by GTPγS (Fig. 3B). The extent of the GTPγS-induced enhancement of $Ca^{2+}$-induced contraction was similar whether neomycin was present or absent.

## Effect of ET-1, GTPγS, and PDBu on Contraction and $MLC_{20}$ Phosphorylation, Without a Change in Their Relation, in Skinned Smooth Muscle

We observed the effects of ET-1, GTPγS, and PDBu on the contraction and $MLC_{20}$ phosphorylation induced by various concentrations of $Ca^{2+}$ in ionomycin-treated β-escin-skinned smooth muscle strips (Fig. 4). ET-1, GTPγS and PDBu each shifted the $Ca^{2+}$–force relation curve to the left with no change in the maximum amplitude of contraction induced by 10 μM $Ca^{2+}$. The threshold concentration of $Ca^{2+}$ needed to produce contraction was slightly lower in the presence of ET-1, GTPγS, or PDBu (each 0.03 μM) than in the control (0.1 μM). The concentration of $Ca^{2+}$ required for half-maximum force ($ED_{50}$) was 0.61 ± 0.18 μM in controls, 0.25 ± 0.04 μM in the presence of ET-1 with GTP, 0.18 ± 0.13 μM in the presence of GTPγS, and 0.14 ± 0.07 μM in the presence of PDBu. The $ED_{50}$ value obtained in the presence of any of the agents was significantly smaller than that in the corresponding control

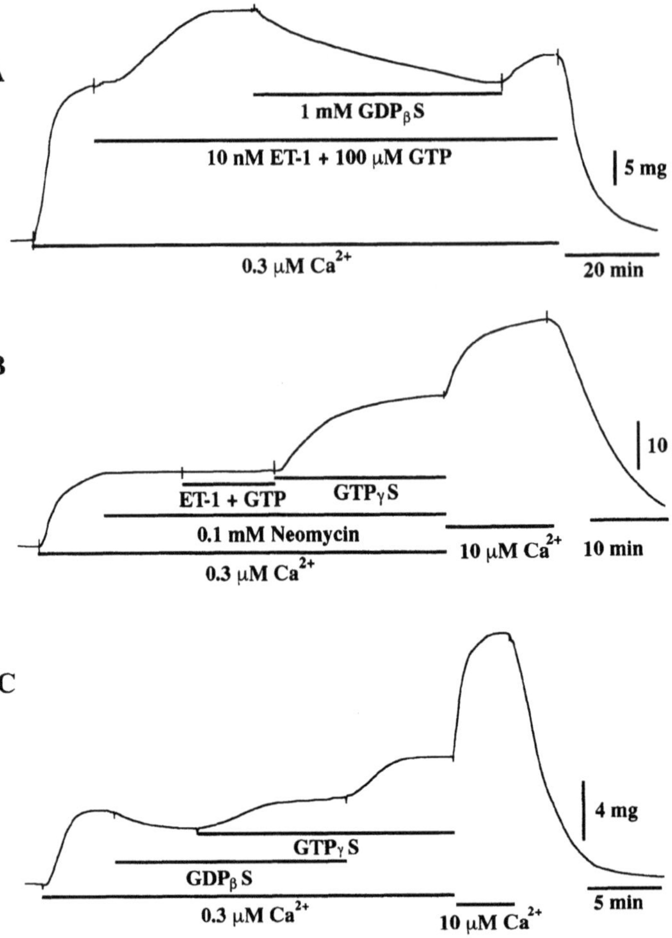


**Fig. 3.** Effects of ET-1 with GTP and of GTPγS alone on contraction induced by 0.3 μM $Ca^{2+}$ in β-escin-skinned smooth muscle strips of the rabbit mesenteric artery. After the muscle strip had been skinned by β-escin, 0.3 μM $Ca^{2+}$ was first applied. **A.** ET-1 10 nM with GTP 100 μM was applied on the steady-state contraction induced by 0.3 μM $Ca^{2+}$. GDPβS 1 mM was then applied in a solution containing 0.3 μM $Ca^{2+}$ and ET-1 with GTP. Each agent was applied as indicated by the bar. **B.** Neomycin 0.1 mM was applied during steady state of the $Ca^{2+}$-induced contraction; 10 nM ET-1 with 100 μM GTP was then applied in the presence of 0.3 μM $Ca^{2+}$ and 0.1 mM neomycin. GTPγS 10 μM was subsequently applied in a solution containing 0.3 μM $Ca^{2+}$ plus 0.1 mM neomycin. Finally, 10 μM $Ca^{2+}$ was applied alone to obtain the maximum $Ca^{2+}$-induced contraction for that strip. **C.** GDPβS (1 mM) was pretreated before application of 10 μM GTPγS in a solution containing 0.3 μM $Ca^{2+}$. GTPγS was then applied in a solution containing 0.3 μM $Ca^{2+}$ with or without GDPβS. Finally, 10 μM $Ca^{2+}$ was applied alone to obtain the maximum $Ca^{2+}$-induced contraction for that strip

($P < 0.05$). The Hill coefficient was 2.2 ± 0.3 in the control, 1.4 ± 0.2 in the presence of ET-1 with GTP, 1.3 ± 0.2 in the presence of GTPγS, and 1.6 ± 0.3 in the presence of PDBu.

Figure 4B shows the effects of ET-1, GTPγS, and PDBu on the $MLC_{20}$ phosphorylation induced by various concentrations of $Ca^{2+}$ (0.03–10.00 μM). In $Ca^{2+}$-free solution containing 4 mM EGTA, $MLC_{20}$ phosphorylation was 0.04 ± 0.04 mol

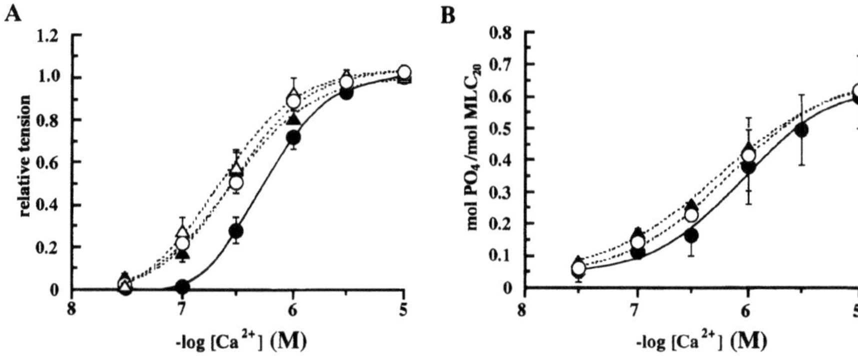


**Fig. 4.** Effects of ET-1, GTPγS, and PDBu on the relations between $Ca^{2+}$–tension (**A**) and $Ca^{2+}$–$MLC_{20}$ phosphorylation (**B**) in β-escin-skinned smooth muscle strips. $Ca^{2+}$ at any given concentration (0.03–10.00 μM) was applied in a relaxing solution in the absence (*closed circles*) or presence of 10 nM ET1 with 100 μM GTP (*open circles*) or 10 μM GTPγS alone (*closed triangles*) or 0.1 μM PDBu alone (*open triangles*). **A.** Maximum amplitude of contraction induced by 10 μM $Ca^{2+}$ was normalized as a relative force of 1.0 for each strip. The curves for the effects of $Ca^{2+}$ on force and $MLC_{20}$ phosphorylation were obtained by fitting the data points to Eq. (1) by a nonlinear least-squares method (see Methods). Results shown are each the mean of six observations, with the standard deviation shown by a vertical bar

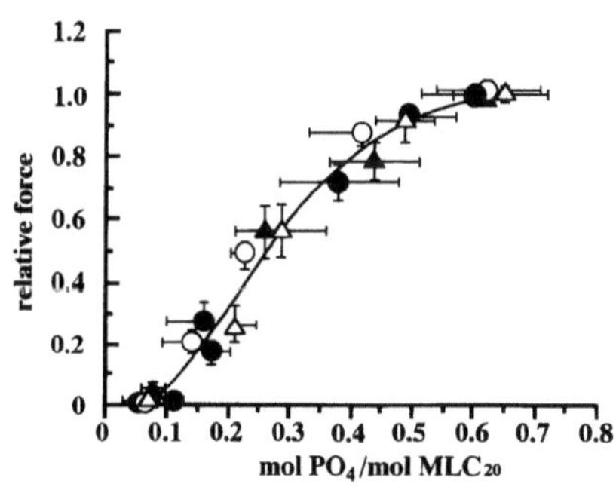


**Fig. 5.** Relation between relative force and $MLC_{20}$ phosphorylation in the presence of various concentrations of $Ca^{2+}$ (0.03–10.00 μM) with or without ET-1 with GTP, GTPγS alone, or PDBu alone in β-escin-skinned smooth muscle strips. *Closed circles*, control; *open circles*, in the presence of 10 nM ET-1 + 100 μM GTP; *closed triangles*, in 10 μM GTPγS; *open triangles*, in 0.1 μM PDBu. The curve was obtained by fitting the data points to Eq. (1) by a nonlinear least-squares method according to the equation: $F/F_0 = (C/K)^N/[1 + (C/K)^N]$, where $N$ is the Hill coefficient, $K$ is the dissociation constant, $C$ and $F/F_0$ represent $MLC_{20}$-phosphorylation (mol $PO_4$/mol $MLC_{20}$) and relative force, respectively. The fitted parameters of $N$ and $K$ were 2.50 and 0.28 mol $PO_4$/mol $MLC_{20}$, respectively ($r = 0.99$)

$PO_4$/mol $MLC_{20}$, and it was concentration-dependently increased by $Ca^{2+}$ (0.03–10.00 μM). Under these conditions, $Ca^{2+}$ (0.03–10.00 μM) mainly increased the monophosphorylated form of $MLC_{20}$ (>95%) whether in the presence or absence of ET-1, GTPγS, or PDBu. ET-1, GTPγS, and PDBu each shifted the relation between $Ca^{2+}$ and $MLC_{20}$ phosphorylation (monophosphorylated form) to the left without affecting the maximum level of $MLC_{20}$ phosphorylation obtained with 10 μM $Ca^{2+}$. The $ED_{50}$ values of $Ca^{2+}$ for the monophosphorylated form of $MLC_{20}$ were 0.88 ± 0.06 μM in controls, 0.63 ± 0.05 μM in the presence of ET-1 with GTP, 0.53 ± 0.08 μM in the presence of GTPγS, and 0.31 ± 0.08 μM in the presence of PDBu. The $ED_{50}$ value obtained in the presence of any of the agents was significantly smaller than that in controls ($P < 0.05$).

Figure 5 shows the relation between relative force and the monophosphyorylated form of $MLC_{20}$ in the presence of various concentrations of $Ca^{2+}$ with and without ET-1 plus GTP, GTPγS alone, or PDBu alone. None of the agents changed this relation.

## Effect of GTPγS and PMA on Contraction, $MLC_{20}$ Phosphorylation, and Shortening Velocity in Skinned Smooth Muscle

The effects of 10 μM GTPγS on the force, maximum shortening velocity ($V_{max}$), and phosphorylation of $MLC_{20}$ were observed in the presence of 0.3 μM $Ca^{2+}$ in saponin-treated skinned smooth muscle. In controls the amplitude of the contraction induced by 0.3 μM $Ca^{2+}$ was 0.33 ± 0.09 times that induced by 10 μM $Ca^{2+}$. The phosphorylation of $MLC_{20}$ was 0.05 ± 0.02 mol $PO_4$/mol $MLC_{20}$ in $Ca^{2+}$-free solution containing 4 mM EGTA, and it was increased to 0.19 ± 0.05 mol $PO_4$/mol $MLC_{20}$ in the presence of 0.3 μM $Ca^{2+}$ or to 0.51 ± 0.09 mol $PO_4$/mol $MLC_{20}$ in 1.0 μM $Ca^{2+}$. $V_{max}$ was

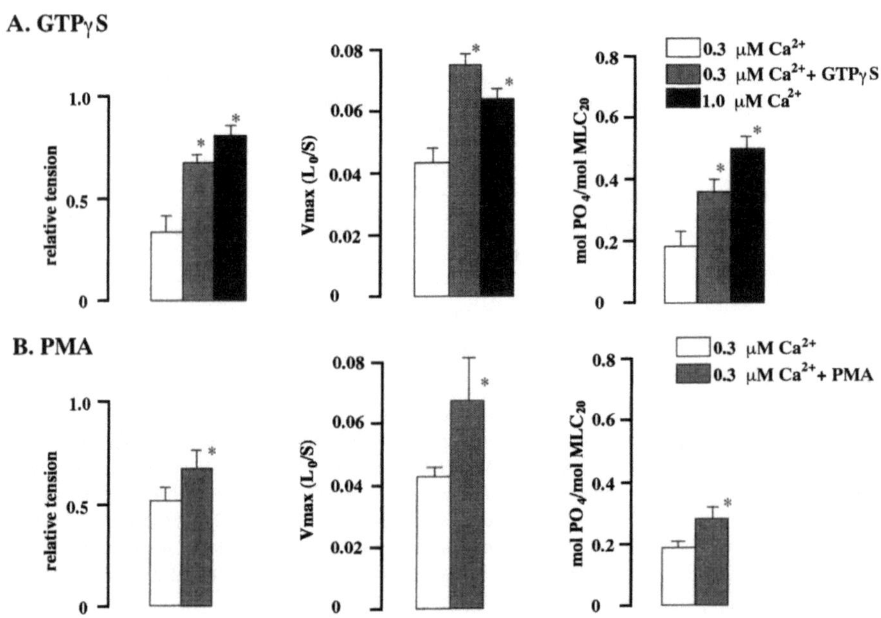


**Fig. 6.** Effects of GTPγS 10 μM alone or PMA 0.1 μM with phosphatidylserine 50 μg/ml on the tension, maximum shortening velocity ($V_{max}$), and $MLC_{20}$ phosphorylation during the application of 0.3 μM $Ca^{2+}$ in ionomycin-treated, saponin-skinned smooth muscles. Slack tests were performed during the steady-state contraction induced by 0.3 μM $Ca^{2+}$. **A.** Effects observed before (controls) and after application of 10 μM GTPγS in the presence of 0.3 μM $Ca^{2+}$ and a comparison with values observed in the presence of 1 μM $Ca^{2+}$ in the absence of GTPγS. **B.** Effects observed before (controls) and 90 min after application of PMA with phosphatidylserine in 0.3 μM $Ca^{2+}$. The amplitude of contraction evoked by 10 μM $Ca^{2+}$ in the absence of GTPγS was normalized as a relative tension of 1.0. *$P < 0.05$ for difference from control

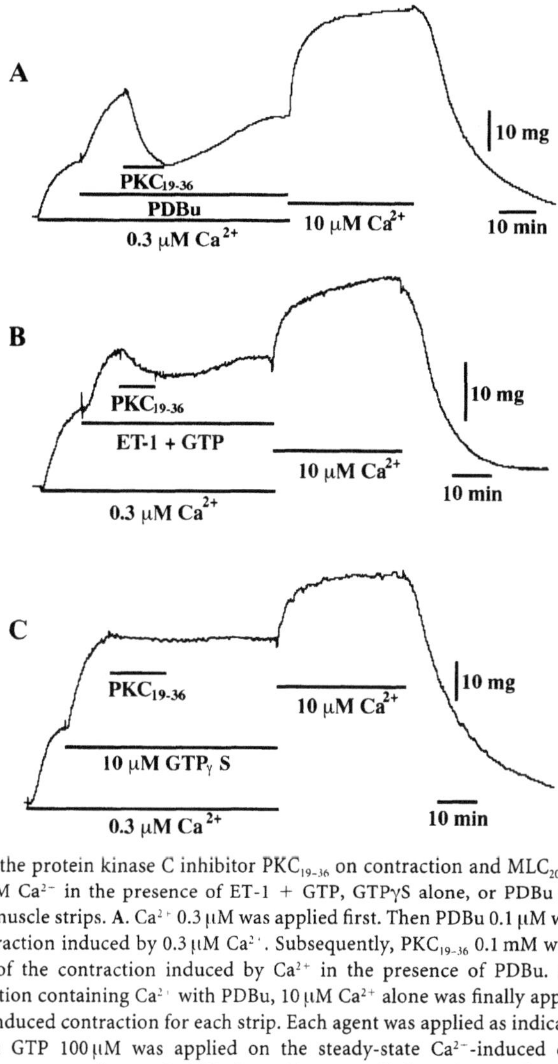


**Fig. 7.** Effects of the protein kinase C inhibitor $PKC_{19-36}$ on contraction and $MLC_{20}$ phosphorylation induced by 0.3 μM $Ca^{2+}$ in the presence of ET-1 + GTP, GTPγS alone, or PDBu alone in β-escin-skinned smooth muscle strips. **A.** $Ca^{2+}$ 0.3 μM was applied first. Then PDBu 0.1 μM was applied on the steady-state contraction induced by 0.3 μM $Ca^{2+}$. Subsequently, $PKC_{19-36}$ 0.1 mM was applied during the tonic phase of the contraction induced by $Ca^{2+}$ in the presence of PDBu. After washout of $PKC_{19-36}$ by a solution containing $Ca^{2+}$ with PDBu, 10 μM $Ca^{2+}$ alone was finally applied to obtain the maximum $Ca^{2+}$-induced contraction for each strip. Each agent was applied as indicated by the bar. **B.** ET-1 10 nM with GTP 100 μM was applied on the steady-state $Ca^{2+}$-induced contraction, with $PKC_{19-36}$ 0.1 mM present or absent. **C.** GTPγS 10 μM was applied on the steady-state $Ca^{2+}$-induced contraction, with $PKC_{19-36}$ 0.1 mM present or absent

measured during the sustained contraction by application of the slack test [29, 30]. With an increase in the concentration of $Ca^{2+}$ from 0.3 μM to 1.0 μM, $V_{max}$ was increased in a concentration-dependent manner (Fig. 6A) [14, 20]. GTPγS 10 μM increased the force, $MLC_{20}$ phosphorylation, and $V_{max}$ in the presence of 0.3 μM $Ca^{2+}$ (Fig. 6A).

Figure 6B shows the effects of 0.1 μMPMA with phosphatidylserine (PS) 50 μg/ml on the force, $V_{max}$, and phosphorylation of $MLC_{20}$ in a solution containing 0.3 μM $Ca^{2+}$ in saponin-treated skinned smooth muscle. PMA with PS was applied during the tonic phase of the $Ca^{2+}$-induced contraction [20]. After application of PMA with PS, the $Ca^{2+}$-induced force was gradually enhanced with a proportional increase in both $MLC_{20}$ phosphorylation and $V_{max}$.

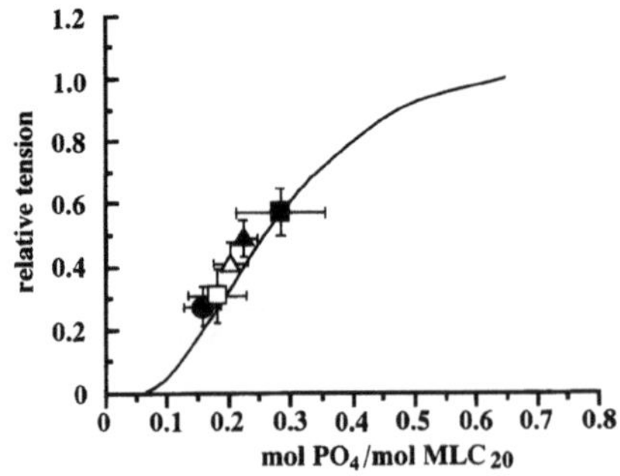


**Fig. 8.** Effects of $PKC_{19\text{-}36}$ on the relation between relative force and $MLC_{20}$ phosphorylation in the presence and absence of either ET-1 with GTP or PDBu alone in β-escin-skinned smooth muscle strips. *Closed circles*, 0.3 μM $Ca^{2+}$ alone (control); *Triangles*, in the presence of 10 nM ET-1 + 100 μM GTP without (*closed triangles*) or with (*open triangles*) 0.1 mM $PKC_{19\text{-}36}$; *Squares*, in 0.1 μM PDBu without (*closed squares*) or with (*open squares*) 0.1 mM $PKC_{19\text{-}36}$. The curve was obtained by fitting the data points by a nonlinear least-squares method using the parameters obtained in Fig. 5 ($r = 0.98$). The maximum amplitude of contraction induced by 10 μM $Ca^{2+}$ was normalized as a relative force of 1.0 for each strip. Results shown are each the mean of four observations; standard deviations are shown by vertical and horizontal bars

## Effect of GTPγS on $Ca^{2+}$-Induced Contraction and $MLC_{20}$ Phosphorylation by a PKC-Independent Mechanism

The peptide inhibitor $PKC_{19\text{-}36}$, a synthetic peptide corresponding to the autoinhibitory domain of PKC, selectively inhibits PKC with a $K_i$ value of 0.2 μM when assayed using a synthetic peptide (Pro-Leu-Ser-Arg-Thr-Leu-Ser-Val-Ala-Ala-Lys-Lys) as a substrate [23]. $PKC_{19\text{-}36}$ 0.1 mM modified neither contraction (1.1 ± 0.2 times control) nor $MLC_{20}$ phosphorylation (0.16 ± 0.06 mol $PO_4$/mol $MLC_{20}$ in controls and 0.17 ± 0.03 mol $PO_4$/mol $MLC_{20}$ in the presence of $PKC_{19\text{-}36}$) induced by 0.3 μM $Ca^{2+}$ alone in β-escin-skinned smooth muscle. The effects of $PKC_{19\text{-}36}$ were then observed on contractions induced by 0.3 μM $Ca^{2+}$ in the presence of ET-1, GTPγS, and PDBu (Fig. 7). $PKC_{19\text{-}36}$ 0.1 mM blocked the PDBu-induced enhancement of both the contraction and $MLC_{20}$ phosphorylation induced by 0.3 μM $Ca^{2+}$ (Figs. 7A, 8). By contrast, $PKC_{19\text{-}36}$ did not inhibit or only partly inhibited the enhancement of the $Ca^{2+}$-induced contraction and MLC phosphorylation induced by the presence of ET-1 (Figs. 7B, 8) or GTPγS (Fig. 7C). $PKC_{19\text{-}36}$ did not modify the relation between relative tension and $MLC_{20}$ phosphorylation in the presence of ET-1 or PDBu (Fig. 8), nor did it modify the relation between tesion and $MLC_{20}$ phosphorylation in the presence of GTPγS (not shown).

# Discussion

## Role of GTP-Binding Proteins in $Ca^{2+}$ Sensitization in Contractile Proteins Induced by ET-1 or GTPγS

Myosin light chain ($MLC_{20}$) phosphorylation by the $Ca^{2+}$–calmodulin–myosin light chain kinase complex plays a pivotal role in the initiation of smooth muscle contraction [38, 39]. It has been found that agonists increase the sensitivity of contraction to

$Ca^{2+}$ in both intact and skinned vascular smooth muscle [10, 11, 40]. In the present experiments, ET-1 shifted the $[Ca^{2+}]_i$–force relation to the left and enhanced the maximum contraction induced by an application of 2.6 mM $Ca^{2+}$ in solutions containing high $K^+$, indicating that ET-1 increases the efficacy with which $[Ca^{2+}]_i$ produces contraction. Agonist-induced $Ca^{2+}$ sensitization of the contractile proteins requires GTP and is inhibited by GDPβS [8, 14, 15]. In the present experiments, ET-1 with GTP and GTPγS alone each enhanced the contraction induced by submaximal $Ca^{2+}$, and the action was greatly attenuated by GDPβS in β-escin-skinned smooth muscle. It is therefore suggested that GTP-binding proteins (G proteins) play a role in this agonist-induced sensitization [8, 15, 34, 35].

It was hypothesized that, in smooth muscle, an agonist acting together with GTP or GTPγS inhibits myosin phosphatases through the activation of unidentified G proteins and thus enhances $Ca^{2+}$-induced $MLC_{20}$ phosphorylation, causing an increase in the $Ca^{2+}$-induced contraction [9, 14, 17]. In β-escin-skinned smooth muscle, ET-1 and GTPγS each increased both the contraction and the $MLC_{20}$ phosphorylation induced by submaximal concentrations of $Ca^{2+}$ (0.03–1.00 μM). The relation between force and $MLC_{20}$ phosphorylation was not modified by ET-1 or GTPγS. These results suggest that GTPγS activates G proteins and consequently increases the $MLC_{20}$ phosphorylation that occurs in response to $Ca^{2+}$, thus causing an increase in the $Ca^{2+}$-induced contraction in β-escin-skinned smooth muscle of the rabbit mesenteric artery.

In the present experiments, the enhancing action of ET-1, but not that of GTPγS, on the $Ca^{2+}$-induced contraction was abolished by neomycin. It is known that neomycin interacts with polyphosphoinositides and inhibits phospholipase C [16, 37]. This action of neomycin is not due to its binding to $IP_3$ [41] because in the present experiments skinned muscle strips were treated with ionomycin in a solution containing relatively high concentration of EGTA (4 mM) to prevent $Ca^{2+}$ release from the storage sites. A direct inhibitory action of neomycin on unidentified G proteins is unlikely, as neomycin did not affect the GTPγS-induced enhancement of the $Ca^{2+}$-induced contraction. Neomycin might be thought to inhibit the activation of PKC through a reduction in DAG production via its inhibitory action on phospholipase C. However, this situation is unlikely because there is little contribution of PKC to ET-1-induced $Ca^{2+}$ sensitization of the contractile proteins in β-escin-skinned strips (see below for details).

## Role of PKC in the G Protein-Mediated Increase in Myofilament $Ca^{2+}$Sensitivity in Skinned Arterial Smooth Muscle

It is known that agonists induce hydrolysis of $PIP_2$ through activation of heterotrimeric G proteins, leading to the generation of $IP_3$ and DAG. The former contributes to the agonist-induced $Ca^{2+}$ release from the $Ca^{2+}$ storage sites, and the latter activates PKC. For the following reasons, it seems likely that ET-1 and GTPγS activate PKC through the production of DAG and then increase the sensitivity of the contractile proteins to $Ca^{2+}$. (1) ET-1 greatly enhances the production of DAG, which activates PKC, through a breakdown of membrane phospholipids in various types of smooth muscle [4, 5, 42, 43]. Moreover, ET-1 induces sustained translocation of PKC from cytosol to the membrane in bovine carotid artery [43]. GTPγS itself produces hydrolysis of $PIP_2$ through activation of heterotrimeric G proteins in vascular smooth muscle [7]. (2) Like GTPγS, the phorbol esters PMA and PDBu, which are activators of

PKC [44], increase contraction, $MLC_{20}$ phosphorylation, and unloaded shortening velocity in the presence of 0.3 μM $Ca^{2+}$ [18–20, 34, 35]. (3) PDBu shifts both the $Ca^{2+}$–force and $Ca^{2+}$–$MLC_{20}$ phosphorylation curves to the left in β-escin-skinned muscle strips [34, 35]. (4) H-7 and staurosporine, inhibitors of PKC, each attenuate the ET-1-induced contraction in intact and α-toxin-treated skinned smooth muscle [6, 13].

However, because staurosporine and H-7 each inhibits PKC at its ATP-binding site, a region with a high degree of sequence homology in most kinases [45, 46], these agents may have some nonspecific actions, especially at high concentrations, such as inhibition of cGMP-dependent protein kinase, cAMP-dependent protein kinase, myosin light chain kinase (MLCK), and other kinases [45, 47, 48]. We previously found that in β-escin-skinned smooth muscle H-7 (at concentrations over 10 μM) concentration-dependently inhibited both contraction and $MLC_{20}$ phosphorylation induced by 10 μM $Ca^{2+}$ without a change in the relation between tension and $MLC_{20}$ phosphorylation [33]. This finding suggested that high concentrations of H-7 may inhibit MLCK as well as PKC. Furthermore, it was reported that when a concentration of staurosporine that selectively inhibits PKC was used this agent failed to alter the amplitude of contraction induced by ET-1 in swine carotid artery [21]. In the present experiments, we found that $PKC_{19-36}$ (a novel peptide inhibitor of PKC) [23] modified neither the contraction nor the $MLC_{20}$ phosphorylation induced by 0.3 μM $Ca^{2+}$, but it completely inhibited the PDBu-induced enhancement of both the $Ca^{2+}$-induced responses. This finding suggests that $PKC_{19-36}$ selectively inhibits PKC but not MLCK in skinned smooth muscle. Under these conditions, $PKC_{19-36}$ did not modify, or only slightly attenuated, the enhancement of both the $Ca^{2+}$-induced contraction and $MLC_{20}$ phosphorylation induced by ET-1 or GTPγS. These results suggest that PKC is unlikely to play a major role in the increase in the sensitivity of contractile proteins to $Ca^{2+}$ seen in the presence of ET-1 or GTPγS. This hypothesis is supported by the following finding: It is known that neomycin inhibits the production of $IP_3$ and DAG through its inhibitory action on phospholipase C [16, 37]. Indeed, under the present experimental conditions, neomycin completely blocked the endothelin-1-induced enhancement of the $Ca^{2+}$-induced contraction [35]. In contrast, neomycin did not modify the GTPγS-induced enhancement of the $Ca^{2+}$-induced contraction, although the agent would be expected to inhibit this action if PKC activated by the synthesized DAG played a role in the GTPγS-induced response in skinned muscle.

## Signal Transduction Mechanism for the G Protein-Mediated Increase in Myofilament $Ca^{2+}$ Sensitivity in Arterial Smooth Muscle

There are two major superfamilies of G proteins involved in signal transduction. One is a superfamily of heterotrimeric G proteins acting as transducers for membrane receptors, and the other is the superfamily including ras p21 and its related small G proteins [49]. It has been suggested that small G proteins, such as rho p21 [50] and ras p21 [51], may contribute to the agonist-induced increase in the sensitivity of the contractile proteins to $Ca^{2+}$. However, it is not yet clear how agonist-receptor signals would couple to these small G proteins. It might be reasonable to speculate that membrane phospholipids that interact with neomycin play a role in the coupling between the ET-1–$ET_A$ receptor complex and unidentified G proteins, which leads to an increase in the sensitivity of the contractile proteins to $Ca^{2+}$. Furthermore, because neomycin did not modify the GTPγS-induced enhancement of the $Ca^{2+}$-induced contraction in β-escin-skinned muscle, the present results also suggest that small G proteins, in addition to the heterotrimeric G proteins, may contribute to the GTPγS-induced increase in myofilament $Ca^{2+}$ sensitivity in skinned smooth muscle.

Steusloff et al. [52] reported that in the α-toxin-skinned longitudinal smooth muscle of guinea pig ileum carbachol plus GTP enhanced the contraction induced by submaximal concentrations of $Ca^{2+}$ with a corresponding increase in tyrosine phosphorylation of unidentified proteins. This result suggests that the other serine/threonine kinases, such as mitogen-activated protein (MAP) kinase, may also play a role in the GTPγS-induced enhancement of $Ca^{2+}$-induced contraction. This point remains to be clarified.

## Mechanism by Which PKC Activation Enhances $Ca^{2+}$-Induced $MLC_{20}$ Phosphorylation in Smooth Muscle

It has been reported that in vitro phosphorylation of heavy meromyosin by PKC does not stimulate actin-activated myosin MgATPase activity but, rather, reduces the rate of phosphorylation of $MLC_{20}$ by MLCK, causing a decrease in the actin-activated MgATPase activity of heavy meromyosin, which had been prephosphorylated by MLCK [53, 54]. Furthermore, in glycerine-treated porcine carotid artery, phosphorylation of $MLC_{20}$ by PKC neither provokes contraction nor modifies the contraction induced by $Ca^{2+}$/calmodulin [55]. These results suggest that $MLC_{20}$ phosphorylation by PKC either has no effect or indeed causes inhibition of $Ca^{2+}$-induced contraction in smooth muscle. In the present experiments, PDBu enhanced both the contraction and $MLC_{20}$ phosphorylation induced by $Ca^{2+}$ in β-escin-skinned muscle. Under these conditions, $Ca^{2+}$ (0.03–1.00 μM) concentration-dependently increased the monophosphorylated, but not the diphosphorylated, form of $MLC_{20}$ in the presence or absence of PDBu. Furthermore, the relation between force and $MLC_{20}$ phosphorylation was the same whether in the presence or absence of PDBu. Furthermore, in the presence of PDBu, the force–$MLC_{20}$ phosphorylation relation was not modified by the novel PKC inhibitor $PKC_{19-36}$. Although we did not try to determine whether PKC directly phosphorylates $MLC_{20}$, all these results suggest that PKC may indirectly activate the phosphorylation of $MLC_{20}$ by MLCK. Phosphorylation of MLCK by PKC causes a decrease in its affinity for the $Ca^{2+}$/calmodulin complex [56, 57], and the level of $MLC_{20}$ phosphorylation is determined by the balance between the effect of MLCK and myosin phosphatase in smooth muscle cells. It may be that PKC activated by PDBu increases $MLC_{20}$ phosphorylation through direct or indirect inhibition of myosin phosphatase.

It was hypothesized that PKC increases $Ca^{2+}$-induced contraction through the phosphorylation of caldesmon [58] or calponin [36, 59], which are thin-filament-linked regulatory proteins in smooth muscle. However, the physiologic role of these proteins in smooth muscle has not yet been clarified, especially in response to agonist stimulation; and their part, if any, in the physiologic actions of PKC is unclear.

*Acknowledgments.* We thank Dr. R.J. Timms for the language editing. Dr. J.T. Stull, at the Department of Physiology and Pharmacology, University of Texas Southwestern

Medical Center, kindly supplied the antibody against bovine tracheal $MLC_{20}$. This work was partly supported by a grant-in-aid from the Ministry of Education of Japan.

## References


1. Yanagisawa M, Kurihara H, Kimura S, Tomobe Y, Kobayashi M, Mitsui Y, Yazaki Y, Goto K, Masaki T (1988) A novel potent vasoconstrictor peptide produced by vascular endothelial cells. Nature 332:411–415
2. Yanagisawa M, Masaki T (1989) Molecular biology and biochemistry of the endothelins. Trends Pharmacol Sci 10:374–378
3. Kasuya Y, Takuwa Y, Yanagisawa M, Kimura S, Goto K, Masaki T (1989) Endothelin-1 induces vasoconstriction through two functionally distinct pathways in porcine coronary artery: contribution of phosphoinositide turnover. Biochem Biophys Res Commun 161:1049–1055
4. Lee T-S, Chao T, Hu K-Q, King GL (1989) Endothelin stimulates a sustained 1,2-diacylglycerol increase and protein kinase C activation in bovine aortic smooth muscle cells. Biochem Biophys Res Commun 162:381–386
5. Griendling KK, Tsuda T, Alexander RW (1989) Endothelin stimulates diacyglycerol accumulation and activates protein kinase C in cultured vascular smooth muscle cells. J Biol Chem 264:8237–8240
6. Ohlstein EH, Horohonich S, Hay DWP (1989) Cellular mechanisms of endothelin in rabbit aorta. J Pharmacol Exp Ther 250:548–555
7. Sasaguri T, Hirata M, Kuriyama H (1985) Dependence on $Ca^{2+}$ of the activities of phosphatidylinositol 4,5-bisphosphate phosphodiesterase and inositol 1,4,5-triphosphate phosphatase in smooth muscles of the porcine coronary artery. Biochem J 231:497–503
8. Nishimura J, Kolber M, van Breemen C (1988) Norepinephrine and GTP-$\gamma$-S increase myofilament $Ca^{2+}$ sensitivity in $\alpha$-toxin permeabilized arterial smooth muscle. Biochem Biophys Res Commun 157:677–683
9. Kitazawa T, Gaylinn BD, Denney GH, Somlyo AP (1991) G-protein-mediated $Ca^{2+}$ sensitization of smooth muscle contraction through myosin light chain phosphorylation. J Biol Chem 266:1708–1715
10. Rembold CM (1990) Modulation of the $[Ca^{2+}]$ sensitivity of myosin phosphorylation in intact swine arterial smooth muscle. J Physiol (Lond) 429:77–94
11. Hori M, Sato K, Sakata K, Ozaki H, Takano-Ohmuro H, Tsuchiya T, Sugi H, Kato I, Karaki H (1992) Receptor agonists induce myosin phosphorylation-dependent and phosphorylation-independent contractions in vascular smooth muscle. J Pharmacol Exp Ther 261:506–512
12. Itoh T, Kajikuri J, Kuriyama H (1992) Characteristic features of noradrenaline-induced $Ca^{2+}$ mobilization and tension in arterial smooth muscle of the rabbit. J Physiol (Lond) 457:297–314
13. Nishimura J, Moreland S, Ahn HY, Kawase T, Moreland RS, van Breemen C (1992) Endothelin increases myofilament $Ca^{2+}$ sensitivity in $\alpha$-toxin-permeabilized rabbit mesenteric artery. Circ Res 71:951–959
14. Fujiwara T, Itoh T, Kubota Y, Kuriyama H (1989) Effects of guanosine nucleotides on skinned smooth muscle tissue of the rabbit mesenteric artery. J Physiol (Lond) 408:535–547
15. Kitazawa T, Kobayashi S, Horiuti K, Somlyo AV, Somlyo AP (1989) Receptor-coupled, permeabilized smooth muscle: role of the phosphatidylinositol cascade, G-proteins, and modulation of the contractile response to $Ca^{2+}$. J Biol Chem 264:5339–5342
16. Kobayashi S, Kitazawa T, Somlyo AV, Somlyo AP (1989) Cytosolic heparin inhibits muscarinic and $\alpha$-adrenergic $Ca^{2+}$ release in smooth muscle. J Biol Chem 264:17997–18004
17. Kubota Y, Nomura M, Kamm KE, Mumby MC, Stull JC (1992) GTP$\gamma$S-dependent regulation of smooth muscle contractile elements. Am J Physiol 262:C405–C410
18. Itoh T, Kanmura Y, Kuriyama H, Sumimoto K (1986) A phorbol ester has dual actions on the mechanical response in the rabbit mesenteric and porcine coronary arteries. J Physiol (Lond) 375:515–534

19. Chatterjee M, Tejada M (1986) Phorbol ester-induced contraction in chemically skinned vascular smooth muscle. Am J Physiol 251:C356–C361
20. Fujiwara T, Itoh T, Kubota Y, Kuriyama H (1988) Actions of a phorbol ester on factors regulating contraction in rabbit mesenteric artery. Circ Res 63:893–902
21. Moreland R, Cilea J, Moreland S (1992) Staurosporine decreases stiffness but not stress in endothelin-1-stimulated arterial muscle. Am J Physiol 262:C862–C869
22. Shimamoto H, Shimamoto Y, Kwan C-Y, Daniel EE (1992) Participation of protein kinase C in endothelin-1-induced contraction in rat aorta: studies with a new tool, calphostin C. Br J Pharmacol 107:282–287
23. House C, Kemp BE (1987) Protein kinase C contains a pseudosubstrate prototype in its regulatory domain. Science 238:1726–1728
24. Gong MC, Fuglsang A, Alessi D, Kobayashi S, Cohen P, Somlyo AV, Somlyo AP (1992) Arachidonic acid inhibits myosin light chain phosphatase and sensitizes smooth muscle to calcium. J Biol Chem 267:21492–21498
25. Itoh T, Kuriyama H, Suzuki H (1983) Differences and similarities in noradrenaline- and caffeine-induced mechanical responses in the rabbit mesenteric artery. J Physiol (Lond) 337:609–629
26. Grynkiewicz G, Poenie M, Tsien RY (1985) A new generation of $Ca^{2+}$ indicators with greatly improved fluorescence properties. J Biol Chem 260:3440–3450
27. Poenie M, Alderton J, Steinhart RA, Tsien RY (1986) Calcium rises abruptly and briefly throughout the cell at the onset of anaphase. Science 233:886–889
28. Itoh T, Kanmura Y, Kuriyama H (1986) Inorganic phosphate regulates the contraction-relaxation cycle in skinned muscles of the rabbit mesenteric artery. J Physiol (Lond) 376:231–252
29. Edman KAP (1979) The velocity of unloaded shortening and its relation to sarcomere length and isometric force in vertebrate muscle fibers. J Physiol (Lond) 291:143–159
30. Arner A, Hellstrand P (1985) Effects of calcium and substrate on force-velocity relation and energy turnover in skinned smooth muscle of the guinea-pig. J Physiol (Lond) 360:347–365
31. Satoh S, Kubota Y, Itoh T, Kuriyama H (1987) Mechanisms of the $Ba^{2+}$-induced contraction in smooth muscle cells of the rabbit mesenteric artery. J Gen Physiol 89:215–237
32. Persechini A, Kamm KE, Stull JT (1986) Different phosphorylated forms of myosin in contracting tracheal smooth muscle. J Biol Chem 261:6293–6299
33. Suzuki A, Itoh T (1993) Effects of calyculin A on tension and myosin phosphorylation in skinned smooth muscle of the rabbit mesenteric artery. Br J Pharmacol 109:703–712
34. Itoh T, Suzuki A, Watanabe Y (1994) Effect of a peptide inhibitor of protein kinase C on G-protein-mediated increase in myofilament $Ca^{2+}$-sensitivity in rabbit arterial skinned muscle. Br J Pharmacol 111:311–317
35. Yoshida M, Suzuki A, Itoh T (1994) Mechanisms of vasoconstriction induced by endothelin-1 in smooth muscle of rabbit mesenteric artery. J Physiol (Lond) 477:253–265
36. Itoh T, Suzuki S, Suzuki A, Nakamura F, Naka M, Tanaka T (1994) Effects of exogenously applied calponin on $Ca^{2+}$-regulated force in skinned smooth muscle of the rabbit mesenteric artery. Pflugers Arch 427:301–308
37. Cockcroft S, Gomperts BD (1985) Role of guanine nucleotide binding protein in the activation of polyphosphoinositide phosphodiesterase. Nature 314:534–536
38. Hartshorne DJ (1987) Biochemistry of the contractile process in smooth muscle. In: Johnson LR (ed) Physiology of the Gastrointestinal Tract, 2nd edn, vol 1. Raven, New York, pp 423–482
39. Itoh T, Ikebe M, Kargacin GJ, Hartshorne DJ, Kemp BE, Fay FS (1989) Effects of modulators of myosin light-chain kinase activity in single smooth muscle cells. Nature 338:164–167
40. Morgan JP, Morgan KG (1984) Stimulus-specific patterns of intracellular calcium levels in smooth muscle of ferret portal vein. J Physiol (Lond) 351:155–167
41. Prentki M, Deeney JT, Matschinsky FM, Joseph SK (1986) Neomycin: a specific drug to study the inositol-phospholipid signalling system? FEBS Lett 197:285–288
42. Sunako M, Kawahara Y, Hirata K, Tsuda T, Yokoyama M, Fukuzaki H, Takai Y (1990) Mass analysis of 1,2-diacylglycerol in cultured rabbit vascular smooth muscle cells. Hypertension 15:84–88
43. Haller H, Smallwood JI, Rasmussen H (1990) Protein kinase C translocation in intact vascular smooth muscle strips. Biochem J 270:375–381

44. Nishizuka Y (1984) The role of protein kinase C in cell surface signal transduction and tumour promotion. Nature 308:693–698
45. Hidaka H, Inagaki M, Kawamoto S, Sasaki Y (1984) Isoquinolinesulfonamides, novel and potent inhibitors of cyclic nucleotide dependent protein kinase and protein kinase C. Biochemistry 23:5036–5041
46. Edelman AM, Blumenthal DK, Krebs EG (1987) Protein serine/threonine kinase. Annu Rev Biochem 56:567–613
47. Tamaoki T, Nomoto H, Takahashi I, Kato Y, Morimoto M, Tomita F (1986) Staurosporine, a potent inhibitor of phospholipid/$Ca^{++}$ dependent protein kinase. Biochem Biophys Res Commun 135:397–402
48. Rüegg UT, Burgess GM (1989) Staurosporine, K-252 and UCN-01: potent but nonspecific inhibitors of protein kinases. Trends Pharmacol Sci 10:218–220
49. Bourne HR, Sanders DA, McCormick F (1990) The GTPase superfamily: a conserved switch for reverse cell functions. Nature 348:125–132
50. Hirata K, Kikuchi A, Sasaki T, Kuroda S, Kaibuchi K, Matsuura Y, Seki H, Saida K, Takai Y (1992) Involvement of rho p21 in the GTP-enhanced calcium ion sensitivity of smooth muscle contraction. J Biol Chem 267:8719–8722
51. Pfitzer G, Satoh S (1993) Modulation of calcium sensitivity of vascular smooth muscle myofilaments. Pflugers Arch 422(suppl 1):R11
52. Steusloff A, Pfitzer G, Kolquist K, Disalvo J (1993) Is tyrosine phosphorylation (TP) involved in the regulation of smooth muscle $Ca^{2+}$ responsiveness? Pflugers Arch 422(suppl 1):R80
53. Nishikawa M, Sellers JR, Adelstein RS, Hidaka H (1984) Protein kinase C modulated in vitro phosphorylation of the smooth muscle heavy meromyosin by myosin light chain kinase. J Biol Chem 259:8808–8814
54. Ikebe M, Hartshorne DJ, Elzinga M (1987) Phosphorylation of the 20 000-dalton light chain of smooth muscle myosin by the calcium-activated, phospholipid-dependent protein kinase. J Biol Chem 262:9569–9573
55. Sutton TA, Haeberle JR (1990) Phosphorylation by protein kinase C of the 20 000-dalton light chain of myosin in intact and chemically skinned vascular smooth muscle. J Biol Chem 265:2749–2754
56. Ikebe M, Inagaki M, Kanamaru K, Hidaka H (1985) Phosphorylation of smooth muscle myosin light chain kinase by $Ca^{2+}$-activated, phospholipid-dependent protein kinase. J Biol Chem 260:4547–4550
57. Nishikawa M, Shirakawa S, Adelstein RS (1985) Phosphorylation of smooth muscle myosin light chain kinase by protein kinase C: comparative study of the phosphorylated states. J Biol Chem 260:8978–8983
58. Walsh MP (1990) Smooth muscle caldesmon. In: Sperelakis N, Wood JD (eds) Frontiers in Smooth Muscle Research. Alan R. Liss, New York, pp 127–140
59. Nakamura F, Mino T, Yamamoto J, Naka M, Tanaka T (1993) Identification of the regulatory site in smooth muscle calponin that is phosphorylated by protein kinase C. J Biol Chem 268: 6194–6201

# Regulation of $Ca^{2+}$-Dependent Phosphorylation of 20-kDa Myosin Light Chain by the Small Molecular Weight G Protein rho p21 in Vascular Smooth Muscle Cells

Yoh Takuwa[1], Masakuni Noda[1], Chikako Yasuda[1], Mamoru Kumada[2], and Kiyoshi Kurokawa[3]

*Summary.* The nonhydroyzable GTP analog guanosine 5'-(3-O-thio)triphosphatase (GTPγS) was previously shown to enhance $Ca^{2+}$-dependent contraction and phosphorylation of 20-kDa myosin light chain ($MLC_{20}$) in skinned smooth muscle preparations. In β-escin-permeabilized cultured aortic smooth muscle cells, GTPγS enhances $Ca^{2+}$-induced phosphorylation of $MLC_{20}$ in a concentration-dependent manner, which is inhibited by the nonhydrolyzable GDP analog guanosine 5'-O-(β-thio)diphosphate (GDPβS). Down-regulation of protein kinase C by pretreatment with phorbol-12,13-dibutyrate does not affect enhancement by GTPγS of $Ca^{2+}$-dependent $MLC_{20}$ phosphorylation, indicating that protein kinase C is not involved in GTPγS-induced enhancement of $MLC_{20}$ phosphorylation. Western blot analysis reveals that the small-molecular-weight G protein rho p21 is abundant in aortic smooth muscle cells. Pretreatment of permeabilized cells with the botulinum exoenzyme $C_3$, which specifically ADP-ribosylates and inactivates the rho family proteins, completely abolishes GTPγS-induced enhancement of $Ca^{2+}$-dependent $MLC_{20}$ phosphorylation. Under the same condition $C_3$ completely ADP-ribosylates cellular proteins with a relative molecular weight of 21 kDa in permeabilized cells. $C_3$ pretreatment of cells does not potentiate thiophosphorylation of $MCL_{20}$ but decreases the rate of dephosphorylation of $MLC_{20}$. These data suggest that rho p21 is involved in the GTPγS-induced $Ca^{2+}$ sensitization of $MLC_{20}$ phosphorylation in aortic smooth muscle cells, and that this effect of rho p21 is due to inhibition of protein phosphatase activity toward $MLC_{20}$.

*Key Words.* Smooth muscle—Myosin—Phosphorylation—Myosin light chain kinase—Phosphatase

## Introduction

Stimulation of smooth muscle by a variety of excitatory agonists causes activation of phospholipase C via a heterotrimeric GTP-binding protein (G protein) and activation of the $Ca^{2+}$ channel to result in a biphasic rise in the intracellular free $Ca^{2+}$ concentra-

[1] Department of Cardiovascular Biology, Faculty of Medicine, 7-3-1 Hongo, Bunkyo-ku, Tokyo, 113 Japan
[2] Department of Physiology, Faculty of Medicine, 7-3-1 Hongo, Bunkyo-ku, Tokyo, 113 Japan
[3] Department of Internal Medicine, Faculty of Medicine, 7-3-1 Hongo, Bunkyo-ku, Tokyo, 113 Japan

tion ($[Ca^{2+}]_i$) [1–6]. A well defined molecular target of the $Ca^{2+}$ message is the calmodulin-dependent enzyme myosin light chain kinase (MLCK) [7]. It is now generally accepted that the resultant phosphorylation of the 20-kDa myosin light chain ($MLC_{20}$) leads to an interaction of actin with myosin, which in turn underlies the initiation of a contractile response. In addition, accumulated evidence suggests that receptor activation by excitatory agonists modulates the $Ca^{2+}$ sensitivity of smooth muscle contraction. Morgan and Morgan initially observed that receptor activation by an agonist induces a larger contraction than KCl depolarization, which produces an increase in $[Ca^{2+}]_i$ similar in magnitude to that induced by agonists [8]. Rembold and Murphy correlated an increase in the $[Ca^{2+}]_i$ with increases in the extent of $MLC_{20}$ phosphorylation and tension in agonist- or KCl-stimulated smooth muscle and found that an agonist produces larger increases in the extent of $MLC_{20}$ phosphorylation and the tension at a given level of the $[Ca^{2+}]_i$ than does KCl depolarization [9]. Subsequently this observation was confirmed by several other laboratories [10, 11]. These findings indicate that agonists somehow enhance the $Ca^{2+}$ sensitivity of $MLC_{20}$ phosphorylation and contraction. It has recently been demonstrated in staphylococcal α-toxin-, β-escin-, or saponin-permeabilized smooth muscle that an agonist causes potentiation of contraction in a GTP-dependent manner at a constant level of $[Ca^{2+}]_i$, which is accompanied by an increase in the phosphorylation level of $MLC_{20}$ [12–17]. This action of an agonist is inhibited by the nonhydrolyzable GDP analog guanosine 5'-O-(β-thio)diphosphate (GDPβS) and that the nonhydrolyzable GTP analog guanosine 5'-(3-O-thio)triphosphate (GTPγS) mimics this effect of an agonist. These findings suggest that an agonist sensitizes $Ca^{2+}$-dependent $MLC_{20}$ phosphorylation by a mechanism that involves a G protein.

The $Ca^{2+}$ sensitivity of $MLC_{20}$ phosphorylation in an agonist- or GTPγS-stimulated permeabilized smooth muscle could be increased by stimulation of MLCK or inhibition of protein phosphatase for $MLC_{20}$. Kitazawa et al. have demonstrated that addition of GTPγS decreased the rate of dephosphorylation of phosphorylated $MLC_{20}$ but did not enhance thio-phosphorylation of $MLC_{20}$ in permeabilized vascular smooth muscle, suggesting that GTPγS inhibited protein phosphatase toward $MLC_{20}$ [18]. In accordance with this report, Kubota et al. showed that GTPγS caused inhibition of protein phosphatase activity toward both $MLC_{20}$ and heavy meromysin but not stimulation of MLCK activity in homogenates of tracheal smooth muscle [16]. Activation of protein kinase C by an active phorbol ester in smooth muscle also increases the $Ca^{2+}$ sensitivity of the myofilaments in permeabilized smooth muscle, as do GTPγS and an agonist plus GTP [19, 20]. Because in permeabilized smooth muscle GTPγS and an agonist may activate phospholipase C via a heterotrimeric G protein to produce 1,4,5-inositol trisphosphate and 1,2-diacylglycerol, an endogenous activator of protein kinase C, there is an possibility that protein kinase C mediates agonist- or GTPγS-induced $Ca^{2+}$ sensitization of the myofilaments [12, 14, 20]. Itoh et al., however, reported (this volume) that it is not true. Gong et al. [21] reported that exogenously added arachidonic acid increases the $Ca^{2+}$ sensitivity of $MCL_{20}$ phosphorylation presumably through inhibition of protein phosphatase activity toward $MLC_{20}$ in permeabilized smooth muscle. They also showed that arachidonic acid in vitro caused dissociation of trimeric protein phosphatase toward $MLC_{20}$ isolated from smooth muscle into the catalytic subunit and the complex of the two regulatory subunits and inhibition of its $MLC_{20}$ phosphatase activity. Moreover, they found that an agonist and GTPγS stimulate arachidonic acid release from smooth muscle. However, whether arachidonic acid plays a physiologic role in the regulation of protein phosphatase activity toward $MLC_{20}$ in smooth muscle remains unclear. The precise molecular mechanisms by which agonist stimulation or GTPγS leads to inhibition of protein phosphatase activity toward $MLC_{20}$ have not been clarified. It also remains unknown which G protein is involved in this phenomenon.

Takai and colleagues [22] have reported that the low-molecular-weight GTP-binding protein rho p21 is involved in GTPγS enhancement of $Ca^{2+}$-induced contraction in permeabilized vascular smooth muscle. The rho p21 family is composed of three members: rho A, B, and C [23]. Studies demonstrated that the rho family proteins are involved in regulation of the actin cytoskeleton and cell adhesion in non-smooth-muscle cells [24]. These observations raise an interesting possibility that rho p21 might be involved in GTPγS-induced $Ca^{2+}$ sensitization of $MCL_{20}$ phosphorylation. By using the bacterial exoenzyme $C_3$, which specifically ADP-ribosylates and inactivates rho p21 [25, 26], we examined the involvement of rho p21 in GTPγS-induced $Ca^{2+}$ sensitization of $MLC_{20}$ phosphorylation in permeabilized aortic smooth muscle cells. Furthermore, if rho p21 was involved, we tried to determine whether rho p21 increased $Ca^{2-}$ sensitivity of $MLC_{20}$ phosphorylation through inhibition of $MLC_{20}$ phosphatase or stimulation of MLCK.

## GTPγS Enhancement of $MLC_{20}$ Phosphorylation in Permeabilized Vascular Smooth Muscle Cells

We tried to establish permeabilized, cultured vascular smooth muscle cells to promote biochemical analysis of the regulation of the $Ca^{2+}$ sensitivity of $MLC_{20}$ phosphorylation. We permeabilized porcine aortic smooth muscle cells employing β-escin. Raising the ambient free $Ca^{2+}$ concentration in these cells causes an increase in the extent of $MLC_{20}$ phosphorylation in a manner dependent on $Ca^{2+}$ concentration. The phosphorylation level of $MLC_{20}$ becomes maximal at 0.3 μM free $Ca^{2+}$. At lower free $Ca^{2+}$ concentrations (≤0.3 μM) only monophosphorylation of $MLC_{20}$ is observed. At higher free $Ca^{2+}$ concentrations (≥1 μM), in addition to monophosphorylated $MLC_{20}$, diphosphorylated $MLC_{20}$ appears. Even at $Ca^{2+} < 0.01$ μM, a small extent (9% of total $MLC_{20}$) of monophosphorylation of $MLC_{20}$ is found consistently. The addition of the MLC kinase inhibitor wortmannin [27] decreases the phosphorylation level of $MLC_{20}$ at both 0.3 μM and 1.0 μM $Ca^{2+}$ to the value found at $Ca^{2+} < 0.01$ μM. However, the small extent of $MLC_{20}$ phosphorylation observed at $Ca^{2+} < 0.01$ μM cannot be suppressed by the same concentration of wortmannin. These results suggest that $Ca^{2+}$-induced phosphorylation of $MLC_{20}$ is mediated by MLCK, but that basal phosphorylation of $MLC_{20}$ at $Ca^{2+} < 0.01$ μM is mediated by another protein kinase.

Addition of either GTP 100 μM or GTPγS 3 μM potently enhances 0.1 μM $Ca^{2+}$-induced $MLC_{20}$ phosphorylation in permeabilized cells (Fig. 1), although GTP is less potent than GTPγS. GTPγS not only increases the level of monophosphorylated $MLC_{20}$ but causes diphosphorylation of $MLC_{20}$. GTPγS 30 μM shifts the $Ca^{2+}$–$MLC_{20}$ phosphorylation curve to the left by approximately one order of magnitude and significantly increases the maximal level of $MLC_{20}$ phosphorylation. The enhancing effect of GTPγS is dose-dependent with the half-maximally and maximally effective concentrations of 5 and 30 μM, respectively. In contrast, both CTP and UTP at 100 μM are totally ineffective, and both nucleotides at 1 mM slightly increase the phosphorylation level of $MLC_{20}$, indicating that the potentiating effect is relatively specific

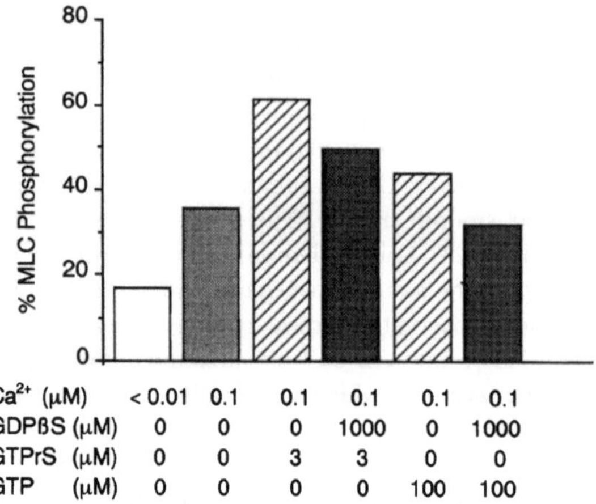


**Fig. 1.** Enhancement of $Ca^{2+}$-induced myosin light chain ($MLC_{20}$) phosphorylation by GTP and GTPγS in β-escin-permeabilized aortic smooth muscle cells. Permeabilized cells were incubated at Ca < 0.01 or 0.1 μM with or without 3 μM GTPγS or 100 μM GTP for 15 min in the presence or absence of 1 mM GDPβS. The level of $MLC_{20}$ phosphorylation was determined

for guanine nucleotides. When GDPβS is added before GTP or GTPγS, it inhibits the enhancing effects of both GTP and GTPγS, suggesting the involvement of a G protein in GTP- or GTPγS-induced $Ca^{2+}$ sensitization of $MLC_{20}$ phosphorylation. These results indicate that β-escin-permeabilized aortic smooth muscle cells have the mechanism for GTPγS-induced $Ca^{2+}$ sensitization of $MLC_{20}$ phosphorylation, like that in skinned vascular smooth muscle tissue.

The addition of phorbol-12,13-dibutyrate (PDBu) barely increases the phosphorylation level of $MLC_{20}$, nor does it enhance $Ca^{2+}$-induced $MLC_{20}$ phosphorylation in permeabilized cells. Pretreatment of intact aortic smooth muscle cells with a higher concentration of PDBu for 24 h, which nearly completely down-regulates protein kinase C, does not inhibit GTPγS-induced enhancement of $MLC_{20}$ phosphorylation at all. Thus consistent with the results presented by Itoh et al. (this volume), these data indicate that protein kinase C is not involved in GTPγS-induced enhancement of $MLC_{20}$ phosphorylation.

## Botulinum Exotoxin $C_3$ Inhibition of GTPγS-Induced Potentiation of $MLC_{20}$ Phosphorylation

Western blotting analysis using anti-rho A p21 antibody demonstrates that aortic smooth muscle cells contain abundant 21-kDa protein to react with this antibody, and that the permeabilization procedure does not cause any detectable loss of the 21-kDa protein from cells. A previous study by Kawahara et al. [28] demonstrated that rho A is present in abundance and accounts for 50% of total GTP binding activity in the cytosol fraction from bovine aortic smooth muscle. Thus it is likely that the 21-kDa protein that reacts with the anti-rho A antibody represents rho A p21, although the possibility that rho B and rho C also exist in aortic smooth muscle is not excluded.

To examine the role of rho p21 in GTPγS-induced enhancement of $MLC_{20}$ phosphorylation, we investigated the effect of pretreatment of cells with botulinum

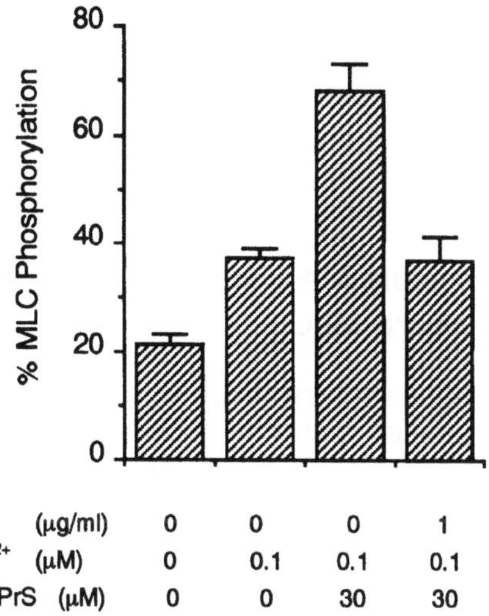


**Fig. 2.** Inhibition by botulinum exoenzyme $C_3$ of GTPγS-induced enhancement of $MLC_{20}$ phosphorylation. Permeabilized cells were pretreated with $C_3$ 1 μg/ml for 20 min or left untreated and then incubated in 0.1 μM $Ca^{2+}$ with or without 30 μM GTPγS for 15 min. The level of $MLC_{20}$ phosphorylation was determined

exotoxin $C_3$, which ADP-ribosylates and inactivates the rho family proteins. As shown in Fig. 2, in cells pretreated with $C_3$ GTPγS enhancement of $Ca^{2+}$-induced $MLC_{20}$ phosphorylation is nearly completely inhibited. However, $C_3$ does not affect $Ca^{2+}$-induced $MLC_{20}$ phosphorylation at all. The effectiveness of this $C_3$ treatment in permeabilized cells is proved by the fact that a 21-kDa protein in permeabilized cells is fully ADP-ribosylated by incubation with $C_3$ and $^{32}P$-NAD.

## Effect of GTPγS on $Ca^{2+}$-Dependent Thiophosphorylation of $MLC_{20}$ and Dephosphorylation of $MLC_{20}$

To evaluate whether GTPγS-induced enhancement of $MLC_{20}$ phosphorylation is mediated through activation of MLCK or inhibition of protein phosphatase toward $MLC_{20}$, we conducted two sets of experiments. First, we studied the effect of GTPγS on thiophosphorylation of $MLC_{20}$. Thiophosphorylated $MLC_{20}$ is a poor substrate for phosphatase [29]. When ATP is used as substrate in the absence of GTPγS, upon addition of 0.1 μM $Ca^{2+}$ the phosphorylation level of $MLC_{20}$ rapidly increases up to about 50% of total $MLC_{20}$ at 2 min and stays at this level for 10 min. Addition of GTPγS enhances $MLC_{20}$ phosphorylation at every time point examined. When ATPγS is used as substrate, the level of $MLC_{20}$ continues to rise for 10 min and reaches higher levels than those in the presence of ATP. GTPγS does not change the thiophosphorylation level of $MLC_{20}$ at each time point. These results indicate that GTPγS does not potentiate the activity of MLC kinase. These results are consistent with the data in skinned vascular smooth muscle tissue reported previously by Kitazawa et al. [18].

In the second set of experiments, we examined whether GTPγS affects the rate of dephosphorylation of phosphorylated $MLC_{20}$ and if $C_3$ pretreament inhibits the GTPγS effect on dephosphorylation of $MLC_{20}$ in permeabilized cells. In $C_3$-pretreated and nonpretreated permeabilized cells, myosin was phosphorylated by introducing $Ca^{2+}$ 0.1 μM alone or together with GTPγS 30 μM. Phosphorylated $MLC_{20}$ started to be dephosphorylated by both removal of $Ca^{2+}$ and addition of the MLC kinase inhibitor wortmannin. In nonpretreated cells the phosphorylation level of $MLC_{20}$ rapidly fell to 32% of the initial phosphorylation level by 10 min. GTPγS decreased the dephosphorylation rate of $MLC_{20}$, and the phosphorylation level was maintained at 55% of the initial level at 10 min. $C_3$ pretreatment completely abolished this effect of GTPγS. These results strongly suggest that GTPγS induces a decrease in the $MLC_{20}$ phosphatase activity through a rho p21-dependent mechanism.

## Possible Mechanism for Regulation of Phosphatase Activity Twoard $MLC_{20}$

The results presented here strongly suggest that the action of rho p21 lies in the regulation of protein phosphatase activity toward $MLC_{20}$. It was recently demonstrated that the major smooth muscle myosin phosphatase belongs to protein phosphatase 1 (PP1) and is composed of three subunits: a 37-kDa catalytic subunit (PP1δ isozyme) and 130- and 20-kDa regulatory subunits [30]. It was also demonstrated that the regulatory components function to target the catalytic subunit of the smooth muscle protein phosphatase to myosin filament and to increase the phosphatase activity toward myosin. In liver and skeletal muscle PP1, with a subunit structure similar to that of smooth muscle myosin phosphatase, is associated with glycogen particles and regulates activities of enzymes of glycogen metabolism by dephosphorylating glycogen particle-associated enzymes [31]. It is known that in addition to its roles in determining the location of PP1 catalytic subunit and enhancing phosphatase activity, the regulatory subunit of glycogen-associated phosphatase confers a mechanism for the regulation of phosphatase activity by covalent modification, such as phosphorylation and dephosphorylation and interaction of allosteric effectors [31, 32]. By analogy, it is an interesting possibility that activated rho p21 causes a change in the phosphorylation state of the regulatory subunit of the smooth muscle myosin-associtate PP1 to alter the phosphatase activity. In this regard, it should be noted that rac, a small-molecular-weight G protein closely related to rho p21, activates a cytosolic serine/threonine protein kinase [33]. Alternatively, it might be possible that a complex of activated rho p21 and downstream effector proteins induces a change in the phosphatase activity by allosteric interaction with the regulatory subunit of the smooth muscle myosin-associated PP1.

In a number of smooth muslce types an agonist enhances $MLC_{20}$ phosphorylation in the presence of GTP, similar to GTPγS [15, 17]. This action of an agonist selectively requires GTP and is inhibited by GDP or GDPβS, indicating the involvement of a G protein [15]. The results of the present study suggest that rho p21 may be a G protein involved in agonist-induced $Ca^{2+}$ sensitization of $MLC_{20}$ phosphorylation. Most agonists bind to membrane receptors and activate effector molecules such as phospholipase C and phospholipase $A_2$ via a heterotrimeric G protein [1]. The activity of ras-like small-molecular-weight G proteins is controlled by at least two kinds of regulatory molecule: guanine nucleotide exchange proteins and GTPase-activation proteins [34, 35]. There are several known regulatory molecules for the rho family members, including smg GDS, rho GDS, rho GDI, and rho GAP. An agonist may alter the activities of these regulatory molecules to bring about a change in rho activity. It remains to be clarified whether—and, if so, how—an agonist activates the rho family G proteins. The cultured smooth muscle cells employed in the present study should provide a useful model for further biochemical investigation about the role of rho p21 in the regulation of $MLC_{20}$ phosphorylation in smooth muscle.

*Acknowledgment.* This work was supported by grants from the Ministry of Education, Science, and Culture of Japan and by funds for cardiovascular research from Tsumura Co. We thank N. Miyamoto, F. Iwata, and E. Kishimoto for technical and secretarial assistance.

## References

1. Rasmussen H, Takuwa Y, Park S (1985) Protein kinase C in the regulation of smooth muscle contraction. FASEB J 1:177–185
2. Takuwa Y, Takuwa N, Rasmussen H (1986) Carbachol induces a rapid and sustained hydrolysis of polyphosphoinositides in bovine tracheal muscle. J Biol Chem 261:14670–14675
3. Morgan JP, Morgan KG (1984) Stimulation-specific patterns of intracellular calcium levels in smooth muscle of ferret portal vein. J Physiol (Lond) 351:155–167
4. Takuwa Y, Rasmussen H (1987) Measurement of cytoplasmic free $Ca^{2+}$ concentration in rabbit aorta using the photoprotein, arquorin. J Clin Invest 80:248–257
5. Takuwa Y, Takuwa N, Rasmussen H (1987) Measurement of cytoplasmic free $Ca^{2+}$ concentration in bovine tracheal muscle using aequorin. Am J Physiol 253:C817–C827
6. Takuwa Y, Kelley G, Takuwa N, Rasmussen H (1988) Protein phosphorylation changes in bovine carotid artery smooth muscle during contraction and relaxation. Mol Cell Endocrinol 60:71–86
7. Kamm KE, Stull JT (1985) The function of myosin and myosin light chain kinase phosphorylation in smooth muscle. Annu Rev Pharmacol Toxicol 25:593–620
8. Morgan JP, Morgan KG (1984) Stimulus-specific patterns of intracellular calcium levels in smooth muscle of ferret protal vein. J Physiol (Lond) 351:155–167
9. Rembold CM, Murphy RA (1988) Myoplasmic $[Ca^{2+}]$ determines myosin phosphorylation in agonist-stimulated swine arterial smooth muscle. Circ Res 63:593–603
10. Suematsu E, Resuick M, Morgan KG (1991) Changes of $Ca^{2+}$ requirement for myosin phosphorylation by prostaglandin $F_{2\alpha}$. Am J Physiol 261:C253–C258
11. Tang D, Stull JT, Kubota Y, Kamm KE (1992) Regulation of the $Ca^{2+}$ dependence of smooth muscle contraction. J Biol Chem 267:11839–11845
12. Nishimura J, Kolber M, van Breemen C (1988) Norepinephrine and GTPγS increase myofilament $Ca^{2+}$ sensitivity in α-toxin permeabilized arterial smooth muscle. Biochem Biophys Res Commun 157:677–683
13. Fujiwara T, Itoh T, Kubota Y, Kuriyama H (1989) Effects of guanosine nucleotides on skinned smooth muscle tissues of the rabbit mesenteric artery. J Physiol (Lond) 408:535–547
14. Kitazawa T, Kobayashi S, Horiuchi K, Somlyo AJ, Somlyo AP (1989) Receptor-coupled, permeabilized smooth muscle. J Biol Chem 264:5339–5342
15. Kitazawa T, Gaylinn BD, Denney GH, Somlyo AP (1991) G-protein-mediated $Ca^{2+}$ sensitization of smooth muscle contraction through myosin light chain phosphorylation. J Biol Chem 266:1708–1715
16. Kubota Y, Nomura M, Komm KE, Mumby MC, Stull JT (1992) GTPγS-dependent regulation of smooth muscle contractile elements. Am J Physiol 262:C405–C410
17. Moreland S, Nishimura J, van Breemen C, Ahn HY, Moreland RS (1992) Transient myosin phosphorylation at constant $Ca^{2+}$ during agonist activation of permeabilized arteries. Am J Physiol 263:C540–C544
18. Kitazawa T, Masuo M, Somlyo AP (1991) G protein-mediated inhibition of myosin light-chain phosphatase in vascular smooth muscle. Proc Natl Acad Sci USA 88:9307–9310
19. Fujiwara T, Itoh T, Kubota Y, Kuriyama H (1988) Actions of a phorbol ester on factors regulating contraction in rabbit mesenteric artery. Circ Res 63:893–902
20. Nishimura J, Khail RF, Drenth JP, van Breemen C (1990) Evidence for increased myofilament $Ca^{2+}$ sensitivity in norepinephrine-activated vascular smooth muscle. Am J Physiol 259:H2–H8
21. Gong MC, Fuglsang A, Alessi D, Kobayashi S, Cohen P, Somlyo AV, Somlyo AP (1992) Arachidonic acid inhibits myosin light chain phosphatase and sensitized smooth muscle to calcium. J Biol Chem 267:21492–21498

22. Hirata K, Kikuchi A, Sasaki S, Kuroda S, Kaibuchi K, Matsuura Y, Seki H, Saida K, Takai Y (1992) Involvement of rho p21 in the GTPγS-enhanced calcium ion sensitivity of smooth muscle contraction. J Biol Chem 267:8719–8722
23. Madaule P, Axel P (1985) A novel ras-related gene family. Cell 41:31–40
24. Narumiya S, Morii N (1993) rho Gene products, botulinum C3 exoenzyme and cell adhesion. Cell Signal 5:9–19
25. Kikuchi A, Yamamoto K, Fujita T, Takai Y (1988) ADP-ribosylation of the bovine brain rho protein by botulinum toxin type C1. J Biol Chem 263:16303–16308
26. Narumiya S, Sekine A, Fujiwara M (1988) Substrate for botulinum ADP-ribosyltransferase, Gb, has an amino acid sequence homologous to a putative rho gene product. J Biol Chem 263:17255–17257
27. Nakanishi S, Kakita S, Takahashi I, Kawahara K, Tsukuda E, Sano T, Yamada K, Yoshida M, Kase H, Matsuda Y, Hashimoto Y, Nonomura Y (1992) Wortmannin, a microbial product inhibitor of myosin light chain kinase. J Biol Chem 267:2157–2163
28. Kawahara K, Kawata M, Sunako M, Araki S, Koide M, Tsuda T, Fukuzaki H, Takai Y (1990) Indentification of a major GTP-binding protein in bovince aortic smooth muscle sytosol as the rho A gene product. Biochem Biophys Res Commun 170:637–683
29. Cassidy P, Hoar PE, Kerrick WGL (1979) Irreversible thiophosphorylation and activation of tension in functionally skinned rabbit ileum strips by [$^{35}$S] ATPγS. J Biol Chem 254:11148–11153
30. Alesso D, Macdougall LK, Sola MM, Ikebe M, Cohen P (1992) The control of protein phosphatase-1 by targetting subunits: the major myosin phosphatase in avian smooth muscle is a novel form of protein phosphatase-1. Eur J Biochem 210:1023–1035
31. Hubbard MJ, Cohen P (1993) On target with a new mechanism for the regulation of protein phosphorylation. Trend Biochem Sci 18:172–177
32. Hubbard MJ, Cohen P (1989) Regulation of protein phosphatase-1G from rabbit skeletal muscle. 1. Phosphorylation by cAMP-dependent protein kinases at site 2 releases catalytic subunit from the glycogen-bound holoenzyme. Eur J Biochem 186:701–709
33. Manser E, Leuing T, Salihuddin H, Zhao Z, Lim L (1994) A brain serine/threonine protein kinase activated by rac1. Nature 367:40–46
34. Takai Y, Kaibuchi K, Kikuchi A, Kawata M (1992) Small GTP-binding proteins. Int Rev Cytol 133:187–654
35. Boguski MS, McCormick F (1993) Proteins regulating ras and its relatives. Nature 336:643–654

# Molecular Dissection of Regulatory Light Chain Function in Vertebrate Smooth Muscle Myosins

John Kendrick-Jones[1], Tony Rowe[1], Ana C. Rasera da Silva[2], and Fernando C. Reinach[2]

*Summary.* The recent elucidation of the crystal structures of the myosin head and regulatory domain allows us to analyze the changes in myosin structure that occur during regulation. The light chains are the regulatory subunits and are bound to the α-helical segment of the heavy chain (regulatory domain), which extends from the motor domain (containing the ATPase and actin-binding sites) to the head–tail junction. We have used a recombinant DNA approach to investigate how the light chains in this location regulate the motor domain. To identify the subdomains/regions important for regulatory function, we have synthesized a series of mutant regulatory light chains (RLCs) and characterized them by their ability to restore $Ca^{2+}$ regulation to the desensitized scallop myosin test system. Assays on chimeric RLCs composed of subdomains derived from vertebrate smooth muscle and skeletal muscle myosin RLCs demonstrate that the origin of the third subdomain specifies the regulatory capability of the RLC. A series of smooth muscle myosin RLC mutants with deletions in the fourth subdomain show that the C-terminal helix in this subdomain is essential for regulation. Although the skeletal and scallop RLCs need an intact $Ca^{2+}/Mg^{2+}$-binding site in the first subdomain for function, the smooth muscle RLC behaves differently and does not require divalent metal binding to this site for regulation. These studies demonstrate the regions of the RLC that are important for regulatory function.

*Key Words.* Myosin—Light chains—Regulation

## Introduction

In vertebrate smooth muscle/nonmuscle myosins (myosin IIs) one regulatory light chain (RLC) and one essential light chain (ELC) are associated with the neck region of each myosin head [1, 2] and are involved in regulating contractile motor activity (Fig. 1). In these myosins "motor" activity is controlled by phosphorylation of the RLCs at a specific serine near the N-terminus by a $Ca^{2+}$/calmodulin-dependent myosin light chain kinase [3–5]. In vitro studies have provided compelling evidence that RLC phosphorylation regulates vertebrate smooth muscle/nonmuscle myosin filament as-

[1]MRC Laboratory of Molecular Biology, Hills Road, Cambridge, CB2 2QH, UK
[2]Department of Biochemistry, Instituto de Química, Universidade de São Paulo, CP 20780, CEP 01498 São Paulo SP, Brazil

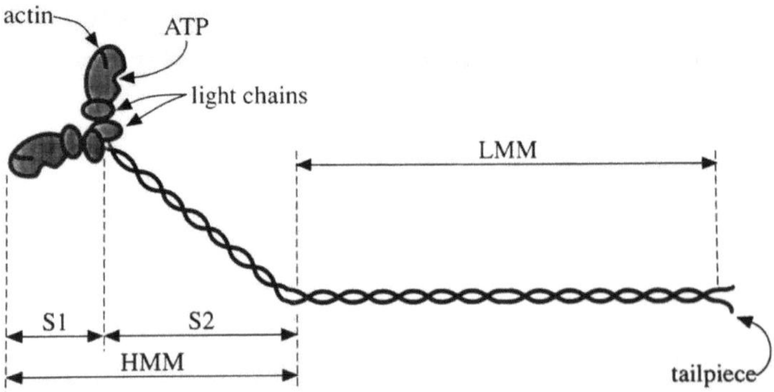


**Fig. 1.** Myosin II molecule showing the two globular heads and the α-helical coiled-coil tail. One essential light chain (ELC) and one regulatory light chain (RLC) are bound to the neck region of the heads. The heads (motor domains) contain the ATPase and actin-binding sites. The proteolytic subfragments of the myosin molecule [1] are shown: *S1*, subfragment 1, myosin head; *S2*, subfragment 2; *HMM*, heavy meromyosin; *LMM*, light meromyosin

sembly as well as myosin interaction with actin. For example, under physiologic conditions in vitro, unphosphorylated myosin filaments disassemble in the presence of MgATP into monomers with folded tails and a sedimentation coefficient of 10S [6–8]. RLC phosphorylation, by unfolding the 10S monomers into transient "assembly-competent" 6S intermediates, promotes filament assembly. In the folded (10S) state the myosin is enzymatically inactive; that is, the products of ATP hydrolysis are effectively "trapped" at the active site and are released slowly [9, 10]. These in vitro studies clearly demonstrate that the RLC under phosphorylation control has a profound effect on the conformation of the whole myosin molecule. However, in smooth muscles stable myosin filaments exist regardless of the phosphorylation state of the myosin [11], demonstrating that myosin filament assembly is not a major regulatory step in smooth muscle in vivo. It is possible that in nonmuscle cells there may exist a cytoplasmic pool of inert 10S monomers that can be recruited into enzymatically active filaments when the RLCs are phosphorylated and assembled with actin to form a contractile apparatus (e.g., in the contractile ring during cytokinesis). The genetic studies of Karess et al. [12] have provided direct proof that the RLC is required for cytokinesis in *Drosophila*.

## Myosin Light Chains

There is now considerable structural information on the myosin light chains. On the basis of sequence analysis, they are homologous to members of the EF-hand family of $Ca^{2+}$-binding proteins such as calmodulin and troponin C [13]. The crystal structures of calmodulin and troponin have been determined [14], and they have served as models for predicting light chain structure. An RLC based on these structures is shown in Figure 2. The characteristic feature of these proteins is that they are composed of four EF-hand subdomains arranged in two domains separated by a long central helix. Each subdomain consists of a 12 residue $Ca^{2+}$-binding loop flanked by α-helices of 10–13 residues. The subdomains are designated 1 to 4 and the helices A to H sequentially from the N- to the C-terminus. In all the RLCs three of the $Ca^{2+}$-

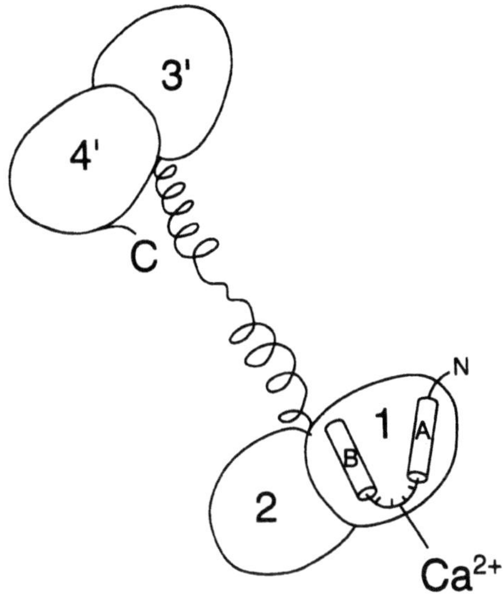


**Fig. 2.** Myosin RLC, based on the crystal structures of troponin C and calmodulin [13] and the recently identified structures of the myosin head [20] and myosin regulatory domain [21]. The subdomains are labeled *1–4* sequentially from the N- to the C-terminus. The subdomains are arranged in pairs 1 + 2 and 3 + 4 to form two domains separated by a central helix. Subdomain 1 is a high affinity $Ca^{2+}/Mg^{2+}$-binding site (helix *A*—$Ca^{2+}$ binding loop—helix *B*), whereas subdomains 2, 3, and 4 have lost divalent metal binding owing to deletions and nonconservative substitutions of the $Ca^{2+}$-coordinating residues [14, 15]. The central region between the two domains, which forms a continuous helix in the troponin C and calmodulin structures [13], is bent in the RLCs and ELCs to allow them to bind to the myosin heavy chain (see Fig. 8)

binding loops have deletions and nonconservative substitutions that would render them incompetent to bind $Ca^{2+}$. Only the first subdomain has retained all the residues necessary to form a competent high-affinity divalent metal-binding site [15, 16]. This site binds $Ca^{2+}/Mg^{2+}$ non-selectively and has a structural role in vertebrate skeletal muscle and scallop myosins [17, 18]. It is distinct from the regulatory $Ca^{2+}$-specific site, unique to molluscan myosins, which is located on the ELC [19, 20]. The recent elucidation of the crystal structures of a vertebrate skeletal muscle myosin head and associated light chains [21] and the regulatory domain of scallop myosin [22] can greatly increase our understanding of the structure of the light chains and how they regulate the myosin head.

The aim of our studies was to understand how the RLCs regulate the activity of the vertebrate smooth muscle myosin motor domain. Two classes of vertebrate regulatory light chains can be distinguished on the basis of their regulatory function: (1) the regulatory light chains in smooth muscle myosins are phosphorylated by a specific $Ca^{2+}$/calmodulin-dependent kinase that regulates myosin interaction with actin and hence the contractile response; whereas (2) the regulatory light chains in striated muscle myosins, although also phosphorylated by a specific $Ca^{2+}$/calmodulin-dependent kinase, play no apparent regulatory role [23]. Skeletal muscle contraction is controlled by $Ca^{2+}$ binding to the thin filament-associated troponin–tropomyosin complex. Using a recombinant DNA approach, we initially focused on identifying the functional regions that distinguish vertebrate smooth RLCs from skeletal muscle

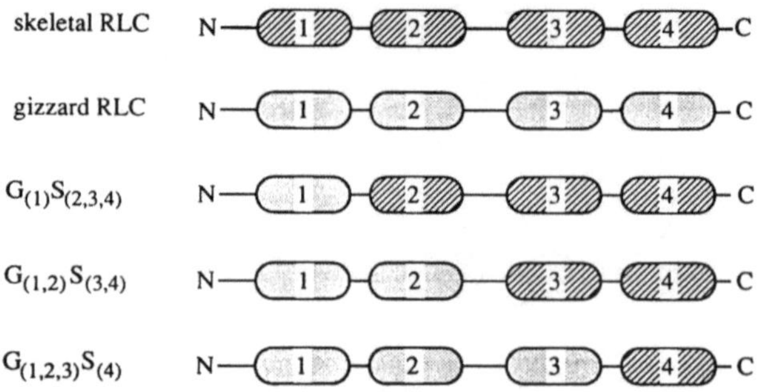


**Fig. 3.** Wild-type RLCs and chimeric mutants. The four subdomains are labeled *1–4*, and the linker regions between them are shown as *single lines*. Subdomains derived from the wild-type gizzard RLC are *stippled*, and those from the skeletal RLC are *hatched*. The chimeras are named according to their subdomain composition. *G* and *S* refer to gizzard and skeletal RLCs, respectively; *numbers in parentheses* are the subdomains, i.e. $G_{(1)}S_{(2,3,4)}$ is gizzard RLC subdomain 1 fused to skeletal RLC subdomains 2, 3, and 4; $G_{(1,2)}S_{(3,4)}$ is gizzard RLC subdomains 1 and 2 fused to skeletal RLC subdomains 3 and 4; and $G_{(1,2,3)}S_{(4)}$ is gizzard RLC subdomains 1, 2, and 3 fused to skeletal RLC subdomain 4

myosin RLCs. One possible approach to identify the specific residues involved in regulatory function was to substitute the amino acids that differed in the sequences of these two types of RLC. However, sufficient sequence divergence exists between these RLCs such that it was difficult to design point mutants intelligently. Therefore an alternative strategy was necessary. We chose to synthesize chimeric mutants with subdomains swapped between RLCs from chicken skeletal and smooth muscle (gizzard) myosins. This approach was feasible, as the protein sequences of these RLCs are highly homologous (overall, 52% of the residues in corresponding positions are identical).

## Chimeric RLCs for Identification of the Subdomain Responsible for Regulatory Function

To construct the chimeric light chain mutants, we used the linker regions between the "EF hand" subdomains as the splice junctions in the gizzard and skeletal RLCs (Fig. 2). These linker regions are located in surface loops in the tertiary structures of troponin C and calmodulin [14, 24] and were assumed to be similarly located in the RLCs. Three chimeras were synthesized (Fig. 3): $G_{(1)}S_{(2,3,4)}$, $G_{(1,2)}S_{(3,4)}$, and $G_{(1,2,3)}S_{(4)}$ (G and S refer to gizzard and skeleton, respectively, and the numbers in brackets represent subdomains 1–4; thus $G_{(1)}S_{(2,3,4)}$ is gizzard RLC subdomain 1 fused to skeletal RLC subdomains 2, 3, and 4). They were synthesized in *Escherichia coli* using the T7 RNA polymerase-directed vector pMW 172 [25] and purified by the protocols described by Rowe and Kendrick-Jones [26].

To characterize the functional properties of these RLC mutants it is essential to be able to reconstitute them with myosin heads. In previous studies scallop myosin was used to test the structural and functional properties of molluscan and vertebrate RLCs [27]. Molluscan myosins are regulated by $Ca^{2+}$-specific binding to a site localized on the ELC [19, 20]. Removal of the native RLCs by treatment with EDTA (desensitization) abolishes $Ca^{2+}$ binding to this regulatory site and $Ca^{2+}$ sensitivity of the actomyosin interaction [28] (Fig. 4). When RLCs are reconstituted with desensitized

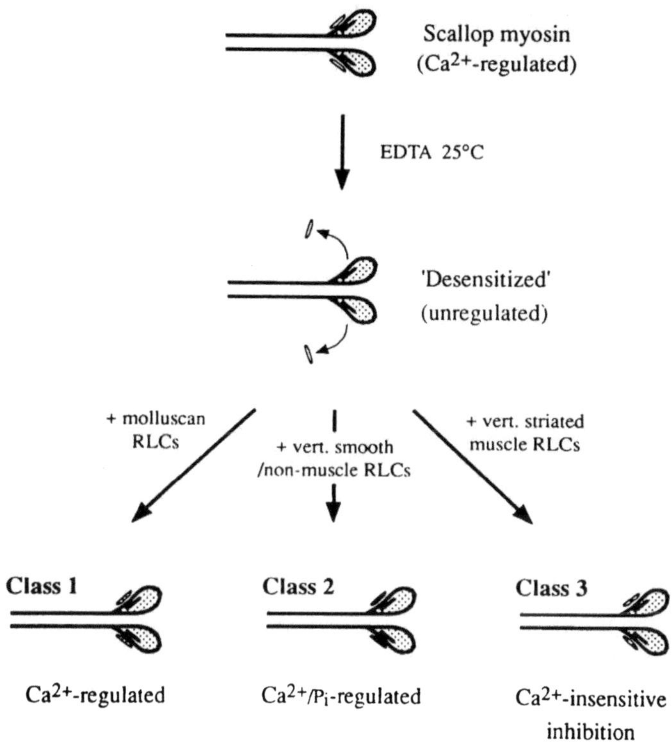


**Fig. 4.** Procedure used to characterize the regulatory Light chain (*RLC*) mutants. Treatment of scallop myosin with EDTA at 25°C releases both RLCs and abolishes inhibition of the actomyosin MgATPase in the absence of $Ca^{2+}$, leading to the "desensitized" state. RLCs from molluscan and vertebrate myosins can be divided into three distinct functional classes based on their ability to restore regulation to the actomyosin MgATPase of desensitized scallop myosin. Molluscan RLCs are able to restore $Ca^{2+}$ regulation (*class 1*); vertebrate smooth muscle/nonmuscle RLCs confer $Ca^{2+}$/phosphorylation-dependent regulation (*class 2*); and vertebrate striated muscle RLCs impose $Ca^{2+}$/phosphorylation-insensitive inhibition (*class 3*)

scallop myosin, their regulatory capability depends on the class of myosin from which they originated [review: 29]. For example, molluscan and vertebrate smooth muscle/nonmuscle RLCs reform the regulatory $Ca^{2+}$-binding site and confer regulation by $Ca^{2+}$, phosphorylation, or both. However, vertebrate striated muscle RLCs fail to restore $Ca^{2+}$ binding and inhibit the actomyosin interaction in a $Ca^{2+}$/phosphorylation-independent fashion.

Because scallop myosin is the only myosin where we can reversibly remove the RLCs, it has been used previously as a model system to test the properties of RLCs/mutants [27]. An obvious question arises: How appropriate is the scallop myosin system for testing the functional capability of these smooth muscle myosin RLC mutants? Several observations have suggested that in molluscan and vertebrate smooth muscle myosins, the regulatory systems mediated by the RLCs operate by similar basic mechanisms, even though the regulatory signals differ. For instance, smooth muscle myosin RLCs are able to mediate $Ca^{2+}$- or phosphorylation-dependent regulation of scallop myosin function [30, 31]. Both types of myosin are able to adopt a folded monomer (10S) conformation where binding of the tail to the neck region of

**Table 1.** Ability of chimeric regulatory light chains (RLCs) to restore $Ca^{2+}$ regulation to desensitized scallop myofibrils.

| Desensitized myofibrils | Actomyosin ATPase ($\mu$mol $H^+$ $min^{-1}$ $mg^{-1}$) $-Ca^{2+}$ | $+Ca^{2+}$ | $Ca^{2+}$ binding at 2.1 $\mu$M $Ca^{2+}$ (mol $Ca^{2+}$/mol myosin) |
|---|---|---|---|
| Control | 0.24 | 0.28 | 0.47 |
| + Scallop RLC | 0.04 | 0.35 | 1.84 |
| + Gizzard RLC | 0.07 | 0.26 | 1.44 |
| + Skeletal RLC | 0.05 | 0.07 | 0.68 |
| + $G_{(1)}$ $S_{(2,3,4)}$ | 0.06 | 0.07 | 0.55 |
| + $G_{(1,2)}$ $S_{(3,4)}$ | 0.04 | 0.04 | 0.62 |
| + $G_{(1,2,3)}$ $S_{(4)}$ | 0.06 | 0.26 | 1.55 |

Actomyosin MgATPase activities were measured at 25°C and pH 7.8; $Ca^{2+}$ binding was measured at a free $Ca^{2+}$ ion concentration of 2.1 $\mu$M at 0°C and pH 7.0.
The RLCs and chimeric RLCs were incubated with the desensitized scallop myofibrils (DMFs) at a 1:1 molar ratio of RLC to myosin head at 4°C for 2 h. Unbound RLCs were removed by washing and the amount of bound RLCs determined by gel densitometry (~1 mol RLC or mutant bound per myosin head). MgATPase activities and $Ca^{2+}$ binding measurements were carried out as described previously [27, 37]. This experiment was repeated three times; typical data are shown [25].

the myosin heads traps the products of ATP hydrolysis at the active site [10, 32]. In both systems, subfragment 1 (a single myosin head or motor domain) is unregulated, whereas heavy meromyosin (two heads attached to a portion of rod) is regulated, implying a crucial role in regulation for the second head domain or the subfragment 2 portion of the rod [33, 34]. The underlying similarity between these regulatory systems suggest that the results obtained with gizzard RLC-desensitized scallop myosin hybrids are relevant to the situation in vertebrate smooth muscle myosins, where the regulatory signal is RLC phosphorylation.

The regulatory capabilities of the chimeric RLCs were assessed when reconstituted with scallop desensitized myofibrils (DMFs) by their ability to mediate $Ca^{2+}$ regulation of the interaction of myosin ATP with actin (as measured by the steady-state MgATPase activity) (Table 1). As expected, removal of the native scallop RLCs (desensitized myofibril control) abolished both high-affinity $Ca^{2+}$-specific binding and inhibition of the actomyosin MgATPase in the absence of $Ca^{2+}$ [29]. All the expressed RLCs, when bound to DMFs, inhibited the actomyosin MgATPase in the absence of $Ca^{2+}$, demonstrating that they refolded into a functional conformation. The gizzard RLC reformed the regulatory $Ca^{2+}$-binding site and restored $Ca^{2+}$ regulation to the actomyosin MgATPase, whereas the skeletal RLC failed to restore $Ca^{2+}$ binding and inhibited the actomyosin MgAPTase in the presence and absence of $Ca^{2+}$ [28]. The chimeric mutants $G_{(1)}S_{(2,3,4)}$ and $G_{(1,2)}S_{(3,4)}$ behaved like the wild-type skeletal RLC. They did not reform the regulatory $Ca^{2+}$-binding site and inhibited the actomyosin MgATPase in a $Ca^{2+}$-independent fashion, indicating that the origin of the C-terminal subdomains specifies the regulatory properties of the RLC. Interestingly, $G_{(1,2,3)}S_{(4)}$ restored $Ca^{2+}$ binding and $Ca^{2+}$ regulation to the actomyosin MgATPase, similar to the wild-type gizzard RLC. Therefore subdomain 3 appears to determine the regulatory function of the RLC. These observations suggest that the ability of the RLC to interact with the scallop myosin head to reform the $Ca^{2+}$-specific binding site and thus regulate the actomyosin MgATPase activity is dependent on the identity of the third subdomain. Thus chimeric RLCs in which this subdomain was derived from the

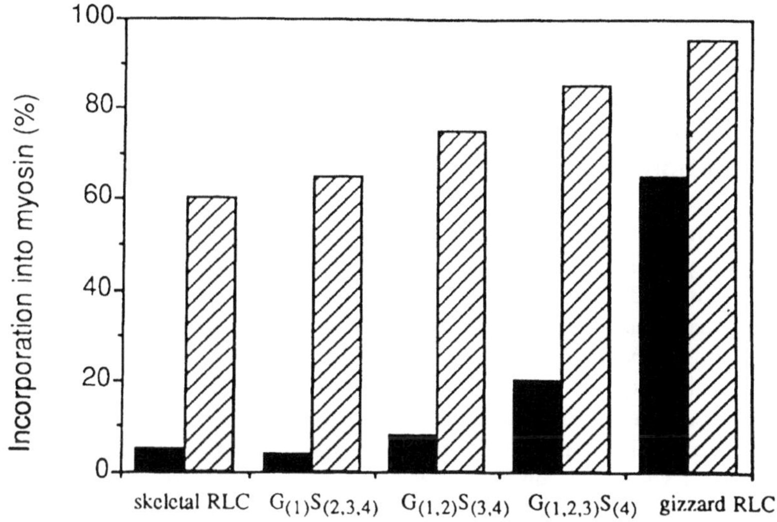


**Fig. 5.** Percentage incorporation of the wild-type RLCs and chimeric mutants into vertebrate smooth muscle myosin (*black bars*) and vertebrate skeletal muscle myosin (*hatched bars*). RLC exchange was carried out by mixing the vertebrate myosins with a 15- to 20-fold molar excess of RLCs or mutants under the conditions described by Morita et al. [35]. The level of incorporation of foreign RLCs and chimeric mutants into the myosins was determined as described in Rowe and Kendrick-Jones [26]

skeletal RLC were nonregulatory, whereas those containing a gizzard RLC subdomain 3 were regulatory.

Although we were unable to selectively remove the RLCs from gizzard or skeletal myosins, under suitable conditons [35] with a large excess of added RLCs we were able to partially exchange the RLCs into these myosins. We found that incorporation of wild-type gizzard RLC into gizzard myosin was at least 60%, whereas only 4% exchange of the skeletal RLC was achieved even with a 15-fold molar excess of exogenous RLC over myosin (Fig. 5). Because of the low level of RLC exchange, we were unable to test the effect of the chimeric RLCs on the regulation by phosphorylation of gizzard myosin-ATP interaction with actin. The values of exchange of $G_{(1)}S_{(2,3,4)}$ and $G_{(1,2)}S_{(3,4)}$ (5–8%) were similar to those observed with skeletal RLC, whereas incorporation of $G_{(1,2,3)}S_{(4)}$ was intermediate between that obtained with the gizzard and the skeletal RLCs. Therefore when subdomains 3 or 4 are derived from gizzard RLC the mutant light chain has a higher affinity for gizzard myosin than when these subdomains are from skeletal RLC. Greater degrees of exchange of all the RLCs into skeletal myosin were observed (Fig. 5). Exchange rates of 60–95% were consistently obtained, with the gizzard RLC being incorporated to the greatest extent and the skeletal RLC to the least extent. We found that all the wild-type RLCs and chimeric mutants had no significant effect on the actin-activated MgATPase activity of skeletal myosin. All the MgATPase rates were roughly the same, and control experiments demonstrated that the MgATPase activites were unaffected by the exchange procedure. The RLCs could be phosphorylated with myosin light chain kinase, but no effect on MgATPase activity was observed. These experiments clearly demonstrate that the wild-type vertebrate smooth or skeletal RLCs and their chimera can be incorporated into skeletal muscle myosin; they also support previous data [23] that whether phosphorylated or nonphosphorylated, the RLCs play no apparent role in the regulation of vertebrate striated muscle myosins.

## The Role of the C-Terminal Helix in RLC Subdomain 4 in Regulation

Our observations that interactions between the C-terminal subdomains of the RLC and the heavy chain appear to be crucial for myosin regulation are consistent with the data of Trybus and Chatman [36], who also constructed chimeric mutants and demonstrated that the identity of the C-terminal domain dictates the ability of the RLC to mediate phosphorylation-dependent control of smooth muscle myosin motor function. To further explore the role of this region of the RLC in myosin regulation, we constructed a series of vertebrate smooth muscle myosin RLC mutants with successive deletions of three amino acids from the C-terminus (Fig. 6). The mutants are denoted as Cd3, Cd6, Cd9, and so on, where 3, 6, 9, and so on residues were deleted from the C-terminus. To check the conformation and heavy chain binding of these deletion mutants, we measured their binding affinities to DMFs. In the presence of $Mg^{2+}$, all native RLCs bind to scallop DMFs with such high affinity that it is impossible, using conventional binding assay techniques, to determine their true binding constants. For instance, they bind to scallop DMFs in a 1:1 molar ratio (RLCs/myosin heads) when added at this equimolar ratio, even in the 10–100 nM range. Therefore to compare the relative affinities of the RLC mutants for scallop DMFs the following procedure was employed: RLC-DMF hybrids (prepared by adding the RLCs to DMFs at a slight molar excess of RLCs over myosin heads) were washed with a $Mg^{2+}$-containing buffer at temperatures of 4°, 12°, and 25°C and analyzed by polyacrylamide gel electrophoresis to determine if the RLCs remained bound to the myosin heads at the expected 1:1 molar ratio. Gel-densitometric analysis revealed that after washing at 4° and 12°C all the expressed RLCs remained bound to scallop DMFs at a molar ratio of about 0.7 mol RLCs or more per mole of myosin heads. After washing at 25°C, the band intensity of Cd39 was reduced to a value of 0.4 mol of RLC per mole of myosin heads, whereas the band intensities of Cd3 to Cd12 were similar to those observed at

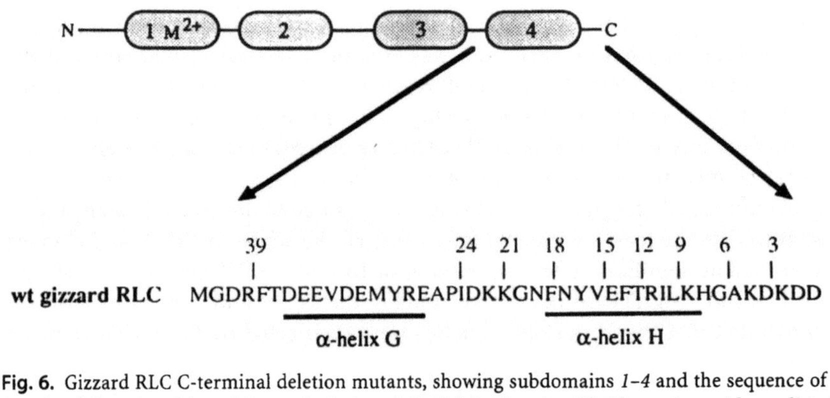


**Fig. 6.** Gizzard RLC C-terminal deletion mutants, showing subdomains *1–4* and the sequence of the fourth subdomain of the wild-type (*wt*) gizzard RLC ($Met^{129}$ to $Asp^{171}$). The amino acids predicted to form the α-helices *G* and *H* are *underlined*, and the numbers above the sequence indicate the positions of the deletions. Thus the mutants Cd3, Cd6, Cd9, and so on were generated with 3, 6, 9, and so on amino acids deleted from the C-terminus

**Table 2.** Ability of RLC C-terminal deletion mutants to restore $Ca^{2+}$ regulation to desensitized scallop myofibrils.

| Desensitized myofibrils | Actomyosin MgATPase ($\mu$mol $H^+$ $min^{-1}$ $mg^{-1}$) $-Ca^{2+}$ | $+Ca^{2+}$ | $Ca^{2+}$ binding at 2.1 $\mu$M $Ca^{2+}$ (mol $Ca^{2+}$/mol myosin) |
|---|---|---|---|
| Control | 0.12 | 0.12 | 0.56 |
| + Scallop RLC | 0.02 | 0.17 | 2.17 |
| + Gizzard RLC | 0.05 | 0.14 | 1.25 |
| + Cd3 | 0.04 | 0.13 | 1.17 |
| + Cd6 | 0.04 | 0.12 | 1.20 |
| + Cd9 | 0.06 | 0.11 | 1.29 |
| + Cd12 | 0.12 | 0.14 | 1.45 |
| + Cd15 | 0.10 | 0.12 | 1.45 |
| + Cd18 | 0.12 | 0.13 | 1.07 |
| + Cd21 | 0.11 | 0.13 | 0.87 |
| + Cd24 | 0.10 | 0.13 | 1.02 |
| + Cd39 | 0.13 | 0.12 | 0.57 |

The MgATPase activities and $Ca^{2+}$ binding measurements were determined as described in Table 1 with the exception that the assays were performed at 12°C because Cd39 was partially released from the DMFs at 25°C.
The DMF-RLC hybrids were prepared as described in Table 1. The experiment was repeated three times; typical data are shown. Similar results were observed when desensitized scallop myosin was used rather than DMFs [39].

the lower temperatures and there was only a slight reduction in Cd15 to Cd24 (<10%) (data not shown). Thus Cd39 (but not Cd3 to Cd24) was partially released from DMFs by washing at 25°C owing to its decreased affinity for myosin heads, indicating that sites on the RLC concerned with binding to the scallop heavy chain–ELC complex were either removed or conformationally altered.

The regulatory capabilities of these C-terminal deletion mutants were assessed by measuring their abilities to restore $Ca^{2+}$-dependent function to scallop DMFs (Table 2). The $Ca^{2+}$-binding affinities of the RLC-DMF hybrids were measured at 2.1 $\mu$M $Ca^{2+}$ in the presence of 2 mM $Mg^{2+}$. Under these conditions, scallop myofibrils bind 2 mol of $Ca^{2+}$ per mole of myosin at the regulatory $Ca^{2+}$-specific sites located on the scallop ELC [19, 22, 37]. The $Ca^{2+}$ binding affinities and steady-state actomyosin MgATPase activities of the RLC-DMF hybrids were determined at 12°C to ensure that a significant amount of Cd39 remained bound to the scallop DMFs during the course of these assays (Table 2). The scallop DMF control sample had a low $Ca^{2+}$-binding affinity, and its actomyosin MgATPase was elevated in the absence and presence of $Ca^{2+}$ [28]. Rebinding the scallop or gizzard RLCs restored both high-affinity $Ca^{2+}$ binding and $Ca^{2+}$ regulation, although the wild-type gizzard RLC-DMF hybrid had a reduced affinity for $Ca^{2+}$ compared to the scallop RLC-DMF hybrid, as previously observed [30]. Mutants Cd3 and Cd6 restored $Ca^{2+}$ sensitivity to DMFs, like the wild-type gizzard RLC, indicating that deletions of up to six residues from the C-terminus had relatively little effect on the regulatory capability of the RLC. Further deletions of 9–15 residues disrupted the inhibitory function of gizzard RLC without affecting its ability to restore $Ca^{2+}$ binding. Obviously the region deleted in these mutants (corresponding to the predicted C-terminal helix H of subdomain 4) is essential for the inhibitory function of the gizzard RLC. The mutants with deletions of 18 to 39 residues, although remaining bound to scallop DMFs during the course of these assays at a molar ratio of about 0.7 mol of RLC per mole of myosin heads, exhibited a progressively decreased ability to restore $Ca^{2+}$ binding. Cd39 appeared to be completely nonfunctional. Because the binding affinities of Cd18 to Cd39 for scallop DMFs were also reduced it is likely that the loss of their abilities to restore $Ca^{2+}$ sensitivity to myosin interactions with actin is due to either the gross disruption of their tertiary structures or the deletion of residues involved in stabilizing the conformation of the C-terminal domain of the RLC and the ELC–heavy chain complex.

We also explored whether it would be possible to exchange these gizzard RLC C-terminal deletions mutants in gizzard myosin in order to study their effect on the phosphorylation-dependent regulation of this myosin. We found that the level of exchange of the deletion mutants Cd3 and Cd6 was at least 90%, implying that their affinities for smooth muscle myosin were comparable to that of the native gizzard RLC. However, further deletions of residues from the C-terminus decreased the ability of the mutants to bind to gizzard myosin; for example, Cd9 was incorporated to a level of about 40%, and no incorporation was detected for Cd12 to Cd39. After RLC exchange, the effects of phosphorylation by $Ca^{2+}$/calmodulin-dependent myosin light chain kinase on the actin-activated MgATPase activities of the gizzard myosins hybrids were measured. All the myosins that were exchanged with the wild-type gizzard RLC or Cd3 to Cd9 could be fully phosphorylated, and they exhibited actomyosin MgATPase activities that were activated by phosphorylation to the same extent. Unfortunately, because of the lower myosin-bindign affinities of Cd12 to Cd39 we were unable to exchange any of these deletion mutants into gizzard myosin and thus were unable to assess their effects on the actin-activated MgATPase of this myosin. These results suggest that the loss of inhibitory function of mutants Cd12 to Cd39, when reconsitituted with desensitized scallop myosin, may be due to the disruption of their tertiary structures.

These site-directed mutagenesis studies have clearly demonstrated the importance of the C-terminus of the smooth muscle myosin RLC for regulation (Fig. 6). Deletions in the helix H of subdomain 4 (mutants Cd12 and Cd15) abolished the ability of the gizzard RLC to switch off the actomyosin MgATPase of scallop DMFs in the absence of $Ca^{2+}$ but did not affect its ability to restore $Ca^{2+}$ binding. Now that we have information on the structure of the light chains [21, 22] we may be able to explain these results in structural terms. The pattern of hydrophobic contacts involved in the interactions between the subdomains observed in the crystal structures of troponin C, calmodulin, and the vertebrate skeletal and scallop myosin RLCs [14, 21, 22] are well conserved in the gizzard RLC, and so one may predict that in this RLC helix H of subdomain 4 forms a hydrophobic interface with helix E of subdomain 3 as well as with helix G in subdomain 4. The hydrophobic interactions stabilizing helices E and H involve conserved Phe residues ($Phe^{106}$ and $Phe^{158}$ in the gizzard RLC; see Fig. 6 where $Phe^{158}$ is in the middle of helix H, 13 residues from the C-terminus). Thus it becomes clear why deletions in helix H destabilize both subdomain 3 and 4 and why this region of the RLC plays a crucial role in $Ca^{2+}$ regulation. Interestingly, Goodwin et al. [18] showed that substitution of the last 11 residues at the C-terminus of scallop RLC with an unrelated sequence of 15 amino acids yielded a mutant that was able to restore $Ca^{2+}$ binding but not $Ca^{2+}$ regulation to scallop DMFs. These results indicate that disruption of helix H with the deletion of the residues involved in hydrophobic interactions with neighboring helices is sufficient to abolish $Ca^{2+}$ regulation. The recent determination of the structure of the regulatory domain of scallop myosin [22] has revealed that the specific $Ca^{2+}$ regulatory binding site is in subdomain 1 of the ELC and is stabilized by linkages involving the region between subdomains 3 and 4 of the RLC (discussed later in more detail).

## Importance of Metal Binding to Subdomain 1 of Gizzard RLC

The primary sequences of more than 20 RLCs have so far been determined by both protein and DNA sequencing [review: 40]. Sequence analysis reveals that the RLCs are homologous to troponin C and calmodulin and have a similar four-subdomain structure [13] (Fig. 2). Sequence alignments show that in all the RLCs only the first subdomain has retained all the residues necessary to form a competent high-affinity divalent metal-binding site [15, 16]. A typical EF hand $Ca^{2+}/Mg^{2+}$-binding site is shown in Fig. 7. Each of the 12 residues of the $Ca^{2+}$-binding loop play an important role in defining its correct conformation for $Ca^{2+}$ binding [14]. In all the crystal structures so far determined [14] (the scallop ELC is an exception, see Fig. 10), five of the six $Ca^{2+}$-coordinating residues (denoted X, Y, Z, −Y, −X, and −Z) are directly involved in providing oxygen ligands for the $Ca^{2+}$ ion (the ligand at the −X position is a water molecule). The $Ca^{2+}$ ion is coordinated by seven oxygen ligands (the acidic residue at the −Z position contributes both of its side-chain oxygen atoms), and the ligand geometry is described as pentagonal bipyramidal. The $Mg^{2+}$ is coordinated by six oxygen ligands; the acidic residue at the −Z position rotates so that only one of its side chain oxygen atoms acts as a ligand. The other residues in the loops are involved in hydrogen-bonding interactions that stabilize the correct conformation of this site for $Ca^{2+}/Mg^{2+}$ binding. The invariant glycine between the residues at the Z and −Y positions is essential to allow the main chain to make a 90° turn so the residues at the −Y, −X, and −Z positions are correctly oriented for $Ca^{2+}/Mg^{2+}$ coordination.

The first subdomain of the RLC binds $Ca^{2+}/Mg^{2+}$ nonselectively with an affinity that is too high to be of any regulatory significance: $K_D$ for $Ca^{2+}$ is $10^{-7}$ to $10^{-8}$ M [41, 42]. However, because of the relative affinities of $Ca^{2+}$ and $Mg^{2+}$ for this site ($K_{Mg}/K_{Ca} = 10$) [41], it is believed to be occupied by a $Mg^{2+}$ ion in resting muscles; and because the $Mg^{2+}$ "off" rate is slow, $Ca^{2+}$ only slowly replaces the $Mg^{2+}$ during contraction [43]. Thus this nonspecific divalent metal-binding is distinct from the regulatory $Ca^{2+}$-specific binding site ($K_D \sim 10^{-6}$ M) located on the ELC on the molluscan myosin head [41]. Site-directed mutagenesis experiments where the acidic residues were substituted with alanines in the first and last $Ca^{2+}$coordinating positions in the binding loop of subdomain 1 of the vertebrate skeletal and scallop myosin RLCs (X and −Z; i.e., 1st and 12th positions in Fig. 7) have shown that metal binding to this site is essential for the proper function of these RLCs [17, 18]. We therefore believed that if we mutated either of these ligating residues in the binding loop of subdomain 1 of the gizzard RLC we would similarly "knock out" $Ca^{2+}/Mg^{2+}$ binding to this site and lose regulation. However, this assumption turned out not to be correct.

As part of a project to investigate the molecular basis for $Ca^{2+}/Mg^{2+}$ discrimination in regulatory proteins, we prepared a number of gizzard RLC mutants with substitutions in the critical ligating positions in the binding loop of subdomain 1. The most interesting mutant in regard to regulation was D12A, where the aspartic acid residue in the 12th position in the binding loop (−Z position) was replaced by alanine. $Ca^{2+}$ binding to this RLC mutant was virtually abolished. (At 1 μM free $Ca^{2+}$ there was no detectable $Ca^{2+}$ binding, whereas at 10 μM $Ca^{2+}$ there was a 10- to 15-fold decrease in

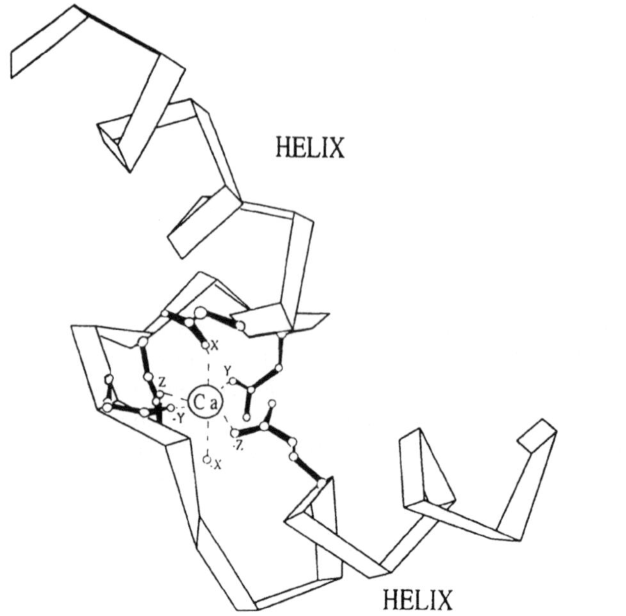


gizzard RLC subdomain 1

x y z -y -x -z

IQEFKEAFNMI **D**Q**N**R**D**G**F**I**D**KE**D**LHDMLASM

α helix — binding loop — α helix

**Fig. 7.** $Ca^{2+}/Mg^{2+}$-binding EF hand subdomain 1 of gizzard RLC. A ribbon representation of the polypeptide backbone based on the $Ca^{2+}/Mg^{2+}$-binding site of troponin C [38] is shown. Each *rectangle* represents the plane of the peptide bond for each amino acid. The side chains are shown only for the $Ca^{2+}$-coordinating residues (except for the −Y position, where the carbonyl oxygen is the ligand, and the −X position, where the ligand is a water molecule). Shown is the sequence of the gizzard RLC subdomain 1 with the α-helical and binding loop regions labeled. It is a typical EF hand motif [12]; the binding loop is made up of 12 residues with oxygen-containing side chains at the positions marked X, Y, Z, −Y, −X, and −Z (at the −Y and −X positions the oxygens are donated by carbonyl and water molecules, respectively). The amino acid residues in these positions are labeled in *bold type*. The geometry of bonding around the $Mg^{2+}$ ion is believed to be approximately octahedral with sixfold coordination, whereas the coordination sphere around the $Ca^{2+}$ ion is arranged in a distorted bipyramid with sevenfold coordination. The extra seventh ligand is provided by the Asp at position −Z, which donates both its side chain oxygens. It is this Asp residue that is changed to an Ala in the mutant D12A, which results in a loss of divalent cation binding

$Ca^{2+}$-binding affinity compared with the wild-type RLC.) However, when this mutant light chain was incorporated into desensitized scallop myosin, it formed a stable complex with the myosin and was still capable of regulating the actomyosin ATPase in a $Ca^{2+}$-dependent manner (Table 3). The same mutation in skeletal muscle myosin RLC or scallop RLC [18] locked the ATPase in the *on* state regardless of the $Ca^{2+}$ concentration, suggesting a role for the $Ca^{2+}/Mg^{2+}$ site in the RLC in the regulation of scallop actomyosin. To exclude the possibility that this $Ca^{2+}/Mg^{2+}$ site was restored in the D12A mutant when it bound to the scallop myosin, the $Ca^{2+}$ capability of the

**Table 3.** Intact $Ca^{2+}/Mg^{2+}$-binding sites in gizzard RLC and regulation of desensitized scallop actomyosin.

| Desensitized myofibrils | Actomyosin ATPase ($\mu$mol $H^+$ $min^{-1}$ $mg^{-1}$) $-Ca^{2+}$ | $+Ca^{2+}$ | $Ca^{2+}$ binding at 1 $\mu$M $Ca^{2+}$ (mol $Ca^{2+}$/mol head) |
|---|---|---|---|
| Control | 0.44 | 0.48 | 0.3 |
| + Gizzard RLC | 0.06 | 0.35 | 1.8 |
| + Gizzard D12A | 0.10 | 0.43 | 0.9 |
| + Skeletal D12A[a] | 0.29 | 0.28 | 0.3 |

Actomyosin MgATPase activities were measured at 25°C (see Table 1). $Ca^{2+}$ binding measurements were performed using a flow dialysis procedure [44] at 20°C in the absence of $Mg^{2+}$. Note that $Ca^{2+}$ binding is expressed as moles of $Ca^{2+}$ per mole of myosin head.

[a] Data for this skeletal mutant were obtained from Goodwin et al. [18].

scallop myosin hybrids were measured (Table 3). The desensitized myosin control showed only background $Ca^{2+}$ binding, whereas the gizzard RLC restored the two $Ca^{2+}$-binding sites to the myosin head ($Ca^{2+}/Mg^{2+}$ site in gizzard RLC and the $Ca^{2+}$ regulatory site on scallop ELC–myosin complex). The mutant D12A restored only one $Ca^{2+}$-binding site, presumably the $Ca^{2+}$-regulatory site required for regulation. These results demonstrate that—provided the RLC is able to interact correctly with the ELC to restore its $Ca^{2+}$-regulatory site—regulation is restored regardless of whether the RLC is capable of binding $Ca^{2+}/Mg^{2+}$. Obviously, smooth muscle RLC behaves different from skeletal muscle and scallop myosin RLCs, which need an intact $Ca^{2+}/Mg^{2+}$-binding site for regulation [17, 18]. Additional experiments are needed to explore the significance of this difference in behavior of these RLCs. For example, an intact $Ca^{2+}/Mg^{2+}$-binding site in the RLC may not be essential for the ATPase activity of the myosin or for its regulation as shown here, but it may be essential for the myosin to generate force and movement.

The ability of smooth muscle RLCs to regulate scallop myosin mediated by $Ca^{2+}$ binding or smooth muscle myosin by phosphorylation indicates that the mechanisms of regulation in vertebrate smooth muscle and molluscan myosins must be similar, even though the regulatory signals differ. At present we have only limited information on the mechanism of light chain regulation.

## Mechanism of Light Chain Regulation: Location of Light Chains on the Myosin Head

The most significant development in the myosin field in recent years has been the x-ray crystal structure determination of the chicken skeletal muscle myosin head by Rayment and colleagues [21] and of the scallop myosin regulatory domain by Cohen, Szent-Györgyi, and colleagues [22]. This structural information, together with known biochemical data, allows us to speculate how the myosin head might develop force and movement and how these activities might be regulated.

We can now pinpoint the exact position of the light chains on the myosin head. In the crystal structure of the myosin head one of the major features is the long (~85 Å) $\alpha$-helix, which extends from the "thick" part of the head, containing the ATP and actin-binding sites, down to the C-terminal end of the heavy chain (Fig. 8). The RLC binds to this $\alpha$-helical heavy chain at a highly hydrophobic sequence (residues 820–

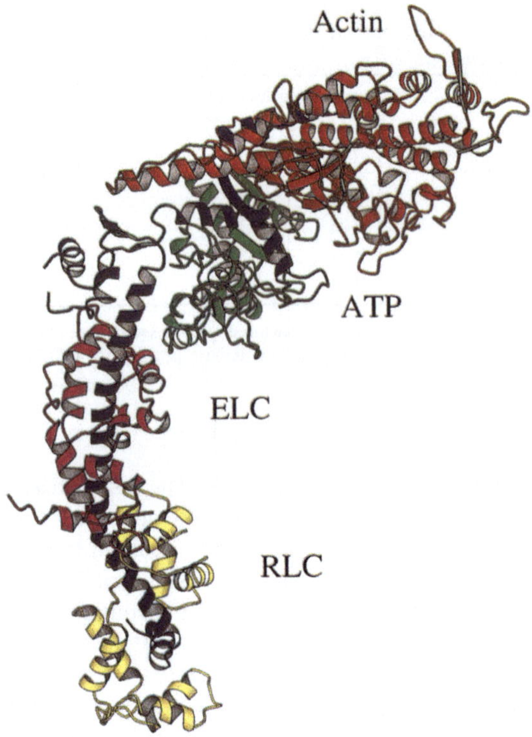


**Fig. 8.** Three-dimensional structure of chicken skeletal muscle myosin subfragment-1 (S1) [20] is presented using the program MOLSCRIPT [45]. S1 contains three polypeptide chains: ELC (*pink*), RLC (*yellow*), and the heavy chain, which is divided into three proteolytic subdomains of N-terminal 25 kDa (*green*), central 50 kDa (*red*) and C-terminal 20 kDa (*blue*). A striking feature is the extended ~8.5 nm α-helical heavy chain at the C-terminus (*blue*), which is stabilized by the binding of the ELC and RLC. The light chain binding sites on this helical heavy chain are constituted by two "IQ motif" sequences (see Fig. 9). The structures of the ELC and RLC bound to the heavy chain are similar to that of calmodulin when bound to its target peptide [23]. In these structures the central helix connecting the N-terminal and C-terminal domains (see Fig. 2) is distorted so the two domains grip the heavy chain. The remainder of the heavy chain constitutes the "bulky" motor domain with the ATPase site (indicated by *ATP*) and the actin-binding region (*Actin*), which spans the large cleft in the end of the head. This diagram was constructed using files provided by Ivan Rayment (acc. no. 1MYS)

840 in Fig. 9) lying close to the invariant proline residue (position 844) that is often chosen as the boundary of the head and tail domains of myosin [54]. The ELC lies slightly farther up the α-helix and abuts but does not overlap the RLC. The position of the light chains are in agreement with previous data obtained using binding assays [56, 57] and electron microscopy [1, 2, 58]. Cheney and Mooseker [55] identified a putative light chain-binding sequence of about 20 amino acids present in the neck region of all myosins. This basic sequence is referred to as the IQ motif because its conserved core usually fits the consensus IQXXXRGXXXR (Fig. 9). Interestingly, the IQ motif is most highly conserved among the "novel" myosins, which may contain up to six copies and are therefore associated with multiple calmodulin subunits. The myosin IIs contain two repeats, one that is well conserved and binds the ELC (positions 794–804) and another that is less well conserved and binds the RLC (positions 820–830) (Fig. 9). The positions of the putative light chain-binding sequences are

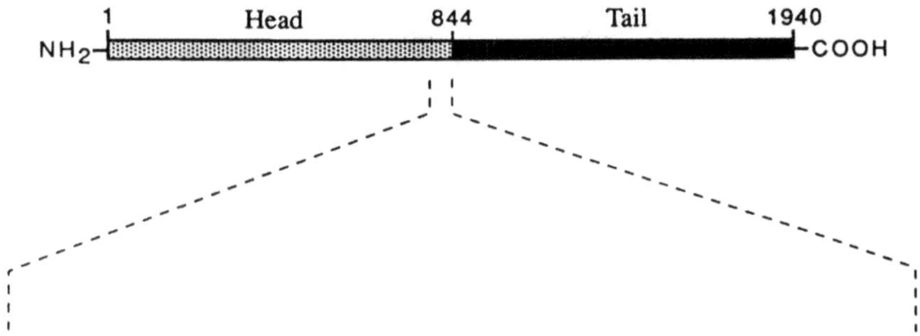


```
          790       800       810       820       830       840
a  RDRTVFSYANKIQRFLRKTAMRKYYYEVKKGGNDALVNKKERRR...........L.SLERP
b  REQAISKMVVSIQAGARAFLARRMYDKMREQTVSAKILQRNIRAWLELKNWAWYQLYVKARP
c  REQRISEIIKAIQAATRGWIARKVYKQAREHTVAARIIQQNLRAYIDFKSWPWWKLFSKARP
d  RDEKLATILTGFQSQIRWHLGLKDRKRRMEQRAGLLIVQRNVRSWCTLRTWEWFKLYGKVKP
e  RDERLSRIITRIQAQARGQLMRIEFKKMVERRDALLVIQWNIRAFMGVKNWPWMKLYFKIKP
f  RDEKLAQLITRTQARCRGFLMRVEFKKMMERRECIFCIQYNVRAFMNVKHWPWMKLFFKIKP
g  RDERLSKIISMFQAHIRGYLIRKAYKKLQDQRIGLSVIQRNIRKWLVLRNWQWWKLYSKVKP
h  RDLKITDVIIAFQAQCRGYLARKAFAKRQQQLTAMKVIQRNCAAYLKLRNWQWWRLFTKVKP
```

**Fig. 9.** Alignment of the amino acid sequences containing the IQ motifs (light chain binding sites) at the C-terminal end of the myosin head domains from: (**a**) *Acanthamoeba* myosin IC [46]; (**b**) *Acanthamoeba* myosin II [47]; (**c**) *Dictyostelium* myosin II [48]; (**d**) nematode myosin heavy chain B [49]; (**e**) rat α-cardiac myosin [50]; (**f**) chicken fast skeletal muscle myosin [51]; (**g**) scallop striated muscle myosin [52]; (**h**) chicken smooth muscle myosin [53]. The amino acid numbering refers that in Molina et al. [51]. The sequences shown are from the conserved arginine residue at position 783 to the invariant proline (position 844), which is thought to mark the boundary of the head and tail domains [54]. The "IQ motifs" (IQXXXRGXXXR) identified by Cheney and Mooseker [55] are thought to be the light chain-binding sites in all these myosins [22]. The conserved residues of these motifs are shown in *bold type*

consistent with site-directed mutagenesis studies showing that both light chains bind to a region within 38 amino acids from the C-terminus of nematode myosin heavy chain B [57]. In addition, a 16-amino-acid region (corresponding to positions 799–814 in Fig. 9) of the rat α cardiac myosin heavy chain has been shown to be required for binding of the ELC [59]. This evidence indicates that the location of the light chains in the α-helical regulatory domain of the myosin head (as observed in the crystal structures) is generally applicable to all the various myosin types.

## Production of Force and Movement by the Myosin Head: Regulation by Light Chains

It is well documented that the two reactive sulfydryl groups (SH1 and SH2) that lie close to the ATP-binding site can be crosslinked by a variety of bifunctional reagents in the presence, but not the absence, of ATP [60, 61]. Crosslinking these thiols traps nucleotide at the active site. In the (nucleotide-free) crystal structure of Rayment et al. [21], the ATP-binding cleft is wide open, and SH1 and SH2 are too far apart to be crosslinked. Rayment et al. [62] suggested that the binding cleft closes when ATP binds, and SH1 and SH2 move close together. They further speculated that the

x y z -y -x -z

normal EF hand f s l f **D** k **D** g **D** g **T** i **T** t k **E** l
1 2 3 4 5 6,7

scallop ELC site f e l f **D** f w **D** **G** r **D** g **A** v d a f
1,2 3,4 5 6 7

--helix-->|<----------loop--------->|<-helix-

**Fig. 10.** Novel $Ca^{2+}$ binding motif in scallop ELC. The sequence of a typical EF hand (subdomain 1 of calmodulin) [13] is compared with that of the $Ca^{2+}$ binding site recently identified in the scallop ELC subdomain 1 by Xie et al. [22]. In the normal EF hand, the $Ca^{2+}$ coordinating residues occur at the positions x, y, z, −y, −x, and −z and are labeled in *bold capital letters*. All these residues donate oxygens from their side chains to coordinate the $Ca^{2+}$, except at the −y position, where the ligand is the backbone carbonyl, and at the x position, where it is a water molecule. The *numbers* below the coordinating residues refer to the ligating groups (note that both oxygens of the side chain in the −z position are involved in ligation). In the scallop ELC $Ca^{2+}$ binding site all the ligating groups (*numbered*) are provided by the residues (in *bold capital letters*) within the nine-residue binding loop; four oxygens are donated by main chain carbonyls (*residues underlined*) and three from the side chains. Note that the Asp residues at positions 1 and 4 in the loop donate both carbonyl and side chain oxygens. The conformation of this unusual $Ca^{2+}$ binding site, where the loop bends tightly around the bound metal ion, is stabilized by cooperative interactions with the RLC and heavy chain

powerstroke might be associated with the opening of this cleft when the products of ATP hydrolysis are released. This change could be transmitted through the globular head region to move the actin-binding site at its tip by about 5 nm relative to the C-terminus. Recovery from the powerstroke might be associated with the closing of the nucleotide-binding cleft again upon ATP binding. These conformational changes in the nucleotide-binding site could be transmitted to the actin-binding regions, thereby switching the myosin between the strong and weak actin-binding states. It should be emphasized that this model of how the myosin head produces force and motion upon interaction with ATP and actin is highly speculative. It does suggest an important structural role for the light chains in stabilizing the α-helical C-terminal region, which may be linked to the mechanism of myosin regulation.

One obvious requirement for the efficient transformation of putative conformational changes of the sort described above into coordinated motion is that the C-terminal single α-helix should be rigid and in all myosins should be stabilized by the light chains. This region, called the regulatory domain [22], may serve as a lever on which the rest of the head pivots during the power stroke. If this notion is true, it raises the possibility that regulation by the light chains may be linked to the control of the flexibility of the heavy chain helix. Hence in the *off* state (which only regulated myosins can achieve) the specific interactions between the heavy and light chains should produce a conformationally tense (rigid) structure for the regulatory domain. In this state, the mobility of the myosin motor domain in the region containing SH1 and SH2 might be restricted, locking the myosin head in the weak actin-binding form with trapped ADP.Pi bound at the active site. Xie and colleagues [22] speculate that some sort of helix–coil transition might provide the means of communication between the regulatory domain and the mobile SH1 and SH2 heavy chain region and hence the ATPase site. During the transition to the *on* state, $Ca^{2+}$ binding or phosphorylation would cause rearrangements in the light chain–heavy chain interactions, producing a more flexible structure for the regulatory domain and relieving the constraints placed on the mobility of the helix. Thus the inhibition of ADP.Pi release would be overcome, and actin-binding would be able to stimulate the cycle of ATP hydrolysis. The structure of the scallop myosin regulatory domain [22] prepared in the presence of $Ca^{2+}$ represents the *on* state. In this myosin the $Ca^{2+}$-binding site is located in subdomain 1 of the ELC. The major surprise is that it has a novel EF hand structure unlike the classic EF hand motif (compare Fig. 7 and Fig. 10). The binding loop in the scallop ELC sharply twists around the bound $Ca^{2+}$ so that four of the ligands are provided by the main chain carbonyls instead of the usual side chain oxygens (Fig. 10). This twisted binding loop is stabilized by hydrogen bond linkage with the RLC and the heavy chain. Xie et al. [22] speculated that $Ca^{2+}$ binding to this site triggers a change in the scallop ELC–RLC–heavy chain interface network, leading to subunit reorganization and possibly a change in the curvature of the long α-helix of the heavy chain. The key residue in the RLC that stabilizes the $Ca^{2+}$-binding loop of the ELC is $Gly^{117}$ of domain III, which is conserved in molluscan and smooth muscle myosin RLCs but is altered in skeletal muscle myosin RLCs. These structural findings confirm our studies with the chimeric RLCs, which demonstrated that the identity of domain III determines the regulatory capability of the RLC.

## Conclusions

Although these three-dimensional structures allow us to speculate how myosin might be regulated, there are many questions that must be resolved before a complete understanding can be achieved. For example, the model does not account for the fact that subfragment 1 (consisting of a single myosin head) is unregulated; that is, it is unable to attain an *off* state under resting conditions [28, 33, 34]. On the other hand, heavy meromyosin (the soluble double-headed proteolytic fragment) is regulated. Taken together, these observations demonstrate that cooperative interactions between the two heads or the S2 part of the rod (or both) are required for regulation. The nature of such interactions is at present unclear. Another intriguing question is, how does phosphoorylation at the N-terminus of the smooth muscle RLC (serine 19 in the gizzard RLC sequence) switch *on* the myosin. This phosphorylation site appears to be some distance away from the RLC-ELC boundary, so it is unclear how it could trigger subunit reorganization to regulate the myosin head. Presumably, within the RLC, interdomain communication between the inhibitory C-terminus and the phosphorylation switch in the N-terminus must occur, possibly via the heavy chain, to change the RLC–ElC interface and switch myosin functions *on* and *off*.

Thus far biochemical and molecular biological studies have provided revealing insights into how the regulatory light chains may function. With knowledge of the structures of both the myosin head and regulatory domain, such studies should begin to reveal further details of the regulatory mechanisms involved.

## References

1. Winkelmann DA, Lowey S (1986) Probing myosin head structure with monoclonal antibodies. J Mol Biol 188:595–612
2. Katoh T, Lowey S (1987) Mapping myosin light chains by immunoelectron microscopy: use of anti-fluorescyl antibodies as structural probes. J Cell Biol 109:1549–1560
3. Adelstein RS, Eisenberg E (1980) Regulation and kinetics of the actin-myosin-ATP interaction. Annu Rev Biochem 49:921–956

4. Hartshorne DJ, Gorecka A (1980) The biochemistry of the contractile proteins of smooth muscle. In: Bohr DF, et al. (eds) Handbook of physiology. American Physiological Society, Bethesda, pp 93–120
5. Small JV, Sobieszek A (1980) The contractile apparatus of smooth muscle. Int Rev Cytol 64:241–306
6. Trybus KM, Huiatt TW, Lowey S (1982) A bent monomeric conformation of myosin from smooth muscle. Proc Natl Acad Sci USA 79:6151–6155
7. Onishi H, Wakabayashi T (1982) Electron microscopic studies of myosin from chicken gizzard muscle: the formation of the intramolecular loop in the myosin tail. J Biol Chem 92:871–879
8. Craig R, Smith R, Kendrick-Jones J (1983) Light-chain phosphorylation controls the confor mation of vertebrate non-muscle and smooth muscle myosin molecules. Nature 302:436–439
9. Cross RA, Cross KE, Sobieszek A (1986) ATP-linked monomer-polymer equilibrium of smooth muscle myosin: the free folded monomer traps ADP·Pi. EMBO J 5:2637–2641
10. Cross RA, Jackson AP, Citi S, Kendrick-Jones J, Bagshaw CR (1988) Active site trapping of nucleotide by smooth and non-muscle myosins. J Mol Biol 203:173–181
11. Somly AV, Butler TM, Bond M, Somlyo AP (1981) Myosin filaments have non-phosphorylated light chains in relaxed smooth muscle. Nature 294:567–569
12. Karess RE, Chang XJ, Edwards KA, Kulkarni S, Aguilera I, Kiehart DP (1991) The regulatory light chain of nonmuscle myosin is encoded by spaghetti-squash, a gene required for cytokinesis in Drosophila. Cell 65:1177–1189
13. Kretsinger RH (1980) Structure and evolution of the calcium-modulated proteins. CRC Crit Rev Biochem 8:119–174
14. Strynadka NCJ, James MNG (1989) Crystal structures of the helix-loop-helix calcium-binding proteins. Annu Rev Biochem 58:951–998
15. Collins JH (1976) Homology of myosin DTNB light chain with alkali light chains, troponin C and parvalbumin. Nature 259:699–700
16. Kendrick-Jones J, Jakes R (1976) Myosin-linked regulation: a chemical approach. In: Ricker G, Weber A, Goodwin J (eds) Myocardial failure: international Boehringer symposium, Munich. Springer-Verlag, Heidelberg New York Tokyo Berlin, pp 28–40
17. Reinach FC, Nagai K, Kendrick-Jones J (1986) Site-directed mutagenesis of the regulatory light-chain $Ca^{2+}/Mg^{2+}$ binding site and its role in hybrid myosins. Nature 322:80–83
18. Goodwin EB, Leinwand LA, Szent-Györgyi AG (1990) Regulation of scallop myosin by mutant regulatory light chains. J Mol Biol 216:85–93
19. Kwon H, Goodwin EB, Nyitray L, Berliner E, O'Neall-Hennessey E, Melandri FD, Szent-Györgyi AG (1990) Isolation of the regulatory domain of scallop myosin: role of the essential light chain in calcium binding. Proc Natl Acad Sci USA 87:4771–4775
20. Kwon H, Melandri FD, Szent-Györgyi AG (1992) Role of gizzard myosin light chains in calcium binding. J Muscle Res Cell Motil 13:315–320
21. Rayment I, Rypniewski WR, Schmidt-Base K, Smith R, Tomchick DR, Benning MM, Winkelmann DA, Wesenberg G, Holden HM (1993) Three-dimensional structure of myosin subfragment-1: a molecular motor. Science 261:50–58
22. Xie X, Harrison DH, Schlichting I, Sweet RM, Kalabokis VN, Szent-Györgyi AG, Cohen C (1994) Structure of the regulatory domain of scallop myosin at 2.8 Å resolution. Nature 368:306–312
23. Persechini A, Stull JT, Cooke R (1985) The effect of myosin phosphorylation on the contractile properties of skinned rabbit skeletal muscle fibres. Proc Natl Acad Sci USA 260:7951–7954
24. Ikura M, Clore GM, Gronenborn AM, Zhu G, Klee CB, Bax A (1992) Solution structure of a calmodulin-target peptide complex of multidimensional NMR. Science 256:632–638
25. Way M, Pope B, Gooch J, Hawkins MG, Weeds AG (1990) Identification of a region in segment 1 of gelsolin critical for actin binding. EMBO J 9:4103–4109
26. Rowe T, Kendrick-Jones J (1992) Chimeric myosin regulatory light chains identify the subdomain responsible for regulatory function. EMBO J 11:4715–4722
27. Kendrick-Jones J, Rasera da Silva AC, Reinach FC, Messer NG, Rowe T, McLaughlin PJ (1991) Recombinant DNA approaches to study the role of the regulatory light chains (RLC) using scallop myosin as a test system. J Cell Sci S14:55–58
28. Szent-Györgyi AG, Szentkiralyi EM, Kendrick-Jones J (1973) The light chains of scallop myosin as regulatory subunits. J Mol Biol 73:170–203

29. Kendrick-Jones J, Scholey JM (1981) Myosin-linked regulatory systems. J Muscle Res Cell Motil 2:347–372
30. Sellers JR, Chantler PD, Szent-Györgyi AG (1980) Hybrid formation between scallp myofibrils and foreign regulatory light-chains. J Mol Biol 144:223–245
31. Kendrick-Jones J, Jakes R, Tooth P, Craig R, Scholey JM (1982) Role of the myosin light chains in the regulation of contractile activity. In: Twarog BM, Levine RJC, Dewey MM (eds) Basic biology of muscles: a comparative approach. Raven Press, New York, pp 255–272
32. Ankrett RJ, Rowe AJ, Cross RA, Kendrick-Jones J, Bagshaw CR (1991) A folded (10S) conformer of myosin from a striated muscle and its implications for regulation of ATPase activity. J Mol Biol 217:323–335
33. Stafford WF, Szentkiralyi EM, Szent-Györgyi AG (1979) Regulatory properties of single-headed fragments of scallop myosin. Biochemistry 24:5273–5280
34. Seidel JC (1980) Fragmentation of gizzard myosin by α-chymotrypsin and papain, the effects on ATPase activity, and the interaction with actin. J Biol Chem 255:4355–4361
35. Morita JI, Takashi R, Ikebe M (1991) Exchange of the fluorescence-labelled 20,000-dalton light chain of smooth muscle myosin. Biochemistry 30:9539–9545
36. Trybus KM, Chatman TA (1993) Chimaeric regulatory light chains as probes of smooth muscle myosin function. J Biol Chem 268:4412–4419
37. Chantler PD, Szent-Györgyi AG (1980) Regulatory light-chains and scallop myosin: full dissociation, reversibility and co-operative effects. J Mol Biol 138:473–492
38. Herzberg O, James MNG (1988) Refined crystal structure of troponin C from turkey skeletal muscle at 2.0 Å resolution. J Mol Biol 203:761–779
39. Rowe T, Kendrick-Jones J (1993) The C-terminal helix in subdomain 4 of the regulatory light chain is essential for myosin regulation. EMBO J 12:4877–4884
40. Collins JH (1991) Myosin light chains and troponin C: structural and evolutionary relationships revealed by amino acid sequence comparisons. J Muscle Res Cell Motil 12:3–25
41. Bagshaw CR, Kendrick-Jones J (1979) Characterization of homologous divalent metal ion binding sites of vertebrate and molluscan myosins using electron paramagnetic resonance spectroscopy. J Mol Biol 130:317–336
42. Bagshaw CR, Kendrick-Jones J (1980) Identification of the divalent metal binding domain of myosin regulatory light chains using spin-labelling techniques. J Mold Biol 140:411–433
43. Bagshaw CR, Reed GH (1977) The significance of the slow dissociation of divalent metal ions from myosin regulatory light chains. FEBS Lett 81:386–390
44. Da Silva ACR, De Araujo AHB, Hertzberg O, Moult J, Sorenson M, Reinach FC (1993) Troponin C mutants with increased calcium affinity. Eur J Biochem 213:599–604
45. Kraulis PJ (1991) MOLSCRIPT: a program to produce both detailed and schematic plots of protein structures. J Appl Cryst 24:946–950
46. Jung G, Korn ED, Hammer JA (1987) The heavy chain of Acanthamoeba myosin IB is a fusion of myosin-like and nonmyosin-like sequences. Proc Natl Acad Sci USA 84:6720–6724
47. Hammer JA, Bowers B, Paterson BM, Korn ED (1987) Complete nucleotide sequence and deduced polypeptide sequence of a non-muscle myosin heavy chain gene from Acanthamoeba: evidence for a hinge in a rod-like tail. J Cell Biol 105:913–925
48. Warrick HM, De Lozanne A, Leinwand LA, Spudich JA (1986) Conserved protein domains in a myosin heavy chain from Dictyostelium discoideum. Proc Natl Acad Sci USA 83:9433–9437
49. Karn J, Brenner S, Barnett L (1983) Protein structural domains in the Caenorhabditis elegans unc-54 myosin heavy chain gene are not separated by introns. Proc Natl Acad Sci USA 80:4253–4257
50. Kraft R, Bravo-Zehnder M, Taylor DA, Leinwand LA (1989) Complete nucleotide sequence of full lenght cDNA for rat β cardiac myosin heavy chain. Nucleic Acids Res 17:7529–7530
51. Molina MI, Kropp KE, Gulick J, Robbins J (1987) The sequence of an embryonic myosin heavy chain gene and isolation of its corresponding cDNA. J Biol Chem 262:6478–6488
52. Nyitray L, Goodwin EB, Szent-Györgyi AG (1991) Complete primary structure of a scallop striated muscle myosin heavy chain. J Biol Chem 266:18469–18476
53. Yanagisawa M, Hamada Y, Katsuragawa Y, Imamura M, Mikawa T, Masaki T (1987) Complete primary structure of vertebrate smooth muscle myosin heavy chain deduced from its complementary DNA sequence. J Mol Biol 198:143–157

54. McLachlan AD, Karn J (1982) Periodic charge distribution in the myosin rod amino acids sequence match cross bridge spacings in muscle. Nature 299:226–231
55. Cheney RE, Mooseker MS (1992) Unconventional myosins. Curr Opin Cell Biol 4:27–35
56. Sellers JR, Harvey EV (1984) Localization of a light chain binding site on smooth muscle myosin by light chain overlay of sodium dodecyl sulfate polyacrylamide electrophoretic gels. J Biol Chem 259:14203–14207
57. Mitchell EJ, Jakes R, Kendrick-Jones J (1986) Localisation of light chain and actin binding sites on myosin. Eur J Biochem 161:25–35
58. Waller GS, Lowey S (1985) Myosin subunit interactions: localization of the alkali light chains. J Biol Chem 260:14368–14373
59. McNally EM, Bravo-Zehnder MM, Leinwand LA (1991) Identification of sequences necessary for the association of cardiac myosin subunits. J Cell Biol 113:585–590
60. Wells JA, Yount RG (1979) Active site trapping of nucleotides by crosslinking two sulfhydryls in myosin subfragment 1. Proc Natl Acad Sci USA 76:4966–4970
61. Wells JA, Knoeber C, Sheldon MC, Werber MM, Yount RG (1980) Crosslinking of myosin S1. J Biol Chem 255:11135–11140
62. Rayment I, Holden HM, Whittaker M, Yohn CB, Lorenz M, Holmes KC, Milligan RA (1993) Structure of the actin-myosin complex and its implications for muscle contraction. Science 261:58–65

# Autoinhibition of Myosin Light Chain Kinase

Satoshi Matsushima[1], Masaaki Ito[2], David J. Hartshorne[3], and Vince Guerriero Jr.[3]

*Summary.* Smooth muscle myosin light chain kinase (MLCK) is a $Ca^{2+}$/calmodulin-activated enzyme that regulates contraction by phosphorylation of the regulatory light chains of myosin. Inactivity of the enzyme in the absence of $Ca^{2+}$/calmodulin is thought to result from interaction of the catalytic core with another region of the enzyme known as the inhibitory region. This inhibitory region resembles the light chain substrate of the kinase, and therefore is called the pseudosubstrate. A partial cDNA for chicken gizzard MLCK was expressed in bacteria and used to make truncation and point mutations to test if the pseudosubstrate sequence was the inhibitory sequence. Truncation at $A^{796}$ removes most of the pseudosubstrate sequence, but the mutant was inactive. Truncation at $K^{793}$ resulted in a constitutively active mutant, indicating that only three amino acids of the pseudosubstrate sequence are needed for inhibition. These results required reevaluation of the pseudosubstrate hypothesis. Based on a model showing the interaction of the pseudosubstrate sequence with the catalytic core, $Y^{794}$ interacts with a hydrophobic pocket in the catalytic core. We propose that this interaction is enough to inhibit the enzyme.

*Key Words.* Myosin light chain kinase—Pseudosubstrate

## Background

Smooth muscle contractile activity is controlled by the intracellular concentration of $Ca^{2+}$, as in other muscle types. The major target for $Ca^{2+}$ binding is calmodulin (CaM); the binding is linked to the contractile apparatus and tension development via myosin light chain kinase (MLCK). The MLCK apoenzyme is inactive, and the ternary complex $Ca^{2-}$/CaM/MLCK, phosphorylates the 20-kDa myosin light chains [1]. Usually S19 on each of the two light chains is phosphorylated, but phosphorylation of T18 may also occur. This phosphorylation of myosin increases actin-activated ATPase activity, and the "activation" of myosin is equated with initiation of the contractile response. When the $Ca^{2+}$ level decreases, CaM dissociates from MLCK and phosphorylation

[1] Department of Cellular and Molecular Pharmacology, Mie University, 2-174 Edobashi, Tsu, Mie, 514 Japan
[2] First Medical Clinic, Mie University, 2-174 Edobashi, Tsu, Mie, 514 Japan
[3] Muscle Biology Group, Shantz Building, University of Arizona, Tucson, AZ 85721, USA

ceases. The myosin is dephosphorylated by a phosphatase, and the contractile apparatus returns to its resting state.

In its simplest scenario the phosphorylation mechanism represents an *on–off* switch. It would thus be predicted that increased levels of phosphorylation would recruit more cross-bridges and lead to increased force. Moreover, the velocity at zero load should be constant for any level of phosphorylation, as this parameter determines the cross-bridge cycling rate. The $Ca^{2+}$ dependence of phosphorylation is determined by the $Ca^{2+}$-binding parameters of CaM and so might be expected to be constant. In general, these predictions are not realized, and it is clear that the simplest interpretation of the phosphorylation scheme is not adequate to describe physiologic behavior.

Are other regulatory mechanisms involved, or can the balance of MLCK and phosphatase be altered? It appears that many of the examples of apparent deviation can be explained by modification of either MLCK or phosphatase, causing a shift in the $Ca^{2+}$–phosphorylation relation. To understand the mechanisms involved in altering either enzymatic activity, it is essential to define the molecular parameters of each system: MLCK and the phosphatase. To date, progress with MLCK is more advanced, and the remainder of this chapter describes some of its molecular properties. (For a discussion of the myosin phosphatase see Ito and Nakano, this volume.)

## Myosin Light Chain Kinase

Myosin light chain kinases are widely distributed and are found in each muscle type and many eukaryotic cells [1]. Skeletal and smooth muscle enzymes are related but distinct; based on limited information it appears that the smooth muscle and nonmuscle kinases are more similar. Several early studies used limited proteolysis as a probe of functional regions. These studies were complemented by determination of a partial cDNA sequence for the gizzard enzyme [2]. The general plan of MLCK was of a central active site flanked on its C-terminal side by the CaM-binding site (Fig. 1). A full-length cDNA was sequenced later [3], and the gizzard MLCK was found to be

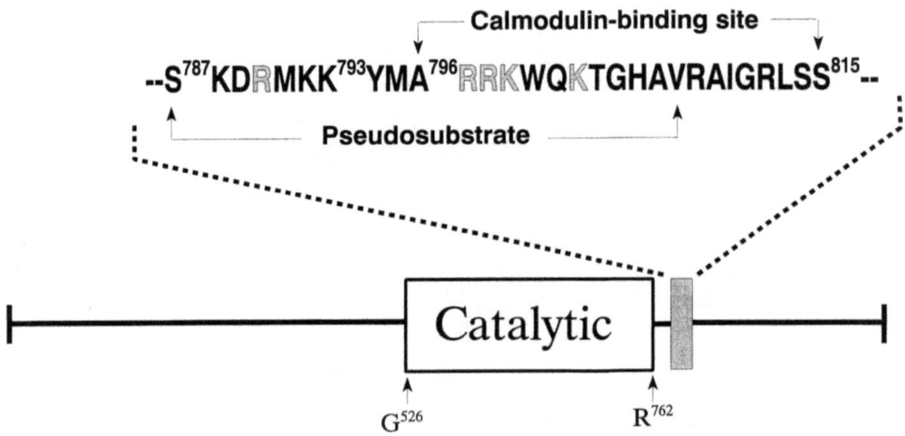


**Fig. 1.** Model of myosin light chain kinase showing calmodulin-binding and pseudosubstrate regions. The enzyme is 972 amino acids in length, and the catalytic domain is located between $G^{526}$ and $R^{762}$. The *shaded area* denotes the regulatory domain, and the amino acid sequence of this region is shown. The *shaded amino acids* are the basic residues that are part of the pseudosubstrate sequence

972 residues long. Based on conserved sequences among kinases [4], the active site was suggested to include $G^{526}$ to $R^{762}$. The CaM-binding site extends from $A^{796}$ to $S^{815}$ (the latter was based on a synthetic peptide analog of a cyanogen bromide fragment [5]). An interesting and still unresolved point is the inhibition of the apoenzyme. Because the apoenzyme is essentially inactive there must be an autoinhibitory mechanism, and several possibilities have been suggested.

One widely accepted hypothesis is that inhibition of MLCK is due to interaction of the active site with a pseudosubstrate region that is dissociated by binding $Ca^{2+}$/CaM [6]. This hypothesis is based on the juxtaposition of basic residues of the light chain phosphorylation site with a stretch of the MLCK sequence, aligning $H^{805}$ of MLCK with $S^{19}$ of the ligth chain. Extension of the light chain sequence to the N-terminus gives a homologous MLCK peptide beginning at $S^{787}$. With this alignment, several basic residues match: For MLCK, residues 802, 799, 797, 792, and 790 line up with light chain residues 16, 13, 12, 11, and 6, respectively. Some of these points are illustrated in Fig. 1.

An important observation was that tryptic proteolysis of gizzard MLCK formed an inactive 64-kDa fragment that on further hydrolysis was converted to an active, CaM-independent 61-kDa fragment [7]. Interpretation of these results was that the 64-kDa fragment had lost the CaM-binding site but retained the autoinhibitory sequence, which was subsequently lost during the conversion to the 61-kDa fragment. Thus the inhibitory domain should be within the 3-kDa difference peptide. From analyses of the N- and C-terminal residues of the 64- and 61-kDa fragments it was determined that the autoinhibitory sequence was at the C-terminal side of the molecule between the active site and the CaM-binding site. Additional studies [8] using limited proteolysis established a maximum sequence for the inhibitory zone, which extended from $K^{779}$ (after tryptic hydrolysis) to $K^{802}$ (after hydrolysis by endoproteinase Lys C). This sequence of 23 residues overlaps with the N-terminal boundary of the CaM-binding site and contains most of the proposed pseudosubstrate sequence $S^{787}$ to $H^{807}$. These results were consistent with the pseudosubstrate hypothesis [9] as the inhibitory mechanism. However, Ikebe et al. [10] claimed that the C-terminal residue of the inactive fragment was $K^{793}$, and this would eliminate the pseudosubstrate sequence as the inhibitory region.

Synthetic peptides based on various sequences of MLCK have also been used to determine the inhibitory sequence. A potent inhibitor of kinase activity is the peptide 783 to 804 [7]. A slightly shorter peptide, 783 to 796, was a less potent inhibitor [7]. Using even shorter peptides, attempts to more closely define the inhibitory region were not successful; it was suggested that multiple inhibitory regions may be required [11]. However, it was pointed out [9] that peptides containing the pseudosubstrate sequence were the most potent inhibitors. A point to be emphasized with respect to the pseudosubstrate hypothesis is that peptides mimicking the light chain substrate (or the light chain itself) are not effective inhibitors compared to peptides based on the kinase sequence. In addition, the serine-containing analog of the pseudosubstrate peptide $R^{797}$ to $V^{807}$ is not phosphorylated [8].

## Mutants of Myosin Light Chain Kinase

To define the inhibitory zone more accurately, a new approach seemed necessary. A logical choice was to design a series of truncation mutants in which the sequence around the suspected inhibitory site was systematically shortened. The expectation was that mutants containing the inhibitory sequence would be inactive; as the inhibitory site is eroded the mutants would become constitutively active. The initial mutants designed by Ito et al. [12] were all initiated at $L^{447}$ but contained varying lengths of C-terminal sequence. As predicted from the proteolysis experiments, a mutant truncated to $T^{778}$ was constitutively active (i.e., CaM-independent) and thus did not contain the inhibitory sequence. These results are in contrast to those of Bagchi et al. [13], who reported the expression of truncated mutants starting at $D^{450}$ and ending at $K^{779}$ or $C^{765}$. These mutants were not constitutively active but could be activated by treatment with trypsin. The authors suggested that the amino-terminus of the protein contained another pseudosubstrate sequence, which had to be digested before constitutive activity could be obtained. The mutants reported by Ito et al. [12] and those reported here contain this same sequence at the amino-terminus, and therefore this explanation is not consistent with these results. It is possible that the mutants reported by Bagchi et al. [13] were not folded properly after expression in bacteria or that a bacterial contaminant inhibited the mutants and was removed by trypsin digestion.

The region initially of interest extended from $Y^{794}$ to $W^{800}$, which contains the tribasic sequence 797 to 799. Truncation mutants were designed, and it was found that truncation to $K^{793}$ resulted in a constitutively active kinase, whereas the kinase truncated at $W^{800}$ had low activity. None of the mutants bound CaM. It was therefore suggested [12] that a critical region of the inhibitory domain is contained within the sequence 794 to 800. These mutants were recovered from the soluble fraction after lysis of the bacteria, and so the yield was low.

Using different expression systems, the same mutants were prepared to confirm these results. For the pET expression vectors it was necessary initially to establish a procedure to recover "active" kinase mutants from the insoluble inclusion bodies. First, inclusion bodies were purified from bacterial lysates and treated with 0.1% Sarkosyl to remove membrane contaminants. The pellet of inclusion body material was then solubilized by the addition of 6 M guanidine HCl. The solubilized kinase was then added to a 50-fold volume of a solution containing 0.8 M NaCl, and this solution was allowed to renature. Renaturation was slow, and optimum activity was recovered only after approximately 5 h. In summary, rapid dilution and slow renaturation were the two key steps for recovery of activity.

This method was tested on native gizzard MLCK by first heating the enzyme in boiling water for 5 min, collecting it by centrifugation, and solubilizing it with guanidine HCl. The solubilized kinase was diluted in a high salt solution, and about 90% of the original activity was recovered after several hours of renaturation. The enzyme retained its absolute dependence on $Ca^{2+}$/CaM, and so it may be assumed that the CaM-binding site also was refolding into a functional conformation.

The results using this procedure on mutants are shown in Table 1. The truncation mutant ending at $W^{800}$ was also prepared in the baculovirus expression system (pBlue.$W^{800}$ in Table 1). The control mutant, pET.E972, contained the full C-terminal sequence from $L^{447}$. Its specific activity is lower than that of native MLCK, but it is considerably higher than previous mutants and approaches the activity of the native enzyme. Truncation at $W^{800}$ generated an inactive kinase. This level of kinase activity is lower than reported previously [12] and indicates clearly that this mutant was inhibited. The W residue is at the C-terminal end of the pseudosubstrate domain ($S^{787}$ to $H^{805}$) and the N-terminal end of the CaM-binding site ($A^{796}$ to $S^{815}$). Consequently these mutants did not bind CaM. Removal of an additional four residues caused a slight increase in activity (see pET.$A^{796}$ in Table 1), but it was probably not significant.

**Table 1.** Activities of truncation mutants of MLCK.

| | Specific activity (nm/min mg) | |
|---|---|---|
| MLCK mutant[a] | $+Ca^{2+}$ | $-Ca^{2+}$ |
| pET.E$^{972}$ | 1480 ± 160 | 1.5 ± 0.8 |
| pET.W$^{800}$ | 2.5 ± 0.7 | 3.5 ± 2.5 |
| pBlue.W$^{800}$ | 5.2 | 1.2 |
| pET.K$^{799}$ | 10 | 11 |
| pET.A$^{796}$ | 41 | 45 |
| pET.K$^{793}$ | 625 ± 289 | 741 ± 251 |
| pET.W$^{800/797-799}$[b] | 19 | 35 |

MLCK, myosin light chain kinase.
[a] Each mutant originated at L$^{4477}$ and ended at the residue shown.
[b] This truncation mutant ended at W$^{800}$, but residues R$^{797}$, R$^{798}$, and K$^{799}$ were mutated to SGE, respectively.

**Table 2.** Point mutations of MLCK.

| | Specific activity (nm min/mg) | |
|---|---|---|
| MLCK mutant[a] | $+Ca^{2+}$ | $-Ca^{2+}$ |
| -YMARRKWQ$^{801}$- | 1480 | 1.5 |
| -YMASGEWQ$^{801}$- | 1180 | 5.4 |
| -YMDRRKWQ$^{801}$- | 1160 | 6.4 |
| -YGARRKWQ$^{801}$- | 2150 | 36.0 |
| -DMARRKWQ$^{801}$- | 815 | 10.0 |
| -YMARRKGQ$^{801}$- | 0.01 | 0.01 |

[a] Each mutant originated at L$^{447}$ and ended at E$^{972}$; the underlined amino acids have been changed.

Similar results were obtained by Yano et al. [14] using truncation mutants that contained the entire amino-terminus of MLCK. Yano et al. [14], however, did not examine shorter mutants to determine the number of amino acids needed for inhibition.

According to the pseudosubstrate hypothesis, three basic residues (R$^{797}$, R$^{798}$, and K$^{799}$) are required for inhibition of the enzyme. However, the truncated mutant pET.A$^{796}$ lacks these three residues but is still inactive. One explanation is that these are truncation mutants and are lacking the carboxyl-terminus of the enzyme. This end of the molecule might be needed for proper folding and activity of the kinase. A new set of point mutants was made that started at L$^{447}$ and included the entire carboxyl-terminus of the molecule. Amino acids in the inhibitory region (794–800) were altered to determine the residues that were important for inhibition (Table 2). Changing the three basic residues to SGE did not affect the activity of the enzyme in the absence of calcium. Similar results were reported for rabbit smooth muscle MLCK [15], rabbit skeletal muscle MLCK [16], and chicken nonmuscle MLCK [17]. The only mutation that dramatically affected activity was changing the W$^{800}$ to G. The W$^{800}$ has been reported to be important for calmodulin activation of MLCK [18], so the results here are consistent with this finding.

Substitution or removal of the basic residues does not produce a constitutively active enzyme, and therefore the pseudosubstrate hypothesis must be reevaluated. An explanation may be found in examination of the model of smooth muscle MLCK that was derived using the crystallographic coordinates of the cyclic AMP-dependent protein kinase [9]. There are 14 amino acids that extend between the inhibitory region and the end of the catalytic region. Addition of these 14 amino acids increases the inhibitory potency of the pseudosubstrate peptide by 30-fold. However, these 14 amino acids alone are ineffective in inhibiting the enzyme. It has been proposed that these linker amino acids are important for aligning the inhibitor peptide and thus strengthening the interaction with the catalytic core [9]. The question then becomes how much of the inhibitory region is needed to inhibit the enzyme. The results presented here indicate that truncation at $K^{793}$ removes the inhibition. However, including the next three amino acids ($Y^{794}$MA) produces an enzyme that is significantly inhibited (Table 1). The model that includes the pseudosubstrate peptide ($S^{787}$ to $V^{807}$) aligns $Y^{794}$ in a hydrophobic pocket of the catalytic core of MLCK ($L^{713}$SPF$M^{717}$). This pocket also accommodates $W^{800}$. Changing $Y^{794}$ to an A in a synthetic peptide increased the $K_i$ twofold but inhibited ATP binding by 75-fold [9]. Foster et al. [11] reported that a 14-amino-acid peptide ending with $Y^{794}$ MA is a good inhibitor and competitive with the peptide substrate ($K_i$ of 2 μM). These results, then, confirm that the truncation mutant ending at $A^{796}$ contains part of the inhibitory sequence. Based on the suggested interaction of $Y^{794}$ with a hydrophobic pocket in the catalytic core, it is likely that $Y^{794}$ within this sequence plays an important role, which would not have been predicted from the pseudosubstrate hypothesis.

Another question is why truncation mutagenesis and proteolysis of MLCK both produce constitutively active mutants, but point mutations that change key residues in the pseudosubstrate region are not effective. Because of electrostatic and hydrophobic interactions it is predicted that the pseudosubstrate is tightly bound. Only a large number of point mutations would be able to disrupt this interaction. Indeed, in one case the nonmuscle MLCK was rendered constitutively active by six point mutations changing basic (R or K) residues to E [17] in the sequence that corresponds to $K^{784}$ to $K^{793}$ of chicken smooth muscle MLCK. Four of these changes would be in the pseudosubstrate region and two ($K^{784}$ and $K^{785}$) in the linker peptide. An explanation may be that because the pseudosubstrate and CaM-binding regions overlap, the region of the pseudosubstrate that does not bind CaM must be important for inhibition of the enzyme and stabilizing interaction with the catalytic core. This point can be illustrated by reexamining the importance of $Y^{794}$ for inhibitory activity. Changing this residue to a D (Table 2) does not produce a constitutively active enzyme, as the remainder of the pseudosubstrate region is present to stabilize the interaction with the catalytic core. Binding of CaM, then, would alter the interaction of the remainder of the pseudosubstrate with the catalytic core and remove inhibition.

The experiments presented here were designed to test the inhibitory activity of the pseudosubstrate sequence in smooth muscle myosin light chain kinase. The pseudosubstrate region has been proposed to contain a large region; therefore many point mutations would be needed to interrupt this interaction to produce a constitutively active enzyme. Truncation mutants were found to be more valuable in determining the smallest region capable of inhibiting the enzyme. Other researchers have shown, using synthetic peptides, that the region on the amino side of the pseudosubstrate does have inhibitory activity, and inclusion of this region with a pseudosubstrate peptide increases inhibitory activity. We have shown here that the three amino acids $Y^{794}MA$ on the end of a truncation mutant are enough to inhibit the enzyme.

## References


1. Hartshorne DJ (1987) Biochemistry of the contractile proteins in smooth muscle. In: Johnson LR (ed) Physiology of the gastrointestinal tract, vol 1. Raven Press, New York, pp 423–482
2. Guerriero V Jr, Russo MA, Olson NJ, Putkey JA, Means AR (1986) Domain organization of chicken gizzard myosin light chain kinase deduced from a cloned cDNA. Biochemistry 25:8372–8381
3. Olson NJ, Pearson RB, Needleman DS, Hurwitz MY, Kemp BE, Means AR (1990) Regulatory and structural motifs of chicken gizzard myosin light chain kinase. Proc Natl Acad Sci USA 87:2284–2288
4. Hanks SK, Quinn AM, Hunter T (1988) The protein kinase family: conserved features and deduced phylogeny of the catalytic domains. Science 241:42–52
5. Lukas TJ, Burges WH, Prendergast FG, Lau W, Watterson DM (1987) Calmodulin binding domains: characterization of a phosphorylation and calmodulin binding site from myosin light chain kinase. Biochemistry 25:1458–1464
6. Kemp BE, Pearson RB, Guerriero V Jr, Bagchi IC, Means AR (1987) The calmodulin binding domain of chicken smooth muscle myosin light chain kinase contains a pseudosubstrate sequence. J Biol Chem 262:2542–2548
7. Ikebe M, Stepinska M, Kemp BE, Means AR, Hartshorne DJ (1987) Proteolysis of smooth muscle myosin light chain kinase. J Biol Chem 260:13828–13834
8. Pearson RB, Ito M, Morrice NA, Smith AJ, Condron R, Wettenhall REH, Kemp BE, Hartshorne DJ (1991) Proteolytic cleavage sites in smooth muscle myosin-light-chain kinase and their relationship to structural and regulatory domains. Eur J Biochem 200:723–730
9. Knighton DR, Pearson RB, Sowadski JM, Means AR, Ten Eyck LF, Taylor SS, Kemp BE (1992) Structural basis of the intrasteric regulation of myosin light chain kinase. Science 258:130–135
10. Ikebe M, Maruta S, Reardon S (1989) Location of the inhibitory region of smooth muscle myosin light chain kinase. J Biol Chem 264:6967–6971
11. Foster CJ, Johnston SA, Sunday B, Gaeta FC (1990) Potent peptide inhibitors of smooth muscle myosin light chain kinase: mapping of the pseudosubstrate and calmodulin binding domains. Arch Biochem Biophys 280:397–404
12. Ito M, Guerriero V Jr, Chen X, Hartshorne DJ (1991) Definition of the inhibitory domain of smooth muscle myosin light chain kinase by site-directed mutagenesis. Biochemistry 30:3498–3503
13. Bagchi IC, Kemp BE, Means AR (1989) Myosin light chain kinase structure function analysis using bacterial expression. J Biol Chem 264:15843–15849
14. Yano K, Araki Y, Hales SJ, Tanaka M, Ikebe M (1993) Boundary of the autoinhibitory region of smooth muscle myosin light-chain kinase. Biochemistry 32:12054–12061
15. Fitzsimons DP, Herring BP, Stull JT, Gallagher PJ (1992) Identification of basic residues involved in activation and calmodulin binding of rabbit smooth muscle myosin light chain kinase. J Biol Chem 267:23903–23909
16. Herring BP (1991) Basic residues are important for $Ca^{2+}$/calmodulin binding and activation but not autoinhibition of rabbit skeletal muscle myosin light chain kinase. J Biol Chem 266:11838–11841
17. Shoemaker MO, Lau W, Shattuck RL, Kwiatkowski AP, Matrisian PE, Guerra-Santos L, Wilson E, Lukas TJ, Van Eldik LJ, Watterson DM (1990) Use of DNA sequence and mutant analysis and antisense oligodeoxynucleotides to examine the molecular basis on nonmuscle myosin light chain kinase autoinihibition, calmodulin recognition and acitity. J Cell Biol 111:1107–1125
18. Bagchi IC, Huang Q, Means AR (1992) Identification of amino acids essential for calmodulin binding and activation of smooth muscle myosin light chain kinase. J Biol Chem 267:3024–3029

# Regulation of Myosin Light Chain Kinase Activity in Smooth Muscle

Kristine E. Kamm, Katherine Luby-Phelps, Malu G. Tansey, Patricia J. Gallagher, and James T. Stull

*Summary.* Phosphorylation of myosin regulatory light chain by $Ca^{2+}$/calmodulin-dependent myosin light chain kinase (MLCK) plays a central role in smooth muscle contractility. The quantitative relation between intracellular $Ca^{2+}$ concentrations and light chain phosphorylation is not fixed but modulated dynamically by various regulatory processes. Most of the calmodulin in smooth muscle cells is tightly bound to cellular elements even in the absence of $Ca^{2+}$. Surprisingly, even the most mobile fraction (8% of the total) has a diffusion coefficient in smooth muscle cells sevenfold lower than a freely diffusible, similar size dextran. The possibility is considered that calmodulin available for MLCK activation is limiting. Furthermore, calmodulin availability may be regulated by unidentified processes. The multifunctional $Ca^{2+}$/calmodulin-dependent protein kinase II phosphorylates MLCK adjacent to the C-terminus of the calmodulin-binding domain. This phosphorylation increases the concentration of $Ca^{2+}$/calmodulin required for activation and hence physiologically increases the $Ca^{2+}$ concentration required for light chain phosphorylation. However, in smooth muscle cells the concentration of $Ca^{2+}$ necessary for MLCK phosphorylation is greater than that required for light chain phosphorylation. Thus MLCK is sensitive to small increases in $Ca^{2+}$ during the initiation of contraction and subsequently becomes desensitized to $Ca^{2+}$ after phosphorylation, thereby limiting light chain phosphorylation.

*Key Words.* Calmodulin—Myosin light chain kinase—Phosphorylation

## Introduction

Myosin light chain kinase (MLCK) phosphorylates the regulatory light chain (RLC) of vertebrate myosin II in a $Ca^{2+}$/calmodulin-dependent manner [1, 2]. Smooth and nonmuscle myosin IIs exist in an enzymatically inactive form that is disinhibited after RLC phosphorylation. Thus the $Ca^{2+}$/calmodulin/MLCK holoenzyme serves as the primary $Ca^{2+}$ effector for promoting actin-activated myosin MgATPase activity and mechanical activation in smooth muscle (Fig. 1). Consistent with this situation, most agents promoting smooth muscle contraction demonstrate a dependence of isometric

Department of Physiology, University of Texas Southwestern Medical Center at Dallas, 5323 Harry Hines Boulevard, Dallas, TX 75235-9040, USA

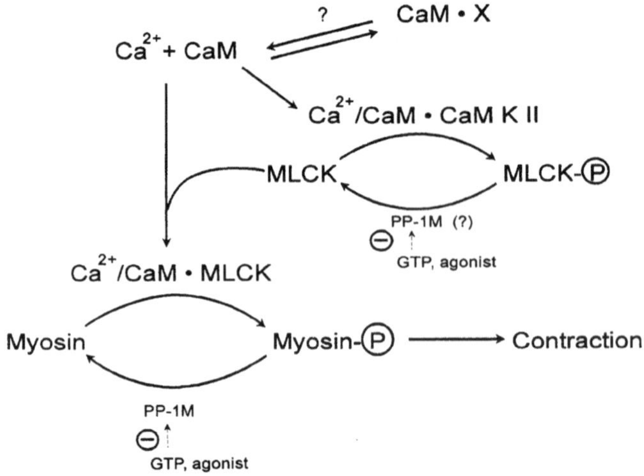


**Fig. 1.** Modulation of $Ca^{2+}$-dependent myosin light chain phosphorylation in smooth muscle. $Ca^{2+}$ entering the cytoplasm binds to calmodulin (CaM), which then binds to and activates myosin light chain kinase (MLCK). The activated kinase ($Ca^{2+}$/calmodulin·MLCK) transfers phosphate (*circled P*) from ATP to the regulatory light chain of myosin, resulting in elevated actomyosin ATPase activity and contraction. Myosin is dephosphorylated by protein phosphatase type 1M (PP-1M), which is inhibited by treatment with agonist or GTP. At higher $Ca^{2+}$ concentrations the multifunctional $Ca^{2+}$/calmodulin-dependent protein kinase II (CaM KII) catalyzes MLCK phosphorylation, thereby desensitizing the enzyme to activation by $Ca^{2+}$/calmodulin. MLCK is dephosphorylated by a protein phosphatase that is also inhibited by GTP and agonist. A fraction of intracellular calmodulin may be immobilized by binding to unidentified intracellular sites (*X*). It is not known if this fraction is in equilibrium with the fraction that activates MLCK or CaM KII

stress on RLC phosphorylation [3]. However, additional regulatory mechanisms not involving RLC phosphorylation are also implicated in smooth muscle contraction [4, 5]. Steady-state values of RLC phosphorylation are attained as a result of the combined rates of phosphorylation by MLCK and dephosphorylation by protein phosphatase type 1M (PP-1M) at a fixed concentration of free cytoplasmic $Ca^{2+}$ ($[Ca^{2+}]_i$) [6–8]. Agents or conditions that alter the activities of either kinase or phosphatase at fixed $[Ca^{2+}]_i$ modulate the $[Ca^{2+}]_i$ dependence of RLC phosphorylation and therefore the contractile stress [9, 10]. We have investigated cellular mechanisms whereby MLCK activity may be limiting in smooth muscle.

Smooth muscle cells and nonmuscle cells express a common form of MLCK [11]. The organization of functional domains of smooth muscle MLCK is shown in Figure 2. It contains a catalytic core in the central portion that is highly homologous to other protein kinases. C-terminal of the catalytic core is a shorter segment that binds calmodulin with high affinity [14]. Calmodulin is a dumbbell-shaped molecule in which a pair of EF $Ca^{2+}$-binding sites are contained in each lobe. A central helix tethers the two lobes, and its flexibility allows them to wrap around the amphiphilic α-helical calmodulin-binding domain of MLCK forming a hydrophobic core. Calmodulin binding results in enzyme activation by removing the autoinhibition that is conferred by a region linking the catalytic core and the calmodulin-binding domain [15]. In addition, smooth muscle MLCK (in contrast to the skeletal muscle form) contains an extended region in the C-terminus that can be expressed as an independent protein,

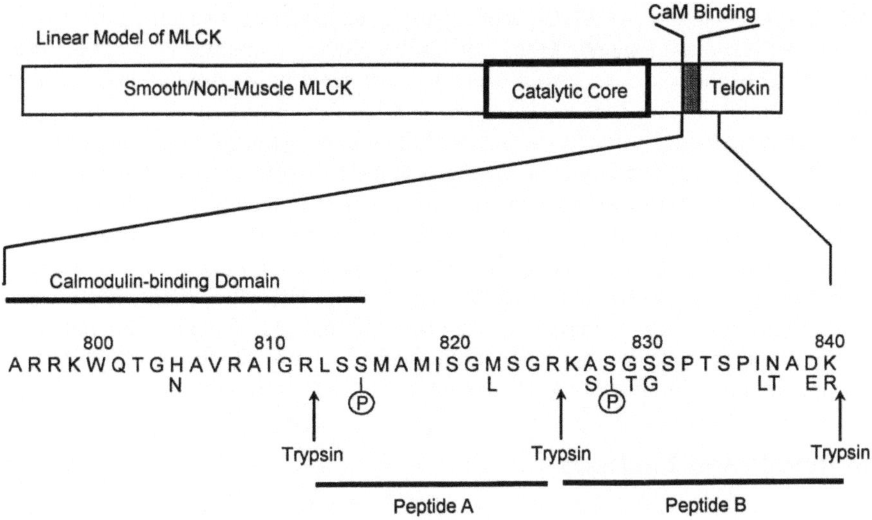


**Fig. 2.** Domain organization and sites of phosphorylation in smooth muscle/nonmuscle MLCK. Amino acid sequence of chicken smooth muscle MLCK is shown for residues 796–840 [12]. Amino acid substitutions found in rabbit and bovine smooth muscle MLCKs are shown below the sequence [11, 13]. Phosphoserine sites designated A and B (*circled P*) are located on tryptic peptides A and B

telokin, in some smooth muscle cells due to an alternate promoter in the MLCK gene [16, 17].

Smooth muscle MLCKs from different animal species are structurally similar [15]. The first 80 residues are highly conserved, and in mammalian kinases there is a tandem repeat of 12 amino acid residues that is not found in chicken smooth muscle MLCK. No function for this tandem repeat is known. MLCK also contains other repeat units, including motif I, which is related to the type III domain of fibronectin, and motif II, which belongs to the CII set of immunoglobulin superfamily. The latter is repeated three times in smooth muscle MLCKs. The β-barrel structure of telokin is similar to the $CH_2$ domain of the immunoglobulin fragment Fc [18]. The molecular function of these repeat units is not clear, although it has been suggested that telokin may be responsible for MLCK binding to myosin [19]. Additionally, biochemical evidence shows an actin-binding site located within the N-terminal sequence containing residues 1–114 [20].

The fraction of MLCK activated at a given $Ca^{2+}$ concentration is predicted from an equilibrium that involves the free concentration of calmodulin as well as the number of $Ca^{2+}$ ions required to bind calmodulin (CaM) to result in activation of MLCK ($n$), the average activation constant for the $n$ apparent $Ca^{2+}$-binding sites on calmodulin ($K_{Ca}{}^{2+}$), and the concentration of $Ca_n$·calmodulin required for half-maximal activation ($K_{CaM}$) [21]. Because the free calmodulin may be less than the total, we considered the possibility that the concentration of calmodulin available to activate MLCK in cells may limit the fraction of active kinase [22]. Although the total tissue calmodulin concentration exceeds MLCK by 10- to 20-fold (40 μM versus 2–4 μM, respectively), we estimate by translational diffusion measurements in smooth muscle cells that only about 8% of the calmodulin is immediately available for activation.

A second variable affecting the fractional activation of MLCK is $K_{CaM}$. Whereas $n$ and $K_{Ca}{}^{2+}$ are not expected to vary, $K_{CaM}$ can be altered upon phosphorylation of

MLCK at a regulatory site. MLCK is phosphorylated in vitro at two sites (sites A and B) in the C-terminal region near the calmodulin-binding domain (Fig. 2). These two sites are phosphorylated by cAMP-dependent protein kinase, protein kinase C, and $Ca^{2+}$/calmodulin-dependent protein kinase II [23–26]. MLCK is desensitized (increase in $K_{CaM}$) to activation by $Ca^{2+}$/calmodulin following phosphorylation of sites A and B; however, phosphorylation at site B alone is insufficient to alter the $Ca^{2+}$/calmodulin activation properties of the enzyme. The predicted physiological consequence of an increased $K_{CaM}$ value is an increase in the $[Ca^{2+}]_i$ necessary for RLC phosphorylation. We show by site-directed mutagenesis that introduction of a negative charge at site A alone is sufficient to increase $K_{CaM}$, whereas a charge substitution at site B has no effect. We have also demonstrated that MLCK is phosphorylated in smooth muscle cells by $Ca^{2+}$/calmodulin-dependent protein kinase II and that phosphorylation diminishes the $Ca^{2+}$ sensitivity of RLC phosphorylation.

## Materials and Methods

### Triton/Glycerol Permeabilization of Tracheal Tissue and Force Measurements

Bovine tracheal muscle strips were skinned as modified from Rüegg et al. [27]. Strips were incubated for 30 min at 4°C in a presoak solution containing 5 mM EGTA buffer, followed by a 30-min incubation in the presoak solution plus 1% Triton X-100, 0.5 mM dithiothreitol, and 0.5 mM leupeptin. After a 15-min rinse, strips were placed in a skinning solution (20 mM imidazole, 4 mM EGTA, 10 mM $MgCl_2$, 7.5 mM ATP, 1 mM $NaN_3$, 0.5 mM dithiothreitol, 0.5 mM leupeptin, pH 6.7, in 50% glycerol) and stored at −20°C for up to 2 weeks. For experiments, strips were washed in relaxing solution containing 6 mM ATP and 6 mM $MgCl_2$. $Ca^{2+}$-buffered solutions for skinned fiber studies were prepared as described by Moreland and Murphy [28]. Strips were immersed in 5 ml of solution at 24°C and were frozen before or after a cycle of contraction–relaxation with a clamp precooled in liquid nitrogen.

### Calmodulin and MLCK Content in Intact and Permeable Tracheal Tissue

Homogenates of frozen tissues prepared in 10% trichloroacetic acid were centrifuged and the supernatant fractions discarded. Pellets were resuspended in a urea sample buffer (24 mM Tris HCl, 266 mM glycine, 8 M urea, 10 mM dithiothreitol, 2 mM EGTA, 5% sucrose). Samples and known amounts of purified calmodulin were subjected to urea/glycerol polyacrylamide gel electrophoresis (PAGE) followed by electrophoretic transfer to nitrocellulose and immunoblotting with a monoclonal antibody raised against calmodulin and a peroxidase-linked goat anti-mouse second-step antibody [29]. Tissue samples were also resuspended in sodium dodecyl sulfate (SDS) sample buffer and electrophoresed on 7.5% SDS PAGE along with known amounts of purified tracheal smooth muscle MLCK, followed by transfer to nitrocellulose and immunoblotting with rabbit antiserum raised against the kinase. Quantitation of calmodulin and MLCK was achieved by densitometric measurements of the stained proteins.

## Bovine Tracheal Cell Culture

Tracheal smooth muscle cells were isolated and prepared for primary culture as previously described [30]. Experiments were conducted 24–48 h after serum deprivation to promote smooth muscle-specific cytodifferentiation [31].

## Labeling of Calmodulin and Co-microinjection with Dextrans

Fluorescence-labeled calmodulin and dextrans of various sizes were co-injected into intact tracheal smooth muscle cells in culture to estimate a membrane pore size following permeabilization with β-escin. Calmodulin was purified from bovine testes and labeled with tetramethylrhodamine isothiocyanate (TRITC) [29]. TRITC-labeled calmodulin was dialyzed against injection buffer and centrifuged for clarification prior to microinjection. Fluorescein-labeled dextrans of various molecular masses (3, 10, 20, or 40 kDa) were resuspended in injection buffer at a concentration of 5 mg/ml and clarified by centrifugation. The calmodulin used for the fluorescence recovery after photobleaching (FRAP) experiments was purified from calf brain and labeled with lissamine rhodamine B (LRB-calmodulin) as previously described [32]. LRB-calmodulin had a dye/protein ratio of 1.4 and was shown to activate smooth muscle MLCK in a $Ca^{2+}$-dependent manner.

Calmodulin analogs and dextrans were co-injected into smooth muscle cells. A specific-size dextran and calmodulin were mixed in a 1:1 ratio and centrifuged for clarification prior to microinjection. It is estimated that the amount injected was ≤10% of the cell volume, resulting in an intracellular concentration of LRB-calmodulin of ≤20 μM. After injection, cells were rinsed three times in complete Dulbecco's modified Eagle's medium (DMEM) without phenol red and allowed to recover for a minimum of 3 h.

### Fluorescence Intensity Measurements

Fourteen-bit fluorescence images were acquired at 0.5 s for each cell at each excitation wavelength (fluorescein excitation 490 nm, emission 520 nm; rhodamine excitation 570 nm, emission 590 nm). All focusing was done with the rhodamine filter set to avoid excess photobleaching of the fluorescein signal. To determine the rate of photobleaching for each fluorophore during image acquisition, a photobleaching curve was generated by acquiring six exposures with each filter set for each of 10 cells and determining the mean fluorescence intensity of the cells after each exposure. Regions of interest with each cell were outlined, and the mean pixel intensity within the region of interest was calculated. The mean value of pixels in a region of comparable area outside the cells was subtracted from each measurement to account for background intensity. Mean intensities were expressed as a percentage of the mean intensity in the first image of each cell.

## Fluorescence Recovery After Photobleaching

The FRAP procedure was performed as described elsewhere [29, 33]. The general format of the experimental approach is shown in Figure 3. A SpectraPhysics 2016 argon ion laser (Mountainview, CA, USA) operated at 250 mW on the 488 nm line or

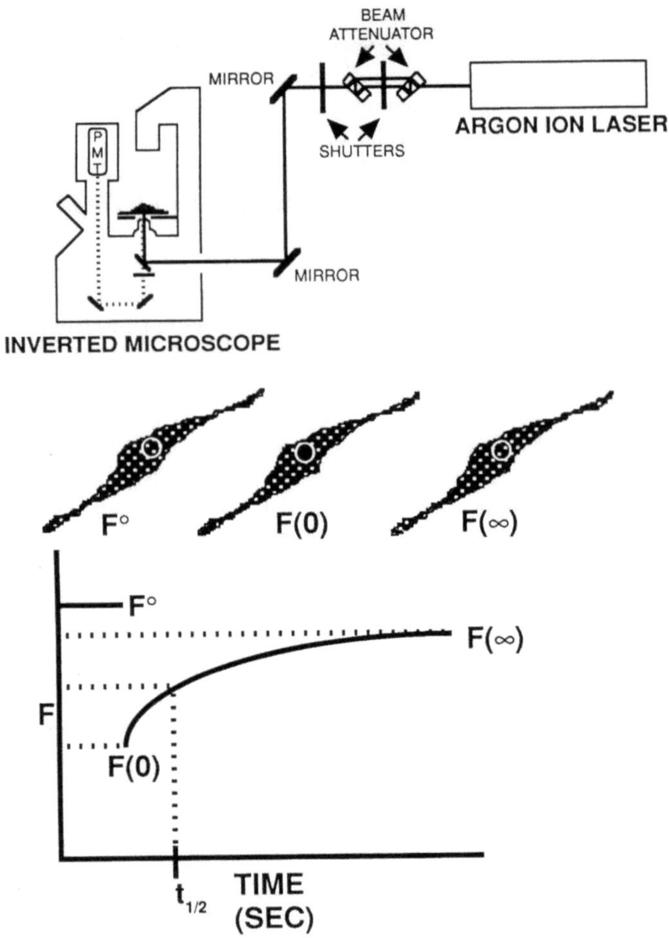


**Fig. 3.** Methods for estimating translational diffusion constants in cytoplasm of living cells by fluorescence recovery after photobleaching (FRAP). **Top. Bottom.** FRAP data record. An initial level of fluorescence intensity (F°) is measured using an attenuated laser beam to excite fluorophores in the sample (shown here as a smooth muscle cell). A brief (millisecond) pulse of unattenuated laser light is used to irreversibly photolyse (bleach) a portion of the fluorophores in the path of the laser, resulting in reduced fluorescence intensity [F(0)]. If fluorophores outside the bleached region are free to move into the bleached spot and exchange with the bleached molecules, a recovery of fluorescence intensity is observed. In the case where transport is by brownian motion, the half-time of the recovery ($t_{1/2}$) is proportional to the diffusion coefficient. The asymptote of the recovery [F(∞)] gives the fraction of fluophores that are mobile on the time scale of the measurement (≤30 s)

700 mW on the 514 nm line was directed into a Zeiss Axiovert 35 (Carl Zeiss Inc., Thornwood, NY, USA) microscope with beam steering optics and focused to a 1.3 μM spot at the specimen plane with a 40 × 0.75 NA objective. Fluorescence was measured with a Thorn EMI (Thorn EMI, Far Rockaway, NJ, USA) 9203A extended red photomultiplier tube cooled to −23°C. Instrument control, data acquisition, and data analysis were performed with Labview software on a Macintosh IIx computer

with 8 megabytes RAM. Bleach times were 10 ms for fluorescein and 400–1000 ms for rhodamine, resulting in a 20–50% bleach. The calculated diffusion coefficients for LRB-calmodulin did not vary significantly with bleaching time, suggesting that recovery during the bleach was not a serious problem despite the relatively long bleaching times. Ten recovery half-lives of data were recorded for each recovery (Fig. 3).

## COS Cell Expression of Rabbit Smooth Muscle MLCK and Oligonucleotide-Directed Mutagenesis

Wild-type and mutant rabbit smooth muscle MLCKs were expressed in COS cells as described elsewhere [11]. Oligonucleotide-directed mutagenesis was performed by the method of Kunkel et al. using mutagenic oligonucleotides designed to produce the desired amino acid substitutions [34]. For each mutant cDNA the desired nucleotide substitutions were verified by DNA sequencing. Kinetic parameters for the expressed wild-type and mutant MLCKs were determined directly in COS cell lysates as described previously [35–37]. COS cell lysates were diluted 50- to 500-fold in the reaction mixture. A 1:10 dilution of mock-transfect COS cells has no detectable MLCK activity. The $Ca^{2+}$/calmodulin-dependent activity of wild-type and mutant MLCKs was measured by $^{32}P$ incorporation into the light chain [21]. Calmodulin activation properties of kinases were performed as described previously in a reaction mixture containing 1 μM calmodulin and various concentrations of free $Ca^{2+}$ as determined by $Ca^{2+}$/EGTA buffers [37, 38].

## MLCK Activity Ratio

To assess changes in the calmodulin activation properties of MLCK ($K_{CaM}$), activities were determined in cell extracts at two $Ca^{2+}$ concentrations as described elsewhere [39–41]. MLCK activities were assayed in cell lysates or tissue homogenates at 1 and 100 μM $Ca^{2+}$ in the presence of 400 nM calmodulin. The MLCK activity ratio decreases quantitatively as the apparent $K_{CaM}$ for MLCK increases. The extent of the changes in $K_{CaM}$ were calculated from the ratio of activities at 1:100 μM $Ca^{2+}$: $[(Vmax/v)' - 1/(Vmax/v) - 1] = (K'_{CaM}/K_{CaM})$, where (v/Vmax) and $K_{CaM}$ represent values obtained under nonstimulated conditions and (v/Vmax)′ and $K'_{CaM}$ represent values obtained under stimulated conditions [37, 39]. An average $K_{CaM}$ value for nonphosphorylated MLCK is 1 nM [42].

## Myosin Light Chain Phosphorylation

Cell lysates or tissue homogenates were processed for urea/glycerol PAGE and immunoblotting of myosin RLC to measure myosin RLC phosphorylation [43].

## $^{32}P$ Labeling and Phosphopeptide Mapping

Muscle strips were prepared as described previously for $^{32}P$ labeling [44]. Strips were mounted isometrically on stainless steel wire holders at near-optimal length, incubated for 4 h at 37°C with $[^{32}P]H_3PO_4$ 125 μCi/ml, and then transferred to experimental solutions. After treatment the strips were frozen by submersion in dichlorodifluoromethane chilled in liquid nitrogen stored at −60°C prior to biochemical analyses.

The MLCK was immunoprecipitated from muscle homogenates containing protein kinase and protein phosphatase inhibitors with a monoclonal antibody raised to bovine tracheal MLCK [45]. Antigen–antibody complexes eluted from Pansorbin were subjected to SDS-PAGE. MLCK was quantified from densitometric scans of Coomassie blue-stained sample protein and standards. Segments of gel containing kinase were excised. After trypsin digestion, lyophilized peptides were resuspended in electrophoresis buffer and subjected to high-voltage thin-layer electrophoresis and chromatography as described elsewhere [44]. Autoradiography was performed to locate [$^{32}$P]phosphopeptides. Labeled spots were scraped and the radioactivity measured. The stoichiometry of MLCK phosphorylation was calculated from total counts perminute per mole of protein divided by counts per minute per mole of intracellular [$\gamma^{32}$P]ATP as assessed from specific activity measurements [43]. The stoichiometry of peptide phosphorylation was calculated as the product of the fraction of total radioactivity in peptide A and the stoichiometry of MLCK phosphorylation [45].

# Results

## Effective Calmodulin Concentration Is Less than Predicted from Total Calmodulin

Permeabilized smooth muscle tissues do not require the addition of MLCK, PP-1M, or calmodulin to effect $Ca^{2+}$-dependent contraction and relaxation, but it has been noted that the addition of calmodulin can potentiate the sensitivity of force to $Ca^{2+}$ [46]. We investigated the calmodulin dependence of force development in tracheal tissue that had been extensively skinned in 4 mM EGTA, 1% Triton X-100, and 50% glycerol. In this permeable smooth muscle preparation addition of calmodulin resulted in no increase in force under relaxing conditions. Increasing the intracellular $Ca^{2+}$ to 2 μM resulted in a small increase in force that was greatly potentiated upon addition of 2 μM calmodulin (Table 1). At high $Ca^{2+}$ concentrations calmodulin had little or no effect (Table 1). Addition of 2 μM calmodulin to the Triton/glycerol skinned trachealis increases the sensitivity of force development to $Ca^{2+}$. Furthermore, because fibers contracted at 2 μM $Ca^{2+}$ without addition of calmodulin, tissue calmodulin was probably not completely extracted.

To determine the extent of calmodulin and MLCK extraction resulting from permeabilization of fibers, tissue contents of both calmodulin and MLCK were evalu-

**Table 1.** Effect of calmodulin on permeable tracheal smooth muscle strips.

| | Force (mg), by free calcium concentration | | |
|---|---|---|---|
| Added CaM (μM) | <0.01 μM | 2 μM | 30 μM |
| 0 | 0 | 30 ± 20 | 430 ± 69 |
| 2.0 | 0 | 300 ± 13 | 507 ± 42 |

The force response was obtained at each free calcium concentration ([$Ca^{2+}$]), in the absence or presence of added 2 μM calmodulin. Permeable strips are estimated to contain 16 μM endogenous calmodulin and 3.4 μM myosin light chain kinase.

ated. The MLCK content in the skinned fibers was similar to that in intact tissues (3.4 ± 0.2 μM versus 3.2 ± 0.4 μM). Calmodulin content, however, decreased only 50% with the extensive skinning procedure in the presence of EGTA (39 ± 2 to 16 ± 2 μM). Additional experiments showed that there was no change in the tissue contents of myosin, actin, filamin, and desmin after skinning. A concentration of 16 μM calmodulin calculated for the permeable fibers is in excess of the 2 μM exogenous calmodulin that was sufficient to elicit a potentiated contraction. Thus it appears that a major portion of the endogenous calmodulin is bound, even in the absence of $Ca^{2+}$. It would be expected that most of this calmodulin fraction is not available to activate MLCK.

## Calmodulin Diffusion Is Restricted in Smooth Muscle Cells

Two experimental approaches were used to investigate calmodulin mobility by fluorescence imaging of labeled calmodulin in cultured tracheal smooth muscle cells. First, we determined if the 16.5 kDa calmodulin diffuses out of cells following permeabilization with β-escin. The integrated fluorescence intensity of cells microinjected with rhodamine-labeled calmodulin was measured by video imaging before and after β-escin permeabilization. Co-injection of rhodamine-labeled calmodulin and various-size fluorescein-labeled dextrans was performed to estimate the cutoff size at which a freely diffusible molecule was lost from the permeable cells. After permeabilization of tracheal cells with β-escin 30 μg/ml in the presence of 4 mM EGTA for 15 min, both calmodulin and the 40 kDa dextran were retained; however, the 3 kDa and 10 kDa dextrans were not. A longer incubation time (45 min) in the presence of β-escin resulted in loss of 40 kDa dextran fluorescence but no significant change in calmodulin. Thus most of the calmodulin appears bound to cellular structures in smooth muscle cells, even in the absence of $Ca^{2+}$.

To test the hypothesis that free calmodulin may limit MLCK activation because binding sites for calmodulin exceed the endogenous amount of calmodulin, FRAP was used to measure the mobility of labeled calmodulin following microinjection into intact tracheal smooth muscle cells in culture. Data for rhodamine-labeled calmodulin were compared with similar measurements for fluorescein-labeled dextran, which does not bind to intracellular components. FRAP recovery curves for labeled calmodulin could not be fitted to a single recovering species. By measuring the rate of recovery during the first 5 s, we obtained an apparent mean diffusion coefficient of $2.2 \times 10^{-9}\ cm^2/s$ for the fastest recovering components (Table 2). In contrast, the full-length recovery curves for fluorescein dextrans fit a single component. One striking feature of these results is the sevenfold lower diffusion coefficient for the fastest mobile fraction of calmodulin compared to a similar-size dextran (Table 2). Another striking feature of the fluorescence recovery curves for LRB-calmodulin was failure to recover to the initial fluorescence intensity. This result indicates that 30%–40% of calmodulin is basically immobile on the time scale of the experiment (30 s). These results are consistent with the hypothesis that calmodulin may be limiting for MLCK activation.

## MLCK Phosphorylation at Site A Decreases Sensitivity to Activation by $Ca^{2+}$/Calmodulin

The predicted effect of site A phosphorylation in MLCK is to decrease the sensitivity of kinase activation to $Ca^{2+}$. This effect is illustrated in Figure 4. Rabbit smooth muscle

**Table 2.** Mobility of calmodulin in smooth muscle cells.

| Substance | Fraction with highest mobility (%) | Diffusion coefficient ($cm^2/s$) |
|---|---|---|
| Dextran | | |
| 20 kDa | 100 | $1.6 \pm 0.2 \times 10^{-8}$ |
| 40 kDa | 100 | $7.0 \pm 1.0 \times 10^{-8}$ |
| Calmodulin (16.5 kDa) | 8 | $2.2 \pm 0.2 \times 10^{-9}$ |

The fastest mobile component of the recovery curve for rhodamine-labeled calmodulin was compared to the monophasic recovery curves of fluorescein-labeled dextrans following fluorescence recovery after photobleaching.

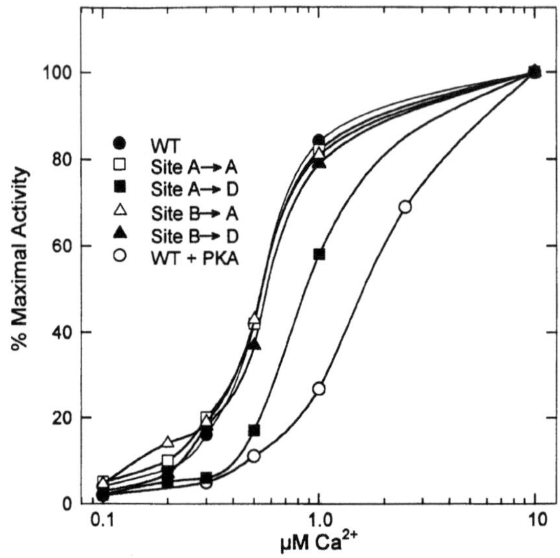


**Fig. 4.** $Ca^{2+}$ activation properties of wild-type and mutant recombinant MLCK. Rabbit smooth muscle MLCKs were expressed in COS cells for measurement of kinase activities in cell lysates at 1 μM calmodulin. MLCK activation was measured at different free $Ca^{2+}$ concentrations, and the data were normalized to the percent maximal activity. *WT*, wild-type MLCK; *WT + PKA*, wild-type MLCK phosphorylated by cAMP-dependent protein kinase. Mutations at the phosphoacceptor serines in site A and site B are shown with arrows pointing to the amino acid substitution (*A*, alanine; *D*, aspartate)

MLCK was expressed in COS cells for measurement of the $Ca^{2+}$/calmodulin dependent activity. After phosphorylation of wild-type MLCK by the catalytic subunit of cAMP-dependent protein kinase, the concentration of $Ca^{2+}$ required for half-maximal activity increased. The calculated change in $K_{CaM}$ increased from 1 nM to 8 nM [39, 45]. To test the hypothesis that site A exerts its effect on calmodulin activation properties independent of phosphorylation of site B, we mutated individually Ser-992 in site A and Ser-1005 in site B to a neutral (alanine) or acidic (aspartate) residue. Whereas neutral substitutions would be expected to have little effect on kinase activation, the acidic group may mimic the negative charge introduced by phosphate incorporation, resulting in an increase in $K_{CaM}$. As seen in Figure 4, substitution of either neutral or acidic amino acids at the phosphoacceptor serine in site B had no effect on calmodulin activation properties of the enzyme. Neither did the neutral substitution for serine in site A; however, the introduction of asparate for the phosphoacceptor at site A resulted in an increase in the concentration of $Ca^{2+}$ required for half-maximal activation, with an increase in $K_{CaM}$ from 1 nM to 4 nM. The aspartate residue appears to partially mimic the effect of phosphate incorporation. This result is consistent with the finding that phosphorylation of site B does not alter calmodulin activation and suggests that the introduction of a negativ charge in site A may be necessary and sufficient for the decreased ability of calmodulin to bind to and activate MLCK.

## MLCK Is Phosphorylated at Site A in Contracting Smooth Muscle

We assessed the phosphorylation of MLCK in intact tracheal smooth muscle tissues using agents known to elevate second messengers that activate protein kinases, which in turn phosphorylate MLCK biochemically (isoproterenol: cyclic AMP-dependent protein kinase; phorbol dibutyrate (PDBu): protein kinase C; carbachol and $K^+$ depolarization: $Ca^{2+}$/calmodulin-dependent protein kinase II). MLCK phosphorylation was assessed by two methods [45]. First, $^{32}P$ incorporation into specific sites was directly measured by phosphopeptide mapping. Second, the functional consequences of kinase phosphorylation were assessed by the MLCK activity ratio. The results of these investigations are summarized in Figure 5, in which the MLCK activity ratio is plotted as a function of phosphate incorporation into peptide A. There was a direct correlation between the MLCK activity ratio and the extent of phosphorylation of peptide A in MLCK; and the relation was nonlinear as expected [39]. This relation was observed with tissues treated with carbachol, KCl, isoproterenol, and PDBu. These results are consistent with observations that quantitative increases in $K_{CaM}$ are determined by the extent of site A phosphorylation. In contrast, the overall stoichiometry of phosphorylation of MLCK was not correlated to the activity ratio. For example,

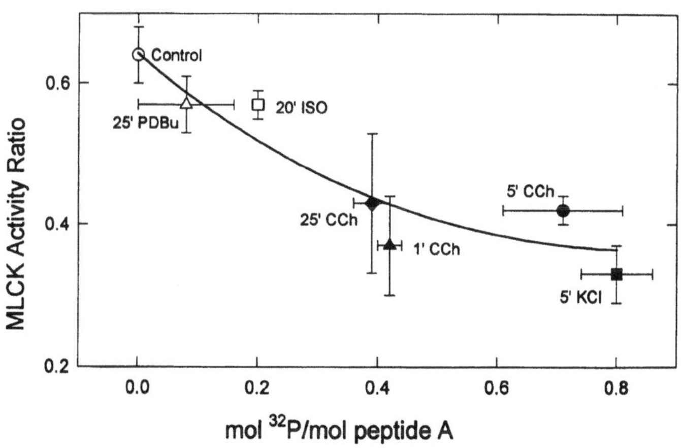


**Fig. 5.** Relation between the myosin light chain kinase (MLCK) activity ratio and the extent of phosphorylation of phosphopeptide A in tracheal smooth muscle tissues. $^{32}P$-labeled muscle strips were treated with 0.1 μM carbachol (CCh) (1, 5, 25 min), 65 mM KCL plus 0.1 μM atropine (5 min), 3 μM isoproterenol (ISO) (20 min), or 1 μM phorbol dibutyrate (PDBu) (25 min) and quick-frozen [45]. MLCK phosphorylation was analyzed for stoichiometry of phosphate incorporation in site A as described in Methods. In parallel experiments, homogenates of tissues treated identically in the absence of $^{32}P$ were assayed for the MLCK activity ratio [45]

when stimulated with PDBu, the total extent of MLCK phosphorylation was 3 mol $^{32}P$ incorporated per mole of kinase, but little phosphate incorporation took place at site A. The striking result of these experiments was that the extent of MLCK phosphorylation at site A was highest in response to agents that elevated intracellular $Ca^{2+}$ and low under conditions that activated cAMP-dependent protein kinase and protein kinase C. These experiments confirm the observations by Miller et al., in which greater decreases in the activity ratio were found in the presence of KCl or carbachol than with isoproterenol [39]. These and other results led us to hypothesize that a $Ca^{2+}$-dependent protein kinase may phosphorylate MLCK in intact cells in response to elevated intracellular $Ca^{2+}$ [45].

## Inhibitors of Calmodulin-Dependent Protein Kinase II Prevent Ca²⁺-Dependent Site A Phosphorylation and Potentiate Myosin RLC Phosphorylation

Tracheal smooth muscle cells in culture were permeabilized briefly with β-escin (15 min) to control cytoplasmic $Ca^{2+}$ concentrations with a $Ca^{2+}$-EGTA buffer system. As seen in Figure 6, both $K_{CaM}$ and myosin RLC phosphorylation increased in a $Ca^{2+}$-dependent manner; however, the concentration of $Ca^{2+}$ required for half-maximal increases in light chain phosphorylation was 250 nM, whereas it was 500 nM for MLCK phosphorylation. Preincubation of the smooth muscle cells with either 1.5 μM KN-62 or a peptide inhibitor CKII of $Ca^{2+}$/calmodulin-dependent protein kinase II inhibited the $Ca^{2+}$-dependent increase in $K_{CaM}$ [29]. Moreover, there was a potentiation of RLC phosphorylation at any given $Ca^{2+}$ concentration, increasing the sensitivity of myosin RLC phosphorylation from 250 to 170 nM $Ca^{2+}$. The low $Ca^{2+}$ sensitivity for MLCK phosphorylation and the inhibition by two types of $Ca^{2+}$/calmodulin-dependent pro-

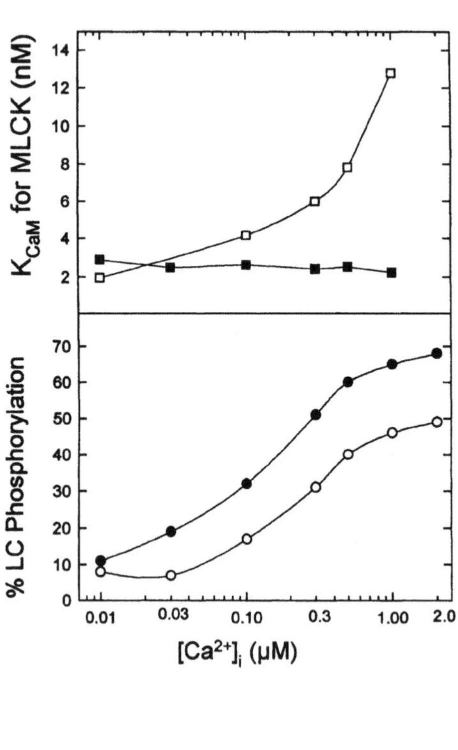


**Fig. 6.** Effect of calmodulin K II inhibitors of $Ca^{2+}$/calmodulin-dependent protein kinase II on the $Ca^{2+}$ dependence of changes in $K_{CaM}$ and myosin light chain (LC) phosphorylation in permeable bovine tracheal smooth muscle cells. Cytosolic [$Ca^{2+}$] was buffered by EGTA-containing solutions. Cells were preincubated 10 min with KN-62 (1.5 μM) or CKII peptide (10 μM) prior to increasing [$Ca^{2+}$] [29]. Data represent averages for all steady-state values obtained with KN-62 and CKII peptide (closed symbols) or control conditions of DMSO alone (open symbols) [29]

**Table 3.** Effect of GTPγS or carbachol on myosin light chain and myosin light chain kinase phosphorylation.

| Condition[a] | MLCK acivity ratio | $PO_4$/LC ratio (mol/mol) |
|---|---|---|
| Relaxed | 0.81 ± 0.04 | 0.02 ± 0.03 |
| $Ca^{2+}$ | 0.67 ± 0.01 | 0.16 ± 0.03 |
| $Ca^{2+}$ + GTPγS | 0.55 ± 0.03 | 0.29 ± 0.07 |
| $Ca^{2+}$ + carbachol | 0.50 ± 0.03 | 0.35 ± 0.05 |

[a] $Ca^{2+}$ at a concentration of 0.3 μM was used in each case.

tein kinase II inhibitors support the notion that the $Ca^{2+}$/calmodulin-dependent protein kinase II phosphorylates MLCK in smooth muscle cells.

## MLCK Site A Is Dephosphorylated by a GTP/Agonist-Sensitive Phosphatase

Little is known about protein phosphatases that dephosphorylate MLCK. Because the kinase associates with both actin and myosin in vitro, actomyosin-associated phosphatase activity (PP-1M?) could play a significant role in the dephosphorylation of MLCK in smooth muscle. PP-1 dephosphorylates both sites A and B in the absence of $Ca^{2+}$/calmodulin [47]. Furthermore, dephosphorylation of the regulatory site (site A) is not inhibited when $Ca^{2+}$/calmodulin is bound to MLCK.

We further investigated the regulation of MLCK phosphorylation by studying the sensitivity of kinase phosphorylation to agents that are known to affect myofibrillar phosphatase activity. Bovine tracheal smooth muscle strips were permeabilized with β-escin, and $Ca^{2+}$ concentrations were buffered by a $Ca^{2+}$-EGTA buffer [48]. There was a $Ca^{2+}$-dependent increase in myosin RLC phosphorylation. As seen with cells, there was a $Ca^{2+}$-dependent decrease in the MLCK activity ratio, indicating that in smooth muscle strips as well as in cells in culture there is $Ca^{2+}$-dependent phosphorylation of MLCK. GTPγS and agonists may potentiate myosin RLC phosphorylation at a fixed $Ca^{2+}$ concentration by inhibiting PP-1M activity [49]. To test the hypothesis that this protein phosphatase, which is bound to smooth muscle actomyosin, may also act on MLCK, we treated β-escin-permeable fibers with GTPγS or carbachol at a fixed $Ca^{2+}$ concentration. As seen in Table 3, individual treatments with GTPγS or carbachol reduced the MLCK activity ratio, indicating that kinase phosphorylation was potentiated at 0.3 μM $Ca^{2+}$ by these treatments. Myosin RLC phosphorylation was potentiated as well. These results support the idea that PP-1M acts to dephosphorylate not only myosin RLC but also MLCK at regulatory site A.

## Discussion

An increase in intracellular free $Ca^{2+}$ concentration is the primary mechanism that initiates smooth muscle contraction [1, 2, 49, 50]. The binding of $Ca^{2+}$ to calmodulin results in rapid association of the $Ca^{2+}$/calmodulin complex to MLCK. Subsequent activation of the kinase results in phosphorylation of Ser-19 on the RLC, allowing an increase in actin-activated myosin ATPase activity and muscle contraction. Although this cascade predicts a simple relation between cytosolic $Ca^{2+}$ concentrations and the extent of RLC phosphorylation, there are important secondary mechanisms for modulating the $Ca^{2+}$ sensitivity of smooth muscle contraction. Increases and decreases in RLC phosphorylation and force at constant $Ca^{2+}$ concentrations result from respective changes in MLCK and protein PP-1M activities. Agonist-stimulated inhibition of PP-1M appears to be a primary mechanism for $Ca^{2+}$ sensitization (i.e., an increase in the extent of light chain phosphorylation at constant $Ca^{2+}$ concentration) [49]. The primary mechanism for $Ca^{2+}$ desensitization involves phosphorylation of MLCK at a specific serine residue at the C-terminus of the calmodulin-binding domain [9]. Our investigations have revealed the physiological importance of MLCK phosphorylation and its primary elements of regulation. Of related importance and interest is the demonstration that calmodulin may be limiting for MLCK activation in smooth muscle cells.

Calmodulin is a regulatory, $Ca^{2+}$-binding protein involved in a variety of cell signaling pathways [51]. Although it binds $Ca^{2+}$ specifically, it also demonstrates an amazingly large number of biochemical functions considering its small size. It regulates the activities of a large number of enzymes, including protein kinases, NAD kinase, phosphodiesterase, $Ca^{2+}$ pumps, and molecular motors involved in various aspects of motility and cell shape changes. Thus it is not surprising to find that the binding of calmodulin to cellular components is multifaceted. The large extent of cellular calmodulin binding in the absence of $Ca^{2+}$ was surprising. Data obtained from skinned tracheal smooth muscle tissue and smooth muscle cells in culture show that approximately 40% of the 40 μM cell calmodulin is immobile and tightly bound in the presence of EGTA. Presumably, this binding occurs to cellular proteins, as the extensive skinning of the smooth muscle fibers with Triton-X 100 and glycerol removes membranes. This high-affinity binding is consistent with previous observations on skinned smooth muscle fibers [22].

The mobile fraction of calmodulin cannot be described by diffusion of a single species, so heterogeneity is presumed. The complexity of the fluorescence recovery curves for calmodulin in the cytoplasm of living smooth muscle cells suggests a multiplicity of intracellular binding sites with a range of affinities for calmodulin. Even the fastest recovering component must be involved in binding interactions because its diffusion coefficient was approximately sevenfold slower than the freely diffusible 20 kDa fluorescein-dextran. The reduced diffusion of the fastest component of calmodulin may indicate either high-affinity binding to slowly diffusing species or transient binding to immobile structures. The estimate for the unbound fraction of calmodulin was 8% of the total cellular calmodulin within smooth muscle cells [29]. These results are consistent in principle with the complex mobility of calmodulin in Swiss 3T3 cells [32]. Because most of the endogenous calmodulin in smooth muscle cells is bound, it could be limiting for the activation of $Ca^{2+}$/calmodulin-dependent enzymes where a reversible binding to $Ca^{2+}$/calmodulin species is necessary for enzyme activity. This observation is particularly pertinent to MLCK.

Specific criteria suggested by Krebs and Beavo for establishing the physiological significance of an enzyme phosphorylation have been satisfied in regard to MLCK (Table 4) [56]. The primary conclusion from these studies is that $Ca^{2+}$/calmodulin-dependent protein kinase II phosphorylates MLCK at a specific serine residue in the C-terminus of the calmodulin-binding domain. This phosphorylation results in desensitization of MLCK to activation by $Ca^{2+}$/calmodulin (Fig. 1). However, there are other important issues in regard to modulation of the $Ca^{2+}$ sensitivity of MLCK. Physiological studies show that sites other than site A are phosphorylated in MLCK under physiological conditions. The functions of these phosphorylations are not

**Table 4.** Criteria for establishing the physiological significance of myosin light chain kinase phosphorylation in smooth muscle.

| Criteria | References |
|---|---|
| 1. Purified myosin light chain kinase is phosphorylated at site A at a significant rate by calmodulin-dependent protein kinase II and dephosphorylated by protein phosphatases (PP1 and PP2A). | 25, 47, 52 |
| 2. A decrease in the affinity of myosin light chain kinase for calmodulin correlates with the extent of phosphorylation of site A. | 23, 39, 45, 53 |
| 3. Myosin light chain kinase is phosphorylated and dephosphorylated in intact cells with accompanying changes in sensitivity to $Ca^{2+}$. | 30, 39, 45, 48, 54, 55 |
| 4. The cellular levels of the protein kinase effector ($Ca^{2+}$) correlates with the extent of myosin light chain kinase phosphorylation. | 40, 48, 54, 55 |

clear; however, data presented herein show that the introduction of a negative charge via site-directed mutagenesis in place of the serine at site A is sufficient to desensitize MLCK activity to $Ca^{2+}$/calmodulin. Introduction of a negative charge at site B, which is also phosphorylated in vivo, is not sufficient for $Ca^{2+}$/calmodulin desensitization. Thus site A phosphorylation appears to be the primary site for modulation of $Ca^{2+}$/calmodulin activation properties.

Another important point raised by these investigations is that site A phosphorylation occurs in response to stimuli that activate primarily the $Ca^{2+}$ signaling pathway, leading to the activation of $Ca^{2+}$/calmodulin-dependent protein kinase II. Activation of other signaling pathways involving other kinases shown to phosphorylate MLCK in vitro are not important in vivo. These pathways include primarily the cAMP-dependent protein kinase and protein kinase C. The results in bovine tracheal smooth muscle have been confirmed in swine carotid arteries [54].

When considering the physiological function of MLCK phosphorylation, it is important to note the complexity involved in a $Ca^{2+}$-dependent kinase phosphorylating another $Ca^{2+}$-dependent enzyme and changing its sensitivity to $Ca^{2+}$ activation (Fig. 1). Although this sequence of events appears to be contradictory in terms of biological efficiency, it is crucial to note the differences in $Ca^{2+}$ sensitivities for the two $Ca^{2+}$-dependent phosphorylations. The $Ca^{2+}$ concentration required for half-maximal MLCK-phosphorylation is approximately 500 nM. When MLCK phosphorylation is inhibited, the $Ca^{2+}$ concentration required for half-maximal RLC phosphorylation is only 170 nM. Thus there is a marked difference in the $Ca^{2+}$ sensitivities of myosin light chain and MLCK phosphorylation. Based on these considerations, the desensitization mechanism is invoked under conditions where cytosolic $Ca^{2+}$ concentrations rise to high levels in cells. This desensitization process is one of several mechanisms that a cell may use to down-regulate the $Ca^{2+}$ signal. Increases in cytosolic $Ca^{2+}$ with the resultant formation of $Ca^{2+}$/calmodulin activates a sarcolemmal $Ca^{2+}$ ATPase that pumps $Ca^{2+}$ out of the cell, thereby limiting the cytosolic $Ca^{2+}$ concentration [57]. $Ca^{2+}$/calmodulin activation of an inositol trisphosphate ($IP_3$) kinase results in the conversion of this second messenger to an inactive metabolite incapable of releasing $Ca^{2+}$ from the sarcoplasmic reticulum [58]. An additional pathway involves the $Ca^{2+}$/calmodulin activation of nitric oxide synthase; and the resultant product, nitric oxide, stimulates cGMP, which acts to decrease cytosolic $Ca^{2+}$ concentrations via activation of the cGMP-dependent protein kinase [59].

Initially, it was difficult to understand how MLCK was phosphorylated in smooth muscle cells under conditions that result in an increase in cytosolic $Ca^{2+}$ concentrations. When $Ca^{2+}$/calmodulin is bound to purified MLCK, phosphorylation by $Ca^{2+}$/calmodulin-dependent protein kinase II (and other protein kinases) at the serine in site A is inhibited. Thus it is predicted that when cytosolic $Ca^{2+}$ concentrations increase in a cell and $Ca^{2+}$/calmodulin binds to the enzyme, MLCK phosphorylation would be inhibited. A possible explanation for these observations is that not all of the MLCK has $Ca^{2+}$/calmodulin bound when cytosolic $Ca^{2+}$ concentrations are increased. Thus the unbound fraction would be available for phosphorylation by calmodulin-dependent protein kinase II. This possibility may be realized if it is considered that calmodulin may be limiting for activation of MLCK, as only a small fraction of the total cellular calmodulin is mobile. Observations on the complex mobility of calmodulin in smooth muscle cells are consistent with this hypothesis.

The MLCK phosphorylation and dephosphorylation occur at rates sufficient to modulate the $Ca^{2+}$ sensitivity of light chain phosphorylation during cycles of contraction and relaxation in smooth muscle [55]. In bovine trachealis, electrical stimulation results in synchronous release of neurotransmitters from the distal terminal neurons, hydrolysis of glycerophosphoinositides, release of $Ca^{2+}$ from internal stores, and generation of the contractile force [60]. In myometrical tissues, electrical depolarization results in synchronous depolarization of smooth muscle cells, influx of extracellular $Ca^{2+}$, and force development. The uniform synchronous activation in these two preparations facilitated investigations of precise temporal relations among $[Ca^{2+}]_i$, protein phosphorylation, and mechanical output. Both MLCK and myosin light chain were phosphorylated rapidly at the initiation of smooth muscle contrac tion consistent with the proposed $Ca^{2+}$/calmodulin dependence of each phosphorylation. Phosphorylation of myosin light chain was more rapid (maximal levels from 500 ms to 2 s) and preceded that of kinase (maximal increase in $K_{CaM}$ from 2 s to 15 s). This delay in MLCK phosphorylation by $Ca^{2+}$/calmodulin-dependent protein kinase II may result from several factors: (1) The $Ca^{2+}$/calmodulin requirement for half-maximal activation of $Ca^{2+}$/calmodulin-dependent protein kinase II (45–150 nM) is greater than that for half-maximal activation of MLCK (1 nM) [21, 61]. (2) The availability of site A for phosphorylation may be a function of the relative amounts of MLCK that are not calmodulin-bound. (3) Localized areas with high $[Ca^{2+}]_i$ may result in increased activation of enzymes in those cellular domains relative to enzyme activation by $[Ca^{2+}]_i$ in the bulk cytoplasm [62].

A comparison of the patterns of alteration in MLCK phosphorylation in the tonic tracheal muscle and the phasic myometrium reveals that maximal phosphorylation of MLCK is delayed in the myometrium (30 s, compared with 5 s in trachealis) [55]. The delay in MLCK phosphorylation may reflect differences in the amounts of MLCK, $Ca^{2+}$/calmodulin-dependent protein kinase II, or the rates of dephosphorylation of MLCK or $Ca^{2+}$/calmodulin-dependent protein kinase II in the two tissues.

To determine if MLCK phosphorylation was coincident with desensitization of light chain phosphorylation to $[Ca^{2+}]_i$, the time courses of phosphorylation and dephosphorylation of MLCK with increases in $[Ca^{2+}]_i$, phosphorylation/dephosphorylation of myosin light chain, and the development of force during the initiation of contraction have been examined [55]. It was surprising that maximal values of light chain phosphorylation developed several seconds prior to maximal values of $[Ca^{2+}]_i$. In both smooth muscles (though using different sources of activating $Ca^{2+}$), the rate of increase in light chain phosphorylation was significantly greater than the rate of increase in $[Ca^{2+}]_i$, and during force development the extent of light chain phosphorylation reached steady-state levels whereas $[Ca^{2+}]_i$ continued to increase. The onset of MLCK phosphorylation coincided with a diminished rate of light chain phosphorylation, and in the myometrium the greatest extent of MLCK phosphorylation occurred at a time when light chain phosphorylation was depressed. The prolonged phosphorylation of MLCK after the decline in the $Ca^{2+}$ transient is consistent with prolonged $Ca^{2+}$-independent activity of autophosphorylated/calmodulin-dependent protein kinase [61].

Taken together with results obtained on permeable smooth muscle cells in culture, MLCK is highly sensitive to changes in $[Ca^{2+}]_i$ during the initiation of contraction. Thereafter the enzyme is phosphorylated by another $Ca^{2+}$-dependent protein kinase ($Ca^{2+}$/calmodulin-dependent protein kinase II) and desensitized to further increases in cytosolic $Ca^{2+}$. The temporal relation between phosphorylation of MLCK and the decreased rate of light chain phosphorylation may explain the reported time-dependent shift in the coupling of $[Ca^{2+}]_i$ and force production during activation of single smooth muscle cells [63].

*Acknowledgment.* The work described was supported in part by grants from the National Institutes of Health (HL26043, HL06296).

## References


1. Kamm KE, Stull JT (1985) The function of myosin and myosin light chain kinase phosphorylation in smooth muscle. Annu Rev Pharmacol Toxicol 25:593–620
2. Hartshorne DJ (1987) Biochemistry of the contractile process in smooth muscle. In: Johnson LR (ed) Physiology of the gastrointestinal tract. Raven, New York, pp 423–482
3. Rembold CM, Murphy RA (1993) Models of mechanism for crossbridge attachment in smooth muscle. J Muscle Res Cell Motil 14:325–333
4. Tansey MG, Hori M, Karaki H, Kamm KE, Stull JT (1990) Okadaic acid uncouples myosin light chain phosphorylation and tension in smooth muscle. FEBS Lett 270:219–221
5. Khalil RA, Morgan KG (1992) Protein Kinase C: a second E-C coupling pathway in vascular smooth muscle? News In Physiological Sciences 7:10–15
6. Kamm KE, Stull Jt (1986) Activation of smooth muscle contraction: relation between myosin phosphorylation and stiffness. Science 232:80–82
7. Shimizu H, Ito M, Miyahara M, Ichikawa K, Okubo S, Konishi T, Naka M, Tanaka T, Hirano K, Hartshorne DJ, Nakano T (1994) Characterization of the myosin-binding subunit of smooth muscle myosin phosphatase. J Biol Chem 269:20407–30411
8. Shirazi A, Iizuka K, Fadden P, Mosse C, Somlyo AP, Somlyo AV, Haystead TAJ (1994) Purification and characterization of the mammalian myosin light chain phosphatase holoenzyme. J Biol Chem 269:31598–31606
9. Stull JT, Gallagher PJ, Herring BP, Kamm KE (1991) Vascular smooth muscle contractile elements: cellular regulation. Hypertension 17:723–732
10. Chamley-Champbell JH, Campbell GR (1981) What controls smooth muscle phenotype? Atherosclerosis 40:347–357
11. Gallagher PJ, Herring BP, Griffin SA, Stull JT (1991) Molecular characterization of a mammalian smooth muscle myosin light chain kinase. J Biol Chem 266:23936–23944
12. Olson NJ, Pearson RB, Needleman DS, Hurwitz MY, Kemp BE, Means AR (1990) Regulatory and structural motifs of chicken gizzard myosin light chain kinase. Proc Natl Acad Sci USA 87:2284–2288
13. Kobayashi H, Inoue A, Mikawa T, Kuwayama H, Hotta Y, Masaki T, Ebashi S (1992) Isolation of cDNA for bovine stomach 155 kDa protein exhibiting myosin light chain kinase activity. J Biochem 112:786–791

14. Meador WE, Means AR, Quiocho FA (1992) Target enzyme recognition by calmodulin: 2.4 Å structure of a calmodulin-peptide complex. Science 257:1251–1255
15. Stull JT, Krueger JK, Zhi G, Gao Z-H (1995) Molecular properties of myosin light chain kinases. In: International Symposium on Regulation of the Contractile Cycle in Smooth Muscle, Japan (in press)
16. Ito M, Dabrowska R, Guerriero V Jr, Hartshorne DJ (1989) Identification in turkey gizzard of an acidic protein related to the C-terminal portion of smooth muscle myosin light chain. J Biol Chem 264:13971–13974
17. Gallagher PJ, Herring BP (1991) The C-terminus of the smooth muscle myosin light chain kinase is expressed as an independent protein, telokin. J Biol Chem 266:23945–23952
18. Holden HM, Ito M, Hartshorne DJ, Rayment I (1992) X-ray structure determination of telokin, the C-terminal domain of myosin light chain kinase, at 2.8 Å resolution. J Mol Biol 227:840–851
19. Shirinsky VP, Vorotnikov AV, Birukov KG, Nanaev AK, Collinge M, Lukas TJ, Sellers JR, Watterson DM (1993) A kinase-related protein stabilizes unphosphorylated smooth muscle myosin minifilaments in the presence of ATP. J Biol Chem 268:16578–16583
20. Kanoh S, Ito M, Niwa E, Kawano Y, Harshorne DJ (1993) Actin-binding peptide from smooth muscle myosin light chain kinase. Biochemistry 32:8902–8907
21. Blumenthal DK, Stull JT (1980) Activation of skeletal muscle myosin light chain kinase by calcium(2+) and calmodulin. Biochemistry 19:5608–5614
22. Rüegg JC, Pfitzer G, Zimmer M, Hofmann F (1984) The calmodulin fraction responsible for contraction in an intestinal smooth muscle. FEBS Lett 170:383–386
23. Conti MA, Adelstein RS (1981) The relationship between calmodulin binding and phosphorylation of smooth muscle myosin kinase by the catalytic subunit of 3′:5′ cAMP-dependent protein kinase. J Biol Chem 256:3178–3181
24. Nishikawa M, Hidaka H, Adelstein RS (1983) Phosphorylation of smooth muscle heavy meromyosin by calcium-activated, phospholipid-dependent protein kinase: the effect on actin-activated MgATPase activity. J Biol Chem 258:14069–14072
25. Hashimoto Y, Soderling TR (1990) Phosphorylation of smooth muscle myosin light chain kinase by $Ca^{2+}$/calmodulin-dependent protein kinase-II: comparative study of the phosphorylation sites. Arch Biochem Biophys 278:41–45
26. Ikebe M, Reardon S, Scott-Woo GC, Zhou Z, Koda Y (1990) Purification and characterization of calmodulin-dependent multifunctional protein kinase from smooth muscle: isolation of caldesmon kinase. Biochemistry 29:11242–11248
27. Rüegg JC, Meisheri K, Pfitzer G, Zeugner C (1983) Skinned coronary smooth muscle: calmodulin, calcium antagonists, and cAMP influence contractility. Basic Res Cardiol 78:462–471
28. Moreland RS, Murphy RA (1986) Determinants of $Ca^{2+}$-dependent stress maintenance in skinned swine carotid media. Am J Physiol 251:C892–C903
29. Tansey MG, Luby-Phelps K, Kamm KE, Stull JT (1994) $Ca^{2+}$-dependent phosphorylation of myosin light chain kinase decreases the $Ca^{2+}$ sensitivity of light chain phosphorylation within smooth muscle cells. J Biol Chem 269:9912–9920
30. Taylor DA, Stull JT (1988) Calcium dependence of myosin light chain phosphorylation in smooth muscle cells. J Biol Chem 263:14456–14462
31. Owens GK, Loeb A, Gordon D, Thompson MM (1986) Expression of smooth muscle-specific α-isoactin in cultured vascular smooth muscle cells: relationship between growth and cytodifferentiation. J Cell Biol 102:343–352
32. Luby-Phelps K, Lanni F, Taylor DL (1985) Behavior of a fluorescent analogue of calmodulin in living 3T3 cells. J cell Biol 101:1245–1256
33. Simon JR, Gough A, Urbanik E, Wang F, Lanni F, Ware BR, Taylor DL (1988) Analysis of rhodamine and fluorescein-labeled F-actin diffusion in vitro by fluorescence photobleaching recovery. Biophys J 54:801–815
34. Kunkel TA, Roberts JD, Zakour RA (1987) Rapid and efficient site-specific mutagenesis without phenotypic selection. Methods Enzymol 154:367–382
35. Herring BP, Stull JT, Gallagher PJ (1990) Domain characterization of rabbit skeletal muscle myosin light chain kinase. J Biol Chem 265:1724–1730
36. Herring BP, Gallagher PJ, Stull JT (1992) Substrate specificity of myosin light chain kinases. J Biol Chem 267:25945–25950

37. Fitzsimons DP, Herring BP, Stull JT, Gallagher PJ (1992) Identification of basic residues involved in activation and calmodulin binding of rabbit smooth muscle myosin light chain kinase. J Biol Chem 267:23903–23909
38. Herring BP (1991) Basic residues are important for $Ca^{2+}$/calmodulin binding and activation but not autoinhibition of rabbit skeletal muscle myosin light chain kinase. J Biol Chem 266:11838–11841
39. Miller JR, Silver PJ, Stull JT (1983) The role of myosin light chain kinase phosphorylation in β-adrenergic relaxation of tracheal smooth muscle. Mol Pharmacol 24:235–242
40. Stull JT, Silver PJ, Miller JR, Blumenthal DK, Botterman BR, Klug GA (1983) Phosphorylation of myosin light chain in skeletal and smooth muscles. Fed Proc 42:21–26
41. Tansey MG, Word RA, Hidaka H, Singer HA, Schworer CM, Kamm KE, Stull JT (1992) Phosphorylation of myosin light chain kinase by the multifunctional calmodulin-dependent protein kinase II in smooth muscle cells. J Biol Chem 267:12511–12516
42. Stull JT, Nunnally MH, Michnoff CH (1986) Calmodulin-dependent protein kinases. In: Krebs EG, Boyer PD (eds) The enzymes. Academic, Orlando, pp 113–166
43. Kamm KE, Hsu L-C, Kubota Y, Stull JT (1989) Phosphorylation of smooth muscle myosin heavy and light chains: effects of phorbol dibutyrate and agonists. J Biol Chem 264:21223–21229
44. Colburn JC, Michnoff CH, Hsu L, Slaughter CA, Kamm KE, Stull JT (1988) Sites phosphorylated in myosin light chain in contracting smooth muscle. J Biol Chem 263:19166–19173
45. Stull JT, Hsu L-C, Tansey MG, Kamm KE (1990) Myosin light chain kinase phosphorylation in tracheal smooth muscle. J Biol Chem 265:16683–16690
46. Sparrow MP, Mrwa U, Hofmann F, Rüegg JC (1981) Calmodulin is essential for smooth muscle contraction. FEBS Lett 125:141–145
47. Nomura M, Stull JT, Kamm KE, Mumby MC (1992) Site-specific dephosphorylation of smooth muscle myosin light chain kinase by protein phosphatases 1 and 2A. Biochemistry 31:11915–11920
48. Tang D-C, Stull JT, Kubota Y, Kamm KE (1992) Regulation of the $Ca^{2+}$ dependence of smooth muscle contraction. J Biol Chem 267:11839–11845
49. Somlyo AP, Somlyo AV (1994) Signal transcution and regulation in smooth muscle. Nature 372:231–236
50. Kamm KE, Stull JT (1989) Regulation of smooth muscle contractile elements by second messengers. Annu Rev Physiol 51:299–313
51. Cohen P, Klee CB (1988) Molecular aspects of cellular regulation. Elsevier, Amsterdam, pp 1–371
52. Pato MD, Adelstein RS (1983) Purification and characterization of a multisubunit phosphatase from turkey gizzard smooth muscle: the effect of calmodulin binding to myosin light chain kinase on dephosphorylation. J Biol Chem 258:7047–7054
53. Nishikawa M, Sellers JR, Adelstein RS, Hidaka H (1984) Protein kinase C modulates in vitro phosphorylation of the smooth muscle heavy meromyosin by myosin light chain kinase. J Biol Chem 259:8808–8814
54. Van Riper DA, Weaver BA, Stull JT, Rembold CM (1995) Myosin light chain kinase phosphorylation in swine carotid artery contraction and relaxation. Am J Physiol (in press)
55. Word RA, Tang D-C, Kamm KE (1994) Activation properties of myosin light chain kinase during contraction/relaxation cycles of tonic and phasic smooth muscles. J Biol Chem 269:21596–21602
56. Krebs EG, Beavo JA (1979) Phosphorylation-dephosphorylation of enzymes. Annu Rev Biochem 48:923–959
57. Carafoli E (1991) Calcium pump of the plasma membrane. Physiol Rev 71:129–153
58. Hand DJ, Taylor CC (1987) Multivariate analysis of variance and repeated measures. Chapman and Hall, New York
59. Forstermann U, Pollock JS, Schmidt HH, Heller M, Murad F (1991) Calmodulin-dependent endothelium-derived relaxing factor/nitric oxide synthase activity is present in the particulate and cytosolic fractions of bovine aortic endothelial cells. Proc Natl Acad Sci USA 88:1788–1792
60. Miller-Hance WC, Miller JR, Wells JN, Stull JT, Kamm KE (1988) Biochemical events associated with activation of smooth muscle contraction. J Biol Chem 263:13979–13982
61. Hanson PI, Schulman H (1992) Neuronal $Ca^{2+}$/calmodulin-dependent protein kinases. Annu Rev Biochem 61:559–601

62. Rizzuto R, Brini M, Murgia M, Pozzan T (1993) Microdomains with high $Ca^{2+}$ close to $IP_3$ sensitive channels that are sensed by mitochondria. Science 262:744–746
63. Yagi S, Becker PL, Fay FS (1988) Relationship between force and $Ca^{2+}$ concentration in smooth muscle as revealed by measurements on single cells. Proc Natl Acad Sci USA 85:4109–4113

# Actin-Binding Property of Myosin Light Chain Kinase and Its Role in Regulating Actin–Myosin Interaction of Smooth Muscle

Li-Hong Ye, Kohichi Hayakawa, Tsuyoshi Okagaki, and Kazuhiro Kohama

*Summary.* Myosin light chain kinase (MLCK), a key regulator of smooth muscle contraction, has a property to bind to actin filaments that is obviously unrelated to the kinase activity innate in MLCK. We have found that this property exerts an inhibitory effect on the ATP-dependent interaction between actin and myosin. The N-terminal, actin-binding fragment was responsible for the inhibition. Calmodulin in the presence of $Ca^{2+}$ ($Ca^{2+}$/CaM) efficiently relieved the inhibition imposed by MLCK. The inhibitory effect was potent enough to explain physiological involvement in terms of the amount of MLCK that was present as a cytoskeletal protein in smooth muscle.

*Key Words.* Myosin light chain kinase—Actin-binding property of MLCK—Actin-binding domain of MLCK—Inhibitory effect of MLCK—Stimulatory effect of MLCK

## 1. Introduction

Contraction of smooth muscle is induced by an ATP-dependent interaction between actin and myosin. This interaction is regulated by $Ca^{2+}$ [1], whose site of action is calmodulin (CaM). CaM activates myosin light chain kinase (MLCK), which then phosphorylates the regulatory light chain of myosin. The phosphorylated myosin is able to interact with actin filaments [review: 2].

This myosin-linked regulatory mode has been well studied biochemically. However, recent technical progress, such as measurement of intracellular $Ca^{2+}$, skinning of cell membranes, and the development of CaM- and MLCK-antagonists has suggested that actual regulation of smooth muscle contraction appears to be much more complex [review: 3]. As is described in a later section, some agents are able to induce contraction of smooth muscle under conditions where myosin phosphorylation does not proceed. Obviously, there may be another important mode of regulation that is not explained by the myosin-linked pathway.

To approach this problem, we were interested in an apparent paradox in smooth

Department of Pharmacology, Gunma University School of Medicine, 3-39-22 Showa-machi, Maebashi, Gunma, 371 Japan

muscle biology; that is, MLCK is associated with actin filaments, not myosin filaments, in vivo [4, 5]. It will be intriguing to determine in vitro if $Ca^{2+}$ regulates the actin–myosin interaction through actin filaments decorated by MLCK and CaM. However, the effect of the activity on the actin–myosin interaction has not been completely analyzed because of the complications due to the kinase activity of MLCK. This difficulty has been overcome [6] by the use of recently developed motility assays [review: 7]. This chapter reports that MLCK binds to actin filaments to inhibit the interaction, which is modulated by CaM in the presence of $Ca^{2+}$. The physiological involvement in regulating the contraction of smooth muscle by this actin-linked mode is discussed.

## 2. Materials and Methods

The MLCK was purified from smooth muscle of chicken gizzard by the method of Adelstein and Klee [8], with slight modifications, and was used for the experiments in this review except that of Fig. 1, where we used MLCK from bovine stomach [9]. Smooth muscle myosin [10], caldesmon [11], calponin [12], and tropomyosin [13] were purified from chicken gizzard. Skeletal muscle myosin was prepared from chicken breast muscle [14]. Actin was purified from acetone powder of chicken breast muscle [14] and used as actin filaments after polymerization. CaM from bovine brain was purchased from Sigma (P-2277, St. Louis, MO, USA).

An ATP-dependent interaction between actin and myosin was monitored by (1) a *Nitella*-based motility assay developed by Shimmen and Yano [15]; (2) a myosin-coated surface assay as described by Okagaki et al. [16]; and (3) measurement of actin-activated ATPase activity [17]. Smooth muscle myosin was used after phosphorylation by MLCK and CaM in the presence of $Ca^{2+}$.

Binding of MLCK to actin filaments was carried out by a centifugation method described elsewhere [17]. In short, MLCK precipitated with actin filaments was subjected to sodium dodecyl sulfate polyacrylamide gel electrophoresis (SDS-PAGE). The amounts of actin and MLCK in the precipitate were quantitated by densitometry.

## 3. Smooth Muscle Contraction Without Myosin Phosphorylation

To phosphorylate smooth muscle myosin, $Ca^{2+}$ is an essential requirement. Therefore it is of interest to determine if the contraction of intact smooth muscle is inducible under conditions that do not allow an increase in the intracellular concentration of $Ca^{2+}$. Uchida and his colleagues were the first to address this question. They allowed oxytocin to induce contraction of uterine smooth muscle in an EGTA-containing, $Ca^{2+}$-free medium [18]. Monitoring of intracellular concentration of $Ca^{2+}$ by loading $Ca^{2+}$-sensitive dye of fura-2 to the muscle failed to detect an increase in $Ca^{2+}$ concentration during the contraction. Accordingly, there are no signs of myosin phosphorylation during the contraction. Similarly, Karaki and his colleagues observed that phorbol ester is able to contract aortic smooth muscle in the $Ca^{2+}$-free medium [19]. During the contraction there was no increase in either $Ca^{2+}$ concentration or myosin phosphorylation.

These examples suggest that myosin phosphorylation is not obligatory for the initiation of smooth muscle contraction and urge us to search for another regulatory pathway of the contraction.

## 4. Actin-Binding Activity of MLCK and Its Regulation of the Actin–Myosin Interaction

To confirm the actin-binding activity of MLCK [20, 21], MLCK from smooth muscle of chicken gizzard was mixed with actin filaments, and then the mixture was centrifuged. Quantitation of the MLCK precipitated with actin filaments showed that MLCK was bound by 0.12 mol/mol monomeric actin with 1.30 μM as an MLCK concentration that allowed half-maximal binding. Binding of MLCK to actin filaments was abolished by $Ca^{2+}$ in the presence of CaM ($Ca^{2+}$/CaM).

How is the ATP-dependent interaction between actin and myosin modified when actin filaments are associated with MLCK? This question was addressed using a *Nitella*-based motility assay that utilized *Nitella*'s actin cables, composed of unidirectional actin filaments [6]. MLCK was introduced to the cables so as to allow MLCK to be bound to them. Then the unbound MLCK, if any, was removed by repeated washing with buffer that did not contain MLCK. Finally, we allowed latex beads (about 2 μm in diameter) coated with skeletal muscle myosin to move along the cables in an

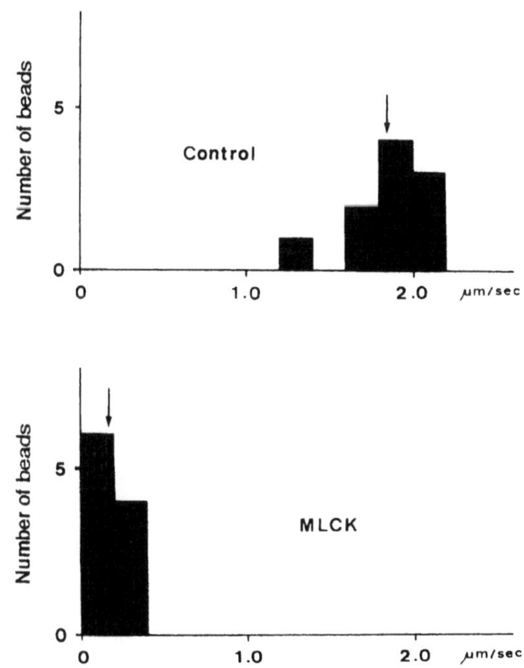


**Fig. 1.** Histogram of the velocity of the ATP-dependent movement of beads coated with skeletal muscle myosin along actin cables of *Nitella* intermodal cells [6]. Actin cables of *Nitella* intermodal cells were exposed (*Control*) and allowed to bind myosin light chain kinase (*MLCK*). Small latex beads coated with skeletal muscle myosin were introduced to the control and MLCK-treated actin cables together with Mg-ATP and EGTA; they were then observed with a phase contrast microscope. *Arrows* indicate mean velocities

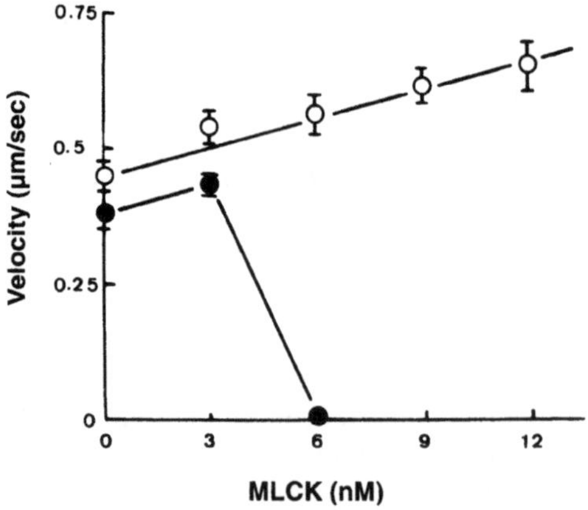


**Fig. 2.** Effect of myosin light chain kinase (*MLCK*) on the movement of actin filaments (3 nM) on a surface coated with phosphorylated myosin of smooth muscle [25]. Fluorescent actin filaments were mounted on a coverslip, the surface of which was coated with smooth muscle myosin that had been phosphorylated by MLCK in the presence of calmodulin (CaM) and $Ca^{2+}$. Movement of the filaments was observed under a fluorescence microscope in the presence of CaM and various concentrations of MLCK. Mean velocities (ordinate) of 20 filaments were plotted against the MLCK concentration (abscissa). *Open circles*, in 0.1 mM $Ca^{2+}$; *closed circles*, in 0.1 mM EGTA

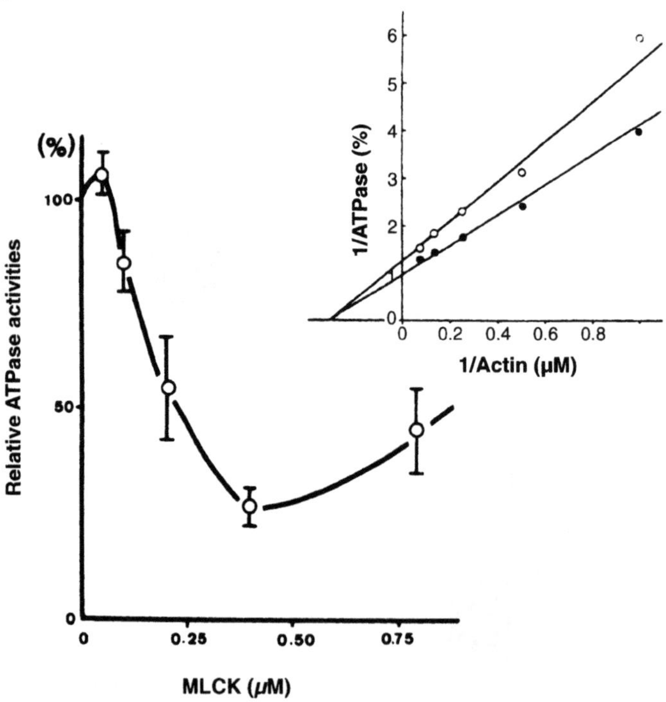


**Fig. 3.** Effect of MLCK on the actin-activated ATPase activity of myosin [24]. Smooth muscle myosin was phosphorylated by MLCK in the presence of CaM and $Ca^{2+}$, then subjected to the ATPase measure in the presence of actin filaments (2 nM)and various concentrations of MLCK. **Inset.** Double reciprocal plots of the actin-activated ATPase activities in the presence (*open circles*) and absence (*closed circles*) of MLCK. Note that the effective concentration of MLCK was different from that of MLCK in Fig. 3, which is indicated in a footnote on the following page.

ATP-dependent manner. The average velocity of the movement was 0.14 μm/s (Fig. 1, bottom panel); however, the same beads moved along the control cables that were not treated with MLCK at an average velocity of 1.90 μm/s (Fig. 1, top panel). Because MLCK was present only in the MLCK-treated cables, the reduced velocity can be related directly to the effect of actin-bound MLCK.

When designing this experiment, we deliberately used skeletal muscle myosin, whose interaction with actin filaments is hardly modified by phosphorylation [22]. Thus it is not likely that the inhibitory effect is brought about by phosphorylating myosin with the MLCK bound to actin cables.

Unlike skeletal muscle myosin, smooth muscle myosin does not interact with actin filaments when it is not phosphorylated [16]. Therefore by using beads coated with phosphorylated smooth muscle myosin, we were able to confirm the inhibitory effect of MLCK (see Fig. 3B in [23]). Further confirmation was achieved by the most popular method, namely, measuring the actin-activated ATPase activity of skeletal muscle myosin (Fig. 3 in [6]) and phosphorylated smooth muscle myosin (Fig. 3), whose activity decreased with an increase in the MLCK concentration[1].

## 5. Inhibitory Effects of MLCK as Examined by the Motility Assay with Myosin-Coated Glass Surface

Despite the advantages of the *Nitella*-based motility assay, the use of *Nitella* actin cables is associated with difficulties quantitating the amount of actin-bound MLCK. Furthermore, conditions that destroy its actin cables (i.e., use of $Ca^{2+}$ and some chemicals) make the experiments difficult. When measuring actin-activated ATPase activity, the property of MLCK that assembles actin filaments [16, 26] also complicates interpretation of the results.

The use of a glass surface coated with myosin [16] overcomes these difficulties[2]. An aliquot of a solution containing fluorescent actin filaments and Mg-ATP was placed on a surface that had been coated with phosphorylated smooth muscle myosin. The motility of actin filaments was then observed. The effect of MLCK on this movement was examined by adding a specified amount of MLCK to the above solution. The complication of assembly of actin filaments by MLCK was avoided by observing the movement of only isolated actin filaments, not the assembled ones. As shown on the ordinate in Figure 2, the mean velocity of their movement in the absence of MLCK was

[1] The concentration of MLCK that inhibits the movement of actin filaments on a myosin-coated surface (Fig. 2) is not identical with the concentration that inhibits the actin-activated ATPase activity of myosin (Fig. 3). The latter is higher than the former. Such a discrepancy in the potency of regulatory proteins according to the assays has also been noted for caldesmon and calponin (see Fig. 7 in [24] for ATPase activity; Figs. 1 and 2 in [25] for the motility assay). This discrepancy can be partly attributed to the difference in the concentrations of actin filaments used in both assays [25a].

[2] Compared with measuring the actin-activated ATPase activity of phosphorylated myosin of smooth muscle, monitoring the movement of actin filaments on a surface coated with phosphorylated myosin of smooth muscle has the following advantages for testing the regulatory role of MLCK: (1) The myosin-coated surface assay directly monitors the actin–myosin interaction in real time. (2) The assay is not complicated by the kinase activity of MLCK (see section 7.2). (3) Because the assay observes isolated actin filaments, we can ignore the movement of assembled actin filaments, if any (see section 5). (4) Similarly, the assay is not complicated by the assembly/disassembly of the myosin. Because the myosin is fixed on a glass surface, there is no room for the myosin to relocate.

0.5 μm/s. In the presence of MLCK, the velocity was reduced as the MLCK concentration increased. MLCK at 6 nM abolished movement. The concentration that inhibited movement half-maximally was about 4 nM.

It must be noted that the concentration of actin filaments in the myosin-coated surface assay was extremely low: 3 nM. Given the dissociation constant of 1.3 μM under conditions comparable to those of the motility assay, only 0.8% of actin filaments were calculated to be in the MLCK-bound form when MLCK was present at 10 nM. Thus we can conclude that only part of the decoration of the actin filaments by MLCK is sufficient for inhibition of their movement.

## 6. Stimulatory Effect of MLCK in the Presence of $Ca^{2+}$/Calmodulin

How does $Ca^{2+}$ modulate this effect of MLCK using CaM as a calcium-binding protein? As shown by the open circles in Figure 2, $Ca^{2+}$ relieved the inhibition imposed by MLCK on movement. Repeated comparisons of the velocity in the presence of MLCK, CaM, and $Ca^{2+}$ (Fig. 2, open circles) with that in their absence (Fig. 2, ordinate) indicated that the former velocity is about 1.5-fold higher than the latter [6, 25]. Although this stimulatory effect was detected with smooth muscle myosin, MLCK was able to stimulate motility on the surface coated with skeletal muscle myosin in the presence of $Ca^{2+}$/CaM (see Fig. 5 in [25]). Because binding of MLCK to actin filaments is abolished by $Ca^{2+}$/CaM (see section 4), the relief itself is readily attributable to actin-binding activity. We do not know how the stimulatory effect is brought about. It is clear, nonetheless, that this stimulatory effect is not mediated by phosphorylation due to the kinase activity of MLCK (see section 7). Obviously, another mechanism to explain the effect must be sought.

## 7. Is Kinase Activity of MLCK Involved in Inhibitory and Stimulatory Effects of MLCK?

To determine if the kinase activity of MLCK is related to the inhibitory and stimulatory effects of MLCK so far described, we carried out the following experiments and concluded that the kinase activity did not cause both effects.

### 7.1 Experiments with Skeletal Muscle Myosin

Skeletal muscle myosin is in an active form regardless of whether it is phosphorylated [22]. In another words, phosphorylation hardly alters its activity. Therefore the regulatory activity of MLCK cannot be ascribed to its kinase activity so long as skeletal muscle myosin is used for assays. The effects of MLCK detected with skeletal muscle myosin were as flows: (1) an inhibitory effect, demonstrated in the *Nitella*-based assay (Fig. 1); (2) an inhibitory effect, demonstrated in the myosin-coated surface assay (see Fig. 5 in [25]); and (3) a stimulatory effect in the presence of $Ca^{2+}$/CaM, shown in the myosin-coated surface assay (see Fig. 5 in [25]).

### 7.2 Use of Glass Surfaces Coated with Smooth Muscle Myosin Preparations with Different Extents of Phosphorylation

The MLCK in question is from smooth muscle, and the discussions about kinase activity of MLCK should be based on the results with smooth muscle myosin. MLCK in the presence of $Ca^{2+}$/CaM phosphorylates the regulatory light chain at Ser 19, the myosin of which is denoted by *My Ser 19*. Phosphorylation then proceeds to Thr 18 (*My Ser 19/Thr 18*) if kinase activity in the reaction mixture is sufficiently high. Actin-activated ATPase activity of *My Ser 19/Thr 18* is higher than that of *My Ser 19* [27].

Concerning the inhibitory effect of MLCK, one would agree that additional phosphorylation is not feasible, because this effect was observed in the absence of $Ca^{2+}$/CaM. On the other hand, during the period when the stimulatory effect was observed in the presence of $Ca^{2+}$/CaM with the surface assay, there is a possibility that phosphorylation may proceed to Thr 18. To address this possibility, we prepared three myosin preparations: *My Ser 19*, *My Ser 19/Thr 18*, and an equimolar mixture of *My Ser 19* and *My Ser 19/Thr 18*. The velocities of the movement were not different among these surfaces (see Fig. 4 in [16]). This control experiment shows that additional phosphorylation, if any occurs, does not modify the result obtained with the surface assay, excluding the involvement of the kinase activity of MLCK during stimulation of the actin-myosin interaction by MLCK.

### 7.3 Avoiding the Possible Complication of Phosphatase Activity in Smooth Muscle Myosin Preparation

The last argument is that the smooth muscle myosin preparation may contain phosphatase activity, which acts to turn over the cycles of phosphorylation and dephosphorylation, thereby producing the inhibitory or stimulatory effects of MLCK. This possibility was ruled out by using smooth muscle myosin that had been throughly freed of phosphatase activity [10]. Furthermore, when we used smooth muscle myosin that had been thiophosphorylated to avoid the effect of phosphatase, we observed inhibitory and stimulatory effects similar to those shown in Fig. 2 (unpublished data).

## 8. Effects of Other Smooth Muscle Actin-Binding Proteins on ATP-Dependent Actin–Myosin Interaction

Caldesmon (CaD), calponin, and tropomyosin are other regulatory proteins of smooth muscle that bind to actin filaments. The effects of these proteins on the actin-myosin interaction have been monitored by motility assays [16, 17, 18–30].

### 8.1 Similarities in Regulatory Ways Between MLCK and Caldesmon

Caldesmon inhibits the actin–myosin interaction, and the inhibition is relieved by $Ca^{2+}$/CaM [reviews: 31, 32]. This mode of action is similar to that of MLCK (Fig. 2). Similarities between CaD and MLCK are also apparent [review: 33] in (1) their actin-binding properties, (2) their competition to bind to actin filaments (Fig. 4), and (3) their primary structures. We [6] can identify the N-terminal portion of MLCK [34], which is highly homologous to the C-terminal portion of CaD [35]; it has been reported to bind to actin filaments and thereby to inhibit the actin–myosin interaction [reviews: 31, 32]. We obtained an actin-binding fragment from MLCK by treating it with CNBr. The fragment consisted of N-terminal, 2-213 residue with a molecular weight of 22 512 [36]. Our preliminary data (Fig. 5) showed that this fragment inhibited the movement of actin filaments on the glass surface coated with smooth muscle

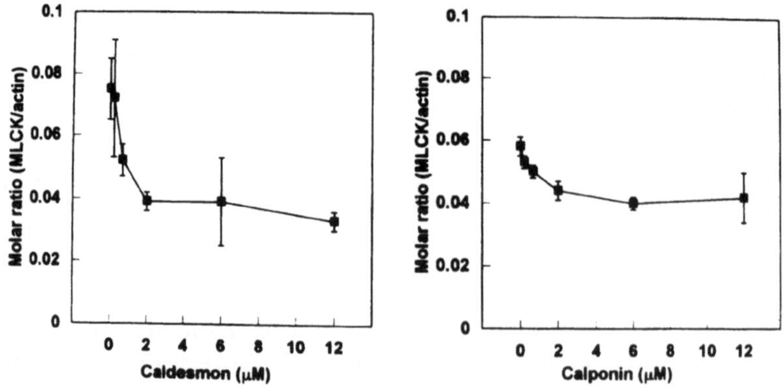


**Fig. 4.** Binding of MLCK to actin filaments in the presence of caldesmon (CaD) or calponin (CaP). **Left panel.** Effect of CaD. MLCK (2 μM) was mixed with actin filaments (12 μM) in the presence of various concentrations of CaD and subjected to centrifugation. The precipitated amounts of MLCK relative to monomeric actin were plotted against the concentration of CaD. **Right panel.** Effect of CaP. Similar to the left panel, MLCK was bound to actin filaments in the presence of various concentrations of calponin

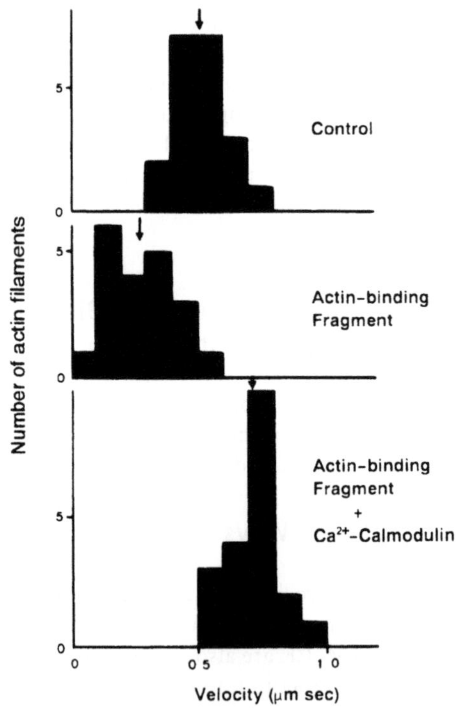


**Fig. 5.** Inhibitory effect of the N-terminal, actin-binding fragment of MLCK on the movement of actin filament [36]. MLCK was subjected to cleavage with CNBr. The N-terminal, actin-binding fragment was then purified. Using the fragment, we carried out the myosin-coated surface assay as described in the legend to Fig. 2, data that are shown by histogram. *Arrows*, mean velocities. **Top panel.** Control (i.e., in the absence of the fragment). **Middle panel.** In the presence of the fragment. **Bottom panel.** In the presence of the fragment and $Ca^{2+}$/CaM

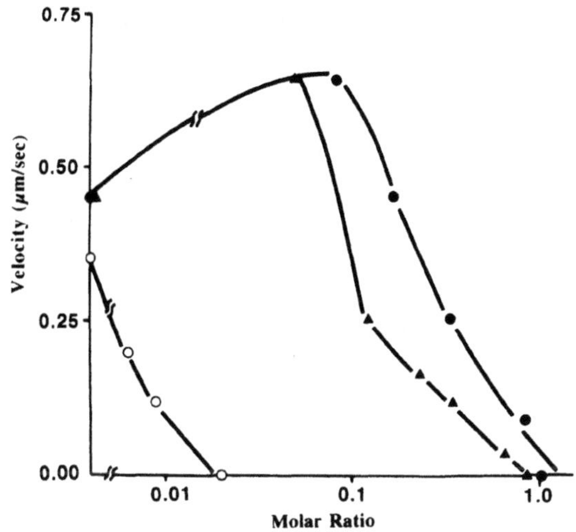


**Fig. 6.** Strength of the inhibitory effect of MLCK compared to those of CaD and CaP [44]. Myosin-coated surface assays were carried out as described in the legend to Fig. 2 in the presence of various concentrations of MLCK (*open circles*), CaP (*triangles*), and CaD (*closed circles*). The velocity of movement of actin filaments on the surface was plotted against the molar ratio relative to CaD

myosin, supporting the idea stated above that an actin-binding domain of MLCK [37] is responsible for the inhibitory effect. Furthermore, the inhibition was relieved by $Ca^{2+}$/CaM (Fig. 5, bottom panel). The CaM-binding site that regulates kinase activity [review: 2] was not included in this fragment; this finding suggests that there is a new site that binds CaM, which regulates MLCK binding to actin filaments.

Similarities in the regulation of actin–myosin interactions between CaD and MLCK are not confined to their inhibitory effects. We [17, 24, 25, 38, 39] and Haeberle et al. [28] detected a stimulatory effect of CaD on this interaction. As shown by the filled circles in Figure 6 (see also Fig. 1 in ref. 25), CaD increased the velocity of movement about 1.4-fold. Further increases in the concentration of CaD changed its effect from stimulatory to inhibitory. To see what portion of CaD is responsible for the stimulatory effect, we purified an N-terminal, myosin-binding fragment from a digest of CaD and found that the fragment stimulated the actin–myosin interaction [39]. We are now interested in obtaining the C-terminal, myosin-binding fragment of MLCK and examining its effect on the interaction.

## 8.2 Similarities in Regulatory Ways Between MLCK and Calponin

Resembling CaD, calponin (CaP) binds to actin filaments and inhibits the actin–myosin interaction. CaP competes with MLCK for binding to actin filaments (Fig. 4). This inhibition is abolished by $Ca^{2+}$/CaM [12, 40–42]. We found that CaP also binds to myosin, and that the ATPase activity of myosin in the absence of actin filaments was stimulated by CaP. When actin filaments were present, the effect of CaP on the activity changed from stimulatory to inhibitory. Therefore we related the stimulatory and inhibitory effects to the myosin-binding and actin-binding properties of CaP, respectively [43].

### 8.3 Regulation of the Actin–Myosin Interaction by Tropomyosin

The effect of smooth muscle tropomyosin on the motility of actin filaments on myosin-coated surfaces was never inhibitory, but always stimulatory [16, 30]. The stimulatory effect of tropomyosin, however, is not potent enough to overcome the inhibitory effect of CaD [16] or MLCK (unpublished data). Because no calcium-binding protein that interacts with tropomyosin is known in smooth muscle, we do not comment on it further.

## 9. Physiological Involvement of the Inhibitory Effect of MLCK in Regulation of the Smooth Muscle Actin–Myosin Interaction

The physiological role of MLCK in regulating the actin–myosin interaction was addressed by the following four experiments.

### 9.1 Inhibitory Strengths of MLCK and CaD

The inhibitory effect of MLCK is brought about by its actions as a cytoskeletal protein, not as an enzyme. The inhibition produced by MLCK has features in common with that by CaD (see section 8), a typical cytoskeletal protein of smooth muscle. Therefore we compared the inhibitory strength of MLCK with that of CaD [44]. As shown in Figure 6, the concentration of CaD that abolishes the movement of actin filaments on a myosin-coated surface was 88 times that of MLCK on a molar basis (mean value from six experiments). Is MLCK present at a molar ratio of $>1/88$ of CaD in smooth muscle cells as a cytoskeletal protein? We subjected myofibrils of smooth muscle prepared by repeated washing to SDS-PAGE. Densitometry indicated that the molar ratio of MLCK to CaD was 1.0:7.1 (mean value from four experiments), suggesting physiological relevance in terms of the concentrations (Table 1).

### 9.2 Effect of Calmodulin Inhibitor

The $Ca^{2+}$/CaM combination reverses the inhibitory effect of MLCK on the movement of actin filaments on a myosin-coated surface (Fig. 2). A CaM inhibitor such as trifluoperazine is expected to antagonize this relief action by $Ca^{2+}$/CaM [review: 45]. Figure 7 shows the effect of trifluoperazine on the movement of actin filaments in the presence of MLCK, $Ca^{2+}$, and CaM [6]. The velocity of the movement was reduced as the trifluoperazine concentration increased. The concentration that produced half-maximal velocity was in the micromolar range, as was the concentration that relaxed smooth muscle tone [46, 47]. This coincidence supports the physiological relevance of the inhibitory effect of MLCK.

One might argue, however, that MLCK is not the sole site of action of $Ca^{2+}$/CaM. CaD and CaP have inhibitory effects on the actin–myosin interaction similar to those of MLCK, and $Ca^{2+}$/CaM relieves this inhibition as well (described above). The amount of CaD required to cause the same inhibition as MLCK is 88 times that of MLCK (Fig. 6). However, the amount of CaD in myofibril is only 7.1 times that of MLCK (Table 1). Similarly, the amount of CaP in myofibrils is insufficient for inhibition to the same extent as MLCK if we take the inhibitory strength of CaP (Fig. 6, triangles) into consideration. Therefore inhibition of the contraction is suggested to

**Table 1.** Amount of MLCK and CaD present in chicken gizzard myofibrils.

| MLCK : CaD (molar ratio) |
|---|
| 1.0 : 6.5 |
| 1.0 : 6.6 |
| 1.0 : 5.7 |
| 1.0 : 9.8 |
| 1.0 : 7.15 ± 1.57 (mean ± SEM) |

From [44], with permission.
MLCK, myosin light chain kinase; CaD, caldesmon.

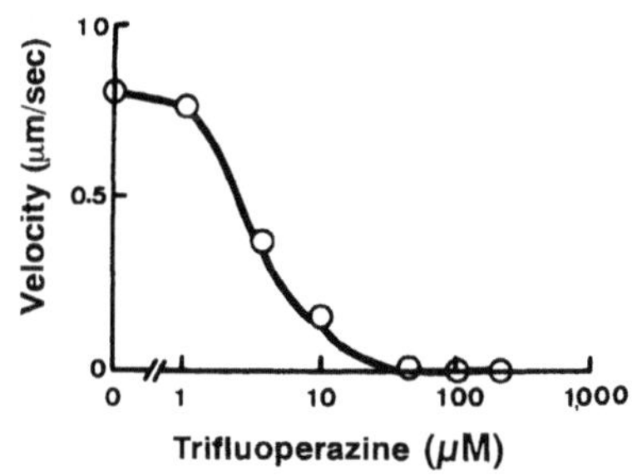


**Fig. 7.** Effect of trifluoperazine as monitored by the myosin-coated surface assay [6, 33]. Motility of actin filament on a myosin-coated surface as described in the legend to Fig. 2 in the presence of MLCK, CaM, $Ca^{2+}$, and various concentrations of trifluoperazine. The mean velocities were plotted against the trifluoperazine concentration

be totally attributable to MLCK, and $Ca^{2+}$/CaM can relieve the inhibition by MLCK, thereby providing a rationale for relating the effect of trifluoperazine to MLCK.

## 9.3 Effect of $Ca^{2+}$

Reasonable concentrations are also evident for $Ca^{2+}$, as shown in Fig. 8. We allowed actin filaments to move on a myosin-coated surface in the presence of MLCK, CaM, and various concentrations of $Ca^{2+}$ [6]. The $Ca^{2+}$ concentration that allowed half-maximal velocity of movement was in the micromolar range and was similar to that required to activate smooth muscle contraction [review: 48].

## 9.4 New Regulatory Protein That Differentiates Kinase Activity of MLCK and Its Inhibitory Effect

The final supporting evidence of the physiological significance of MLCK is provided by a novel regulatory protein (Table 2). We purified a 65-kDa protein on SDS-PAGE that binds to both CaD and MLCK [23]. Although $Ca^{2+}$/CaM modulates both the inhibitory effect and the kinase activity of MLCK, the 65-kDa protein abolishes the inhibitory effect of MLCK but does not alter its kinase activity. The occurrence in smooth muscle cells of a protein that can differentiate these two properties of MLCK support the hypothesis that the actin-linked inhibition of the actin–myosin interac-

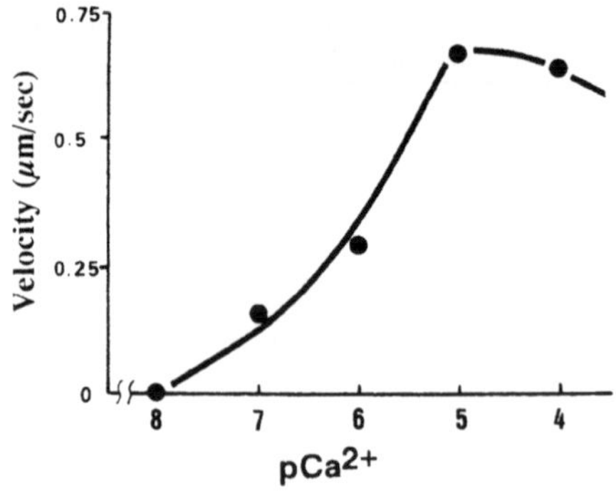


**Fig. 8.** Relation between $pCa^{2+}$ and the velocity of the movement of actin filaments. The myosin-coated surface assay was carried out in the presence of MLCK, CaM, and various concentrations of $Ca^{2+}$ as described in the legend to Fig. 2. $pCa^{2+} = -\log[Ca^{2+}]M$. Note that this figure confirms the findings of Figure 1 (bottom panel) of Kohama et al. [6] using MLCK from chicken gizzard

**Table 2.** Modulator proteins of MLCK.

| Activity | 65 kDa Protein[a] | CaM[b] |
|---|---|---|
| Kinase activity | − | + |
| Inhibitory effect | + | + |

+, effective; −, not effective.
[a] Data from [23].
[b] Data from [2, 6].
CaM, calmodulin.

tion by MLCK plays a physiological role in the regulation of smooth muscle contraction.

## 10. Perspectives

The regulatory role of $Ca^{2+}$ in the actin–myosin interaction was reported for smooth muscle contraction as early as 1966 by Ebashi and his colleagues [1]. They extended their work to advocate a new regulatory protein called leiotonin, which regulates the interaction by binding to actin filaments [review: 49]. Our understanding of the role of MLCK in myosin phosphorylation in the presence of $Ca^{2+}$/CaM for regulating the actin–myosin interaction has brought forth the question of whether leiotonin is related to the kinase activity of MLCK [9]. Our data offer a clue to the answer: As shown in Figure 8, MLCK together with CaM can regulate the interaction via actin filaments in a $Ca^{2+}$-dependent manner under conditions in which kinase activity does not work.

It must be noted that the myosin used in this review has been in the phosphorylated form. Therefore the new effect must be confirmed with unphosphorylated myosin. However, because unphosphorylated myosin is readily phosphorylated by the kinase activity during experiments in the presence of $Ca^{2+}$/CaM, such a confirmation is not feasible. As an approach to this problem, we are now interested in the modification of cDNA for MLCK [50] that is devoid of kinase domain so as to express it as a recombinant protein.

## References


1. Ebashi S, Iwakura H, Nakajima H, Nakamura R, Ooi Y (1966) New structural proteins from dog heart and chicken gizzard. Biochem Z 345:201–211
2. Kamm KE, Stull JT (1985) The function of myosin and myosin light chain kinase phosphorylation in smooth muscle. Annu Rev Pharmacol Toxicol 25:583–620
3. Kohama K, Saida K (eds) (1995) Smooth muscle contraction: new regulatory modes. Karger, Basel
4. Guerriero V Jr, Rowley DR, Means AR (1981) Production and characterization of an antibody to myosin light chain kinase and intracellular localization of the enzyme. Cell 27:447–458
5. De Lanerolle P, Adelstein RS, Feramisco JR, Burridge K (1981) Characterization of antibodies to smooth muscle myosin kinase and their use in localizing myosin kinase in nonmuscle cells. Proc Natl Acad Sci USA 78:4738–4742
6. Kohama K, Okagaki T, Hayakawa K, Lin Y, Ishikawa R, Shimmen T, Inoue A (1992) A novel regulatory effect of myosin light chain kinase from smooth muscle on the ATP-dependent interaction between actin and myosin. Biochem Biophys Res Commun 184:1204–1211
7. Warrick HM, Spudich JA (1987) Myosin structure and function in cell motility. Annu Rev Cell Biol 3:379–421
8. Adelstein RS, Klee CB (1981) Purification and characterization of smooth muscle myosin light chain kinases. J Biol Chem 256:7501–7509
9. Kuwayama H, Suzuki M, Koga R, Ebashi S (1988) Preparation of protein components exhibiting myosin light chain kinase activities from bovine aorta: discrepancies between its enzyme activity and actomyosin activating effect. J Biochem 104:862–866
10. Nakamura S, Nonomura Y (1984) A simple and rapid method to remove light chain phosphatase from chicken gizzard myosin. J Biochem 96:575–578
11. Bretsher A (1984) Smooth muscle caldesmon: rapid purification and F-actin cross-linking properties. J Biol Chem 259:12837–12880
12. Abe M, Takahashi K, Hiwada K (1990) Simplified co-purification of vascular smooth muscle calponin and caldesmon. J Biochem 107:507–509
13. Ishikawa R, Yamashiro S, Matsumura F (1989) Annealing of gelsolin-served actin fragments by tropomyosin in the presence of $Ca^{2+}$: potentiation of the annealing process by caldesmon. J Biol Chem 264:16764–16770
14. Kohama K (1980) Heterogeneity of amino acid incorporation rate in adult skeletal muscle actin. J Biochem 87:997–999
15. Shimmen T, Yano Y (1984) Active sliding movement of latex beads coated with skeletal muscle myosin on Chara actin bundles. Protoplasma 121:132–137
16. Okagaki T, Higashi-Fujime S, Ishikawa R, Takano-Ohmuro H, Kohama K (1991) In vitro movement of actin filaments on gizzard smooth muscle myosin: requirement of phosphorylation of myosin light chain and effects of tropomyosin and caldesmon. J Biochem 109:858–866
17. Ishikawa R, Okagaki T, Higashi-Fujime S, Kohama K (1991) Stimulation of the interaction between actin and myosin by Physarum caldesmon-like protein and smooth muscle caldesmon. J Biol Chem 266:21784–21790
18. Oishi K, Takano-Ohmuro H, Minakawa-Matsuo N, Suga O, Karibe H, Kohama K, Uchida MK (1991) Oxytocin contracts rat uterine smooth muscle in $Ca^{2+}$-free medium without any phosphorylation of myosin light chain. Biochem Biophys Res Commun 176:122–128
19. Sato K, Hori M, Ozaki H (1992) Myosin phosphorylation-independent contraction induced by phorbol ester in vascular smooth muscle. J Pharmacol Exp Ther 261:497–505
20. Sellers JR, Pato MD (1984) The binding of smooth muscle myosin light chain kinase and phosphotase to actin and myosin. J Biol Chem 259:7740–7746
21. Yamazaki K, Ito K, Sobue K, Mori T, Shibata N (1987) Purification of caldesmon and myosin light chain (MLCK) kinase from arterial smooth muscle: comparizons with gizzard caldesmon and MLCK kinase. J Biochem 101:1–9

22. Morgan M, Perry SV, Ottaway J (1976) Myosin light-chain phosphatase. Biochem J 157:687–697
23. Lin Y, Ishikawa R, Kohama K (1992) A novel regulatory protein that affect the function of caldesmon and myosin light chain kinase. Biochem Biophys Res Commun 184:1212–1218
24. Lin Y, Ishikawa R, Kohama K (1993) Role of myosin in the stimulatory effect of caldesmon on the interaction between actin, myosin and ATP. J Biochem 114:279–283
25. Ye L-H, Hayakawa K, Lin Y, Okagaki T, Fujita K, Kohama K (1994) The regulatory role of myosin light chain kinase as an actin-binding protein. J Biochem 116:1377–1382
25a. Sato M, Ye L-H, Kohama K (1995) Myosin light chain kinase from vascular smooth muscle inhibits the ATP-dependent interaction between actin and myosin by binding to actin. J Biochem 118:1–3
26. Hayakawa K, Okagaki T, Higashi-Fujime S, Kohama K (1994) Bundling of actin filaments by myosin light chain kinase from smooth muscle. Biochem Biophys Res Commun 199:786–791
27. Ikebe M, Hartshorne D (1986) Identification, phosphorylation, and dephosphorylation of a second site for myosin light chain kinase on the 20,000-dalton light chain of smooth muscle myosin. J Biol Chem 261:36–39
28. Haeberle JR, Trybus KM, Hemric ME, Warskaw DM (1992) The effects of smooth muscle caldesmon on actin filament motility. J Biol Chem 267:23001–23006
29. Shirinsky VP, Biryukov KG, Hettasch JM, Scllcrs JM (1992) Inhibition of the relative movement of actin and myosin by caldesmon and calponin. J Biol Chem 267:15886–15892
30. Umemoto S, Benzur AR, Sellers JR (1989) Effect of multiple phosphorylations of smooth muscle and cytoplasmic myosins on movment in an in vitro motility assay. J Biol Chem 264:1431–1436
31. Sobue K, Sellers JR (1991) Caldesmon, a novel regulatory protein in smooth muscle and nonmuscle actomyosin system. J Biol Chem 266:12115–12118
32. Marston SB, Redwood CS (1991) The molecular anatomy of caldesmon. Biochem J 279:1–16
33. Okagaki T, Kohama K (1995) Characterization of myosin light chain kinase as an actin-binding protein that regulates the ATP-dependent interaction between actin and myosin of smooth muscle. In: Kohama K, Saida K (eds) Smooth muscle contraction, new regulatory modes. Karger, Basel, pp 111–130
34. Olson NJ, Person RB, Needleman DS, Hurwitz M, Kemp BE, Means AR (1990) Regulatory and structural motifs of chicken gizzard myosin light chain kinase. Proc Natl Acad Sci USA 87:2284–2288
35. Bryan J, Imai M, Lee R, Moore P, Cook RG, Lin W (1989) Cloning and expression of a smooth muscle caldesmon. J Biol Chem 264:13873–13879
36. Hayakawa K, Okagaki T, Ye L-H, Takagi T, Kohama K (1994) Regulatory properties of actin-binding fragment of smooth muscle myosin light chain kinase. Jpn J Pharmacol 64(suppl I):217P
37. Kanoh S, Ito M, Niwa E, Kawano Y, Hartshorne DJ (1993) Actin-binding peptide from smooth muscle myosin light chain kinase. Biochemistry 32:8902–8907
38. Ishikawa R, Okagaki T, Kohama K (1992) Regulation by $Ca^{2+}$-calmodulin of the actin-binding activity of Physarum 210-kDa protein. J Muscle Res Cell Motil 13:321–328
39. Lin Y, Ishikawa R, Okagaki T, Ye L-H, Kohama K (1994) Stimulation of the ATP-dependent interaction between actin and myosin by a myosin-binding fragment of smooth muscle caldesmon. Cell Motil Cytoskeleton 29:250–258
40. Abe M, Takahashi K, Hiwada K (1990) Effect of calponin on actin-activated myosin ATPase activity. J Biochem 108:835–838
41. Winder SJ, Walsh MP (1990) Smooth muscle calponin: inhibition of actomyosin MgATPase and regulation by phosphorylation. J Biol Chem 265:10148–10155
42. Takahashi K, Nadal-Ginard B (1991) Molecular cloning and sequence analysis of smooth muscle calponin. J Biol Chem 266:13284–13288
43. Lin Y, Ye L-H, Ishikawa R, Fujita K, Kohama K (1993) Stimulatory effect of calponin on myosin ATPase activity. J Biochem 113:643–645
44. Ye L-H, Ishikawa R, Kohama K (1994) Actin-linked inhibition of the act-myosin interaction of smooth muscle by myosin light chain kinase: a consideration for the physiologic role of the inhibition. Jpn J Pharmacol 64(suppl I):217P
45. Van Belle H (1984) The effect of drugs on calmodulin and its interaction with phosphodiesterase. Adv Cyclic Nucleotide Protein Phosphorylation Res 17:557–567

46. Kreye VAW, Rüegg JC, Hofmann F (1983) Effect of calcium-antagonist and calmodulin-antagonist drugs on calmodulin-dependent contractions of chemically skinned vascular smooth muscle. Naunyn Schimiedebergs Arch Pharmacol 323:85–89
47. Asano M (1990) Effects of the calmodulin antagonist W-7 on isometric tension development and myosin light chain phosphorylation in bovine tracheal smooth muscle. Jpn J Pharmacol 52:471–481
48. Saida K (1995) $Ca^{2+}$ sensitization of smooth muscle in relation to small GTP-binding protein. In: Kohama K, Saida K (eds) Smooth muscle contraction: new regulatory modes. Karger, Basel, pp 103–112
49. Ebashi S (1980) The Croonian Lecture, 1979: the regulation of muscle contraction. Proc R Soc Lond [Biol] 207:257–286
50. Kobayashi H, Inoue A, Mikawa T, Kuwayama H, Hotta Y, Masaki T, Ebashi S (1992) Isolation of cDNA for bovine stomach 155 kDa protein exhibiting myosin light chain kinase activity. J Biochem 122:786–791

# Pharmacomechanical Coupling Through Regulation of Myosin Light Chain Phosphatase

A.P. Somlyo, M. Gong, K. Iizuka, T. Haystead, and A.V. Somlyo

*Summary.* The two major components of pharmacomechanical coupling, inositol 1,4,5-trisphosphate-induced $Ca^{2+}$ release and modulation of $Ca^{2+}$ sensitivity, are described. Desensitization of phasic smooth muscles to $Ca^{2+}$ is demonstrated through time-dependent divergence, during relaxation, between the level of $Ca^{2+}$ and the decline in $MLC_{20}$ phosphorylation and force. Possible mechanisms of desensitization include down-regulation of myosin light chain kinase and up-regulation of myosin light chain phosphatase. Sensitization to $Ca^{2+}$ is shown to be mediated by a G protein-coupled system that inhibits the (type 1) protein phosphatase responsible for dephosphorylation of the regulatory smooth muscle myosin light chain ($MLC_{20}$). The diacylglycerol/kinase C system and arachidonic acid are identified as potential messengers mediating the inhibition of $MLC_{20}$ phosphatase. The inhibitory effects of arachidonic acid on myosin light chain phosphatase, through dissociation of the catalytic from the targeting subunits, are illustrated, and studies showing the inhibition of voltage-gated $Ca^{2+}$ current by arachidonic acid are summarized. It is suggested that G protein-coupled protein phosphatase inhibition plays a significant role in the regulation of nonmuscle functions in which phosphorylation and dephosphorylation of nonmuscle myosins have functionally important effects.

*Key Words.* Smooth muscle—Myosin—G-protein—Arachidonic acid—Calcium

## Introduction

Pharmacomechanical coupling was defined a quarter of a century ago [1] as a complex of processes that can regulate contractility of smooth muscle without a necessary change in surface membrane potential. In recent years information has started to emerge about some of its molecular mechanisms. Through the use of inhibitors (GDPβS, neomycin, heparin), measurement of inositol 1,4,5-trisphosphate ($InsP_3$) and the kinetics of $Ca^{2+}$ release, it has been established that the $Ca^{2+}$-release component of pharmacomechanical coupling is operated largely through the phosphatidylinositol cascade and its $Ca^{2+}$-releasing messenger $InsP_3$ [2–8; reviews: 9, 10]. We now also know that, as originally anticipated [1], pharmacomechanical coupling can be mediated not only through changes in $[Ca^{2+}]_i$ but also

Department of Molecular Physiology and Biological Physics, University of Virginia, Health Sciences Center Box 449, Jordan Hall, Charlottesville, VA 22908, USA

through modulation of the sensitivity of the contractile regulatory apparatus to $[Ca^{2+}]$.

$Ca^{2+}$ sensitivity is modulated largely through regulation of the level of myosin light chain 20 ($MLC_{20}$) phosphorylation. Because phosphorylation of serine 19 of $MLC_{20}$ permits activation of myosin ATPase by actin, and hence contraction [review: 11], we had proposed that modulation of the $[Ca^{2+}]_i$ sensitivity of contraction is the result of changes in the $Ca^{2+}$ sensitivity of $MLC_{20}$ phosphorylation and is regulated by processes that activate or inhibit myosin light chain kinase and the phosphatase(s) responsible for dephosphorylation of $MLC_{20}$ [12]. Accordingly, we refer to $Ca^{2+}$ desensitization and sensitization as the processes that, respectively, reduce or increase the level of $MLC_{20}$ phosphorylation without a *necessary* change in $[Ca^{2+}]_i$. It should be emphasized that, just as pharmacomechanical and electromechanical coupling can operate simultaneously, $Ca^{2+}$ sensitivity can be modulated simultaneously with changes in $[Ca^{2+}]_i$. Indeed, it is likely that under physiological circumstances the two processes operate together. However, recent studies suggest that ATP hydrolysis products can affect contractility at the cross-bridge level through mechanisms other than $MLC_{20}$ phosphorylation [13–15].

## Desensitization to $Ca^{2+}$

Two major lines of evidence point to the existence of desensitization of the contractile response of phasic smooth muscles to $[Ca^{2+}]_i$. First, the use of $Ca^{2+}$ sensitive fluorescent indicators in intact smooth muscle revealed that in phasic smooth muscle the decline in force and $MLC_{20}$ phosphorylation during a $K^+$ contracture cannot be accounted for by a proportionate decrease in $[Ca^{2+}]_i$[16], and the contractile response to gradual increases in $[Ca^{2+}]_i$ is disproptionately small [17]. The second and conclusive line of evidence showed that in permeabilized phasic smooth muscle a step increase in $[Ca^{2+}]_i$ "clamped" with EGTA also results in a phasic contractile response: the tension and $MLC_{20}$ phosphorylation decline, while $[Ca^{2+}]_i$ remains constant [18]. This type of response is true desensitization to $Ca^{2+}$ and should not be confused with the behavior of some intact muscles in which a high level of depolarization can lead to a decline in $[Ca^{2-}]_i$ and consequently, a decrease in force [19].

A third line of evidence of desensitization identified at least one of its potential mechanisms: the inhibitory phosphorylation of myosin light chain kinase (MLCK) by another kinase [20]. In intact muscle, elevation of $[Ca^{2+}]_i$ due to depolarization with high $K^+$ activates calmodulin/kinase II, which phosphorylates MLCK, decreasing its affinity for $Ca^{2+}$/calmodulin and hence MLCK activity at a given $[Ca^{2+}]_i$ [21, 22]. This demonstrated mechanism of desensitization does not, however, exclude the operation of additional mechanisms, such as activation of MLC phosphatase by, for example, the cyclic GMP-regulated kinase. This possibility is suggested, because phosphorylation of MLCK by this kinase is not inhibitory [23], although G kinase does cause relaxation of permeabilized muscle in the presence of constant $[Ca^{2+}]$ [24]. Theoretically, inhibition of the activity of a putative MLC phosphatase inhibitor could also lead to activation of the phosphatase itself, resulting in desensitization (see below). However, myosin phosphatases isolated from smooth muscle are only weakly sensitive to inhibitors I and II, the known inhibitors of type I protein phosphatases [25–29].

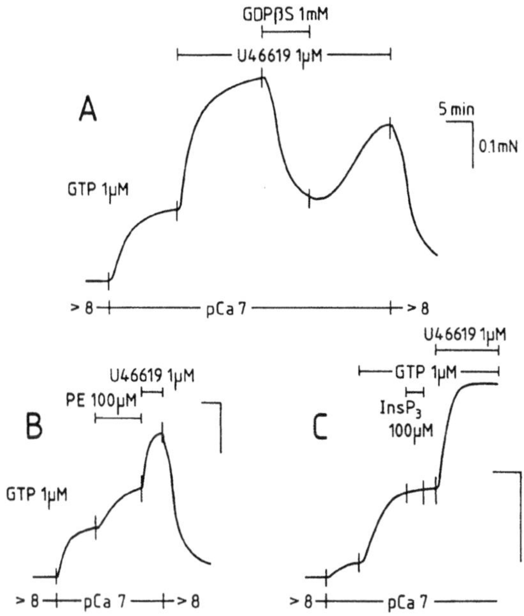


**Fig. 1.** Sensitization of the contractile response to $Ca^{2+}$ by U46619 (A), phenylephrine (PE) and GTP (B) but not by inositol 1,4,5-trisphosphate ($InsP_3$) (C) Rabbit pulmonary artery was permeabilized with staphylococcal α-toxin and treated with A 23187. Each muscle from different animals was contracted submaximally with pCa7 solution buffered with 10 mM EGTA. GTP 1 μM was present in the experiments in **A** and **B** throughout. The addition of U46619 1 μM (**A**, **B**, **C**) produced additional force in the presence of constant $Ca^{2+}$; phenylephrine 100 μM had a significant but lesser effect (**B**), and $InsP_3$ was ineffective (**C**). Added 1 μM GTP superimposed an additional increase in force on the initial $Ca^{2+}$-induced contraction (**C**). GDPβS 1 mM reversibly inhibited the potentiation of the contraction by U46619 (**A**). **B.** Note that the maximal $Ca^{2+}$-sensitizing effect of U46619 in this smooth muscle is much greater than the effect of phenylephrine. (From [31], with permission)

## $Ca^{2+}$ Sensitization

The increase in $[Ca^{2+}]_i$ produced in intact smooth muscle by a given excitatory agent is not necessarily proportional to the accompanying increases in $MLC_{20}$ phosphorylation and force: a relatively small increase in $[Ca^{2+}]_i$ can be associated with a relatively large force response and vice versa [30–32]. Force, under most circumstances, is proportional to the level of $MLC_{20}$ phosphorylation [32], indicating that under physiologic conditions in intact smooth muscle $Ca^{2+}$ sensitization is regulated largely through the level of $MLC_{20}$ phosphorylation. Furthermore, in receptor-coupled, permeabilized smooth muscle, excitatory agonists and GTPγS can increase $MLC_{20}$ phosphorylation and force in the presence of clamped $Ca^{2+}$ [18, 33–37]. The maximal $Ca^{2+}$-sensitizing effect of various agonists can be different even in permeabilized smooth muscle (Fig. 1) [1]. These and other results obtained with $Ca^{2+}$-sensitive fluorescent indicators in intact smooth muscle indicate that unequal increases in both $[Ca^{2+}]_i$ and $Ca^{2+}$ sensitization can contribute to the unequal maximal responses to different agonists characteristic of smooth muscle [1].

$Ca^{2+}$ sensitization, like pharmacomechanically coupled $Ca^{2+}$ release, is mediated by a G protein-coupled cascade(s). $Ca^{2+}$ sensitization can be mimicked by GTPγS

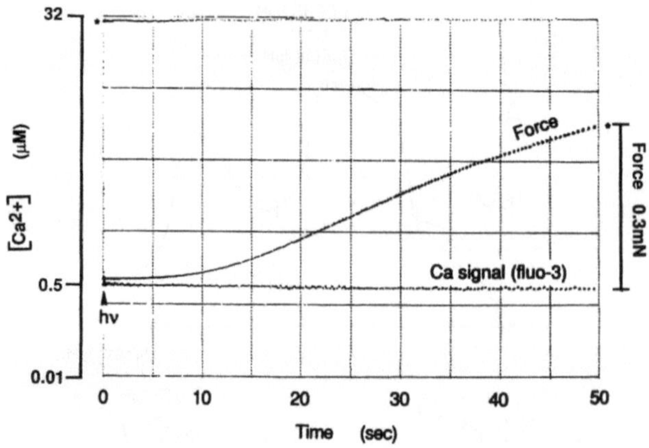


**Fig. 2.** Modulation of force by GTPγS at constant $Ca^{2+}$. Force development and $Ca^{2+}$ were monitored in a small strip of portal vein mounted in a 0.3 mm O.D. quartz capillary. It was permeabilized with α-toxin and incubated with 77 μM caged GTPγS, 50 μM fluo-3 pCa 6.3, and 10 mM EGTA for 3 min prior to photolysis with a 2 ms pulse (*arrowhead*) of 360 nm light from a flash lamp. Note the approximately 10 s delay after photolytic release of GTPγS before the force rises, and that there is no detectable change in $Ca^{2+}$. (From [39], with permission)

[33–37] and inhibited by GDPβS [31, 34]. The G protein(s) involved, whether trimeric, monomeric, or both, are yet to be definitively identified.

The two pharmacomechanical pathways, $InsP_3$-induced $Ca^{2+}$ release and $Ca^{2+}$ sensitization, can be dissociated. Verapamil, in concentrations sufficient to inhibit agonist-induced $Ca^{2+}$ release from the sarcoplasmic reticulum, does not significantly inhibit $Ca^{2+}$ sensitization in permeabilized smooth muscle [38], and the GTPγS-induced $Ca^{2+}$ sensitization is preceded by a longer delay than $Ca^{2+}$ release. Photolysis of caged phenylephrine initiates contraction with a delay of about 1.5 s at room temperature [9], whereas the delay in the $Ca^{2+}$-sensitizing effect of GTPγS is approximately 10 s (Fig. 2) [39]. The fact that verapamil inhibits agonist- and GTPγS-induced but not $InsP_3$-induced $Ca^{2+}$ release suggests that inhibition takes place at a step that precedes the generation of $InsP_3$ by phospholipase C (PLC). The most likely possibility is that verapamil directly inhibits PLC or uncouples it from the relevant G protein (?$G_q$). In this case, verapamil would also interfere with the production of diacylglycerol, a potential messenger of $Ca^{2+}$ sensitization. This action would seemingly exclude diacylglycerol generated from phosphatidylinositol biphosphate by PLC as an essential component of the $Ca^{2+}$-sensitizing cascade, but it does not exclude a $Ca^{2+}$-sensitizing role of diacylglycerol generated through other pathways, such as phosphatidylcholine hydrolysis.

## Myosin Light Chain Phosphatase and $Ca^{2+}$ Sensitization

The major mechanism of $Ca^{2+}$ sensitization is, as suggested originally [12], the result of increased $MLC_{20}$ phosphorylation mediated by inhibition of $MLC_{20}$ phosphatase [35, 36]. Stimulation with an α-adrenergic agonist in the presence of a low concentration of GTPγS inhibits myosin light chain phosphatase, as demonstrated by the slowing of dephosphorylation of $MLC_{20}$ [36], and the increase in $MLC_{20}$ phosphorylation induced by an agonist can be inhibited by GDPβS [35, 36]. These findings indicate the physiologically significant role of a G protein-coupled mecahnism of phosphatase inhibition.

Several enzymes that can dephosphorylate $MLC_{20}$ have been isolated from a variety of smooth muscles [e.g., 25–29, 40], but only some meet the criterion that a phosphatase that participates in physiologic relaxation should dephosphorylate intact myosin, not only isolated $MLC_{20}$. It is now generally agreed that enzymes of the type 1 protein phosphatase family can dephosphorylate intact smooth muscle myosin and mediate $MLC_{20}$ dephosphorylation during physiologic relaxation. Several type 1 protein phosphatases, differing in their subunits (or perhaps the same protein phosphatase and several of its proteotolytic degradation products), have been isolated from smooth muscles [reviews: 27, 29]. The major characteristics of some of these phosphatases are that they are bound to myosin [26, 28, 29, 41], are much more sensitive to inhibition by tautomycin than by okadaic acid [27, 42], are only weakly inhibited (unlike some other type 1 protein phosphatases) by protein phosphatase inhibitors 1 and 2, can dephosphorylate intact myosin, and contain a 37-kDa catalytic subunit (although variously reported as 34–40 kDa) that has low sensitivity, compared to type 2 protein phosphatases, to inhibition by okadaic acid [25–27, 29]. Smooth muscle myosin phosphatase also dephosphorylates in vitro calponin phosphorylated by kinase C [43]. Interestingly, dephosphorylation of MLCK is also inhibited by GTPγS, presumably as the result of inhibition of the protein phosphatase, perhaps $MLC_{20}$ phosphatase, that dephosphorylates MLCK [44].

A trimeric protein phosphatase isolated from gizzard smooth muscle contains, in addition to the 37-kDa catalytic subunit, 130- and 20-kDa subunits as well [26, 27]; one or both of the latter subunits are thought to target the catalytic subunit to myosin. A similar protein phosphatase containing the 37-kDa catalytic and 130- and 20-kDa subunits has also been isolated from relaxed pig bladder smooth muscle; the putative 130-kDa myosin targeting subunit enhances the phosphatase activity of the 37-kDa catalytic subunit by two orders of magnitude [45].

## Diacylglycerol and Arachidonic Acid: Potential $MLC_{20}$ Phosphatase Inhibitory Messengers

Given that G protein-coupled $Ca^{2+}$ sensitization is mediated by inhibition of the dephosphorylation of $MLC_{20}$, several questions remain unanswered: (1) the identity of the G protein(s) involved, alluded to earlier; and (2) the identity of the messenger or cascade that conveys the information from a membrane-bound receptor–G protein complex to the protein phosphatase bound to myosin filaments.

The fact that activation of phospholipase C by excitatory agonists leads to the production of diacylglycerol as well as $InsP_3$, whereas phorbol esters, activators of kinase C, can also cause $Ca^{2+}$ sensitization raised the possibility that the phospholipase C–diacylglycerol–kinase cascade also mediates $Ca^{2+}$ sensitization [46; review: 47]. However, phosphorylation of $MLC_{20}$ at the kinase C sites does not activate and, if anything, inhibits actomyosin ATPase [48]. Furthermore, brain kinase C has either no effect on or relaxes, rather than contracts, permeabilized smooth muscle [49–51]. Neither of these objections is unsurmountable, as kinase C could exert its effect indirectly by increasing phosphorylation of the MLCK site [52] through phosphorylation of some other regulatory site, such as a soluble protein phosphatase inhibitor or myosin light chain phosphatase itself. To account for the absence of

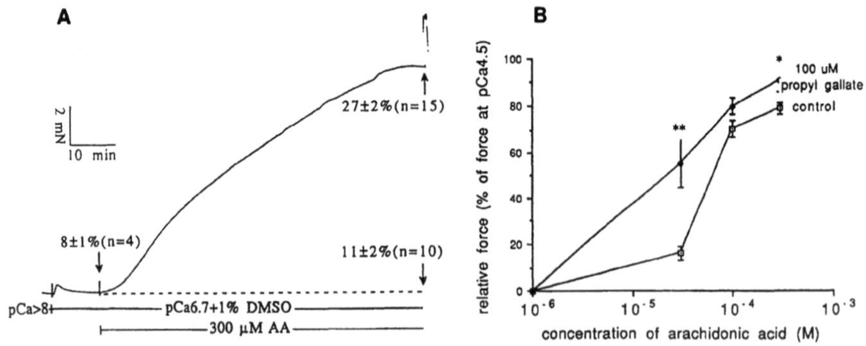


**Fig. 3.** Arachidonic acid (AA)-induced $Ca^{2+}$ sensitization of force and myosin light chain phosphorylation. **A.** AA 300 μM was added to the α-toxin-permeabilized rabbit femoral artery strip in pCa 6.7 solution. The numbers (%) indicate the level of $MLC_{20}$ phosphorylation in similarly treated strips at the time indicated. Note that AA increased both the force and the level of $MLC_{20}$ phosphorylation. **B.** Dose-response curve of AA-induced $Ca^{2+}$ sensitization and the effect of the antioxidant AA-metabolism inhibitor propyl gallate 100 μM. Strips were incubated for 15 min with 100 μM propyl gallate prior to addition of AA. (From [53], with permission)

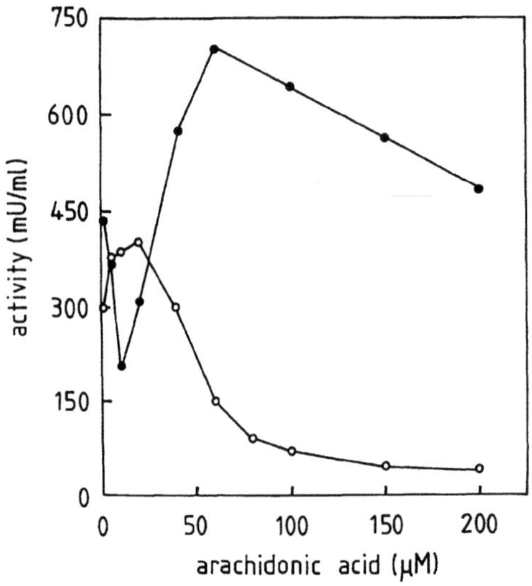


**Fig. 4.** Inhibition by arachidonic acid (AA) of the heavy meromyosin phosphatase activity of the trimeric type 1 myosin-bound protein phosphatase isolated from gizzard smooth muscle. Heavy meromyosin dephosphorylation was assayed at the indicated concentrations of AA for 5 min at 0.02 mU myosin phosphatase activity per milliliter (*open circles*) and phosphorylase dephosphorylation for 10 min at 0.3 mU phosphorylase phosphatase activity per milliliter (*closed circles*). Note that the inhibition of heavy meromyosin dephosphorylation is associated with an increase in phosphorylase phosphatase activity. (From [53], with permission)

the $Ca^{2+}$-sensitizing effect of kinase C, it would have to be assumed that this putative inhibitor is lost during permeabilization or that the brain kinase C used in the cited studies did not contain sufficient concentrations of the appropriate kinase isoform.

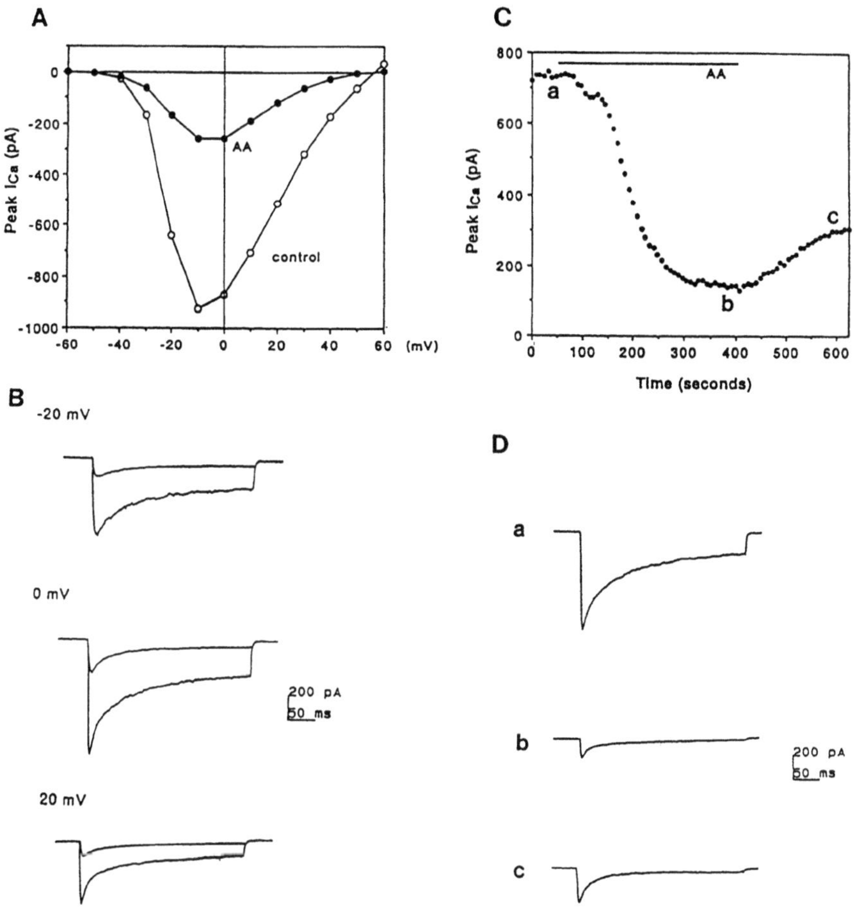


**Fig. 5.** Effect of arachidonic acid (AA) 10 μM on Ba currents through Ca channels ($I_{Ca}$) in rabbit ileum smooth muscle cells. **A, B.** Note the difference beween control responses and after treatment with AA at differing command potentials. Depolarizing pulses (300 ms) from −60 to +60 mV were applied from a holding potential of −60 mV; and the peak amplitudes of $I_{Ca}$ in the absence and presence of AA were plotted. **B.** Traces of individual responses at different potentials. **C, D.** Time course of AA-induced inhibition of $I_{Ca}$. Note the gradual increase in the inhibition of the peak $I_{Ca}$, which reaches a maximum after approximately 6 min of exposure to AA. The three traces in **D** were obtained at the time points indicated in **C** (*a*, *b*, *c*). (Modified [54], with permission)

Arachidonic acid is another potential messenger of $Ca^{2+}$ sensitization. It is known to be released by several agonists and is generated through several pathways in a variety of cell systems [review: 53]. Arachidonic acid can cause $Ca^{2+}$ sensitization of permeabilized smooth muscle by increasing $MLC_{20}$ phosphorylation (Fig. 3) [53]. In solution, arachidonic acid dissociates the trimeric gizzard $MLC_{20}$ phosphatase and inhibits its specific activity toward heavy meromyosin while increasing its activity for phosphorylase (Fig. 4) [53]. However, the concentrations of exogenous arachidonic acid required for $Ca^{2+}$ sensitization are relatively high: The $IC_{50}$ for inhibition of the trimeric phosphatase in solution is 60 μM, and maximal $Ca^{2+}$ sensitization of permeabilized smooth muscle requires about 100 μM. It is possible that the endogenous fatty acid, perhaps transported by fatty acid-binding proteins, is sufficiently active at much lower, less than critical micellar concentrations. In recent studies we have found that in permeabilized smooth muscle at constant [$Ca^{2+}$] both diacylglycerol and arachidonic acid are increased following stimulation with $Ca^{2+}$-sensitizing agonists and GTPγS [54]. It remains to be established whether the concentration of arachidonic acid reached and the time course of its production are sufficient to account for, or at least contribute to, $Ca^{2+}$ sensitization that furthermore may be mediated by multiple pathways, including both kinase C and arachidonic acid.

Arachidonic acid, in addition to its possible involvement in $Ca^{2+}$ sensitization, may also modulate the membrane potential, and consequently electromechanical coupling, through inhibition of voltage-gated $Ca^{2+}$ current (Fig. 5) [55]. The concentration of exogenous arachidonic acid (10 μM) that inhibits $I_{Ca}$ [55] is significantly lower than that required for $Ca^{2+}$ sensitization. Furthermore, endogenous arachidonic acid is generated at the plasma membrane where the ion channels are located. Therefore the effect of arachidonic acid generated in response to agonists may account for some of the time-dependent electrophysiologic effects of agonists and in particular for the delayed inhibition of voltage-gated $Ca^{2+}$ current with its potential consequences on excitation–contraction coupling.

In conclusion, a G protein-coupled pathway that modulates $MLC_{20}$ phosphorylation and contraction through regulation of a type 1 protein ($MLC_{20}$) phosphatase is a major mechanism of regulation in smooth muscle. In view of the ubiquity and similar regulation of nonmuscle myosins by phosphorylation, we believe that such regulation will also be shown to play a major role in the function of nonmuscle systems.

*Acknowledgments.* This work was supported by National Institutes of Health grants 5P01-HL19242 and 1P01-HL48807 and a Human Frontiers for Science grant. We gratefully acknowledge the expertise of Barbara Nordin during preparation of the manuscript and Mary Alice Spina for technical assistance and preparation of the illustrations.

## References

1. Somlyo AV, Somlyo AP (1968) Electromechanical and pharmacomechanical coupling in vascular smooth muscle. J Pharmacol Exp Ther 159:129–145
2. Suematsu E, Hirata M, Hashimoto T, Kuriyama H (1984) Inositol 1,4,5-trisphosphate releases $Ca^{2-}$ from intracellular store sites in skinned single cells of porcine coronary artery. Biochem Biophys Res Commun 120:481–485
3. Somlyo AV, Bond M, Somlyo AP, Scarpa A (1985) Inositol-trisphosphate ($InsP_3$) induced calcium release and contraction in vascular smooth muscle. Proc Natl Acad Sci USA 82:5231–5235
4. Kobayashi S, Somlyo AP, Somlyo AV (1988) Guanine nucleotide and inositol 1,4,5-trisphosphate-induced calcium release in rabbit main pulmonary artery. J Physiol (Lond) 403:601–619
5. Kobayashi S, Somlyo AV, Somlyo AP (1988) Heparin inhibits the inositol 1,4,5-trisphosphate-dependent, but not the independent, calcium release induced by guanine nucleotide in vascular smooth muscle. Biochem Biophys Res Commun 153:625–631
6. Miller-Hance WC, Miller JR, Wells JN, Stull JT, Kamm KE (1988) Biochemical events associated with activation of smooth muscle contraction. J Biol Chem 263:13979–13982
7. Kobayashi S, Kitazawa T, Somlyo AV, Somlyo AP (1989) Cytosolic heparin inhibits muscarinic and adrenergic $Ca^{2+}$ release in smooth muscle. J Biol Chem 264:17997–18004
8. Somlyo AV, Horiuti K, Trentham DR, Kitazawa T, Somlyo AP (1992) Kinetics of $Ca^{2+}$-release and contraction induced by photolysis of caged D-myo-inositol 1,4,5-trisphosphate in smooth muscle: the effects of heparin, procaine and adenine nucleotides. J Biol Chem 267:22316–22322

9. Somlyo AP, Walker JW, Goldman TE, Trentham DR, Kobayashi S, Kitazawa T, Somlyo AV (1988) Inositol trisphosphate, calcium and muscle contraction. Philos Trans R Soc Lond [Biol] 320:399–414
10. Somlyo AP, Somlyo AV (1990) Flash photolysis studies of excitation-contraction coupling, regulation, and contraction in smooth muscle. Annu Rev Physiol 52:857–874
11. Hartshorne DJ (1987) Biochemistry of the contractile process in smooth muscle. In: Johnson LR (ed) Physiology of the gastrointestinal tract, 2nd edn. Raven, New York, pp 423–482
12. Somlyo AP, Kitazawa T, Himpens B, Matthijs G, Horiuti K, Kobayashi S, Goldman YE, Somlyo AV (1989) Modulation of $Ca^{2+}$-sensitivity and of the time course of contraction in smooth muscle: a major role of protein phosphatases? Adv Protein Phosphatases 5:181–195
13. Nishiye E, Somlyo AV, Török K, Somlyo AP (1993) The effects of MgADP on cross-bridge kinetics: a laser flash photolysis study of guinea-pig smooth muscle. J Physiol (Lond) 460:247–271
14. Fuglsang A, Khromov A, Török K, Somlyo AV, Somlyo AP (1993) Flash photolysis studies of relaxation and cross-bridge detachment: higher sensitivity of tonic than phasic smooth muscle to MgADP. J Muscle Res Cell Motil 15:666–673
15. Khromov A, Somlyo AV, Somlyo AP (1994) The differential effect of MgADP on relaxation from isometric contraction by photolysis of Diazo 2 in tonic and phasic smooth muscle. Biophys J 66:A411
16. Himpens B, Matthijs G, Somlyo AV, Butler TM, Somlyo AP (1988) Cytoplasmic free calcium, myosin light chain phosphorylation, and force in phasic and tonic smooth muscle. J Gen Physiol 92:712–729
17. Himpens B, Matthijs G, Somlyo AP (1989) Desensitization to cytoplasmic $Ca^{2+}$ and $Ca^{2+}$-sensitivities of guinea-pig ileum and rabbit pulmonary artery smooth muscle. J Physiol (Lond) 413:489–503
18. Kitazawa T, Somlyo AP (1990) Desensitization and muscarinic re-sensitization of force and myosin light chain phosphorylation to cytoplasmic $Ca^{2+}$ in smooth muscle. Biochem Biophys Res Commun 172:1291–1297
19. Lydrup ML, Himpens B, Droogmans C, Hellstrand P, Somlyo AP (1992) Paradoxical decrease in cytosolic calcium with increasing depolarization by potassium in guinea-pig mesotubarium smooth muscle. Pflugers Arch 420:428–433
20. Adelstein RS, Conti MA, Hathaway DR (1978) Phosphorylation of smooth muscle myosin light chain kinase by the catalytic subunit of adenosine 3′:5′-monophosphate-dependent protein kinase. J Biol Chem 253:8347–8350
21. Stull JT, Hsu L-C, Tansey MG, Kamm KE (1990) Myosin light chain kinase phosphorylation in tracheal smooth muscle. J Biol Chem 265:16683–16690
22. Tansey MG, Luby-Phelps K, Kamm KE, Stull JT (1994) $Ca^{2+}$-dependent phosphorylation of myosin light chain kinase decreases the $Ca^{2+}$-sensitivity of light chain phosphorylation within smooth muscle cells. J Biol Chem 269:9912–9920
23. Nishikawa M, de Lanerolle P, Lincoln TM, Adelstein RS (1984) Phosphorylation of mammalian myosin light chain kinases by the catalytic subunit of cyclic AMP-dependent protein kinase and by cyclic GMP-dependent protein kinase. J Biol Chem 259:8429–8436
24. Pfitzer G, Hofmann F, DiSalvo J, Rüegg JC (1984) cGMP and cAMP inhibit tension development in skinned coronary arteries. Pflugers Arch 401:277–280
25. Tulloch AG, Pato MD (1991) Turkey gizzard smooth muscle myosin phosphatase-III is a novel protein phosphatase. J Biol Chem 266:20168–20174
26. Alessi D, MacDougall LK, Sola MM, Ikebe M, Cohen P (1992) The control of protein phosphatase-1 by targeting subunits: the major myosin phosphatase in avian smooth muscle is a novel form of protein phosphatase-1. Eur J Biochem 210:1023–1035
27. Gong MC, Cohen P, Kitazawa T, Ikebe M, Masuo M, Somlyo AP, Somlyo AV (1992) Myosin light chain phosphatase activities and the effects of phosphatase inhibitors in tonic and phasic smooth muscle. J Biol Chem 267:14662–14668
28. Mitsui T, Inagaki M, Ikebe M (1992) Purification and characterization of smooth muscle myosin-associated phosphatase from chicken gizzards. J Biol Chem 267:16727–16735
29. Okubo S, Erdödi F, Ito M, Ichikawa K, Konishi T, Nakano T, Kawamura T, Brautigan DL, Hartshorne DJ (1993) Characterization of a myosin-bound phosphatase from smooth muscle. Adv Protein Phosphatases 7:295–314

30. Bradley AB, Morgan KG (1987) Alterations in cytoplasmic calcium sensitivity during porcine coronary artery contractions as detected by aequorin. J Physiol (Lond) 385:437–448
31. Himpens B, Kitazawa T, Somlyo AP (1990) Agonist dependent modulation of $Ca^{2+}$-sensitivity in rabbit pulmonary artery smooth muscle. Pflugers Arch 417:21–28
32. Rembold CM, Murphy RA (1988) Myoplasmic [$Ca^{2+}$] determines myosin phosphorylation in agonist-stimulated swine arterial smooth muscle. Circ Res 63:593
33. Fujiwara T, Itoh T, Kubota Y, Kuriyama H (1989) Effects of guanosine nucleotides on skinned smooth muscle tissue of the rabbit mesenteric artery. J Physiol (Lond) 408:535
34. Kitazawa T, Kobayashi S, Horiuti K, Somlyo AV, Somlyo AP (1989) Receptor-coupled, permeabilized smooth muscle: role of the phosphatidylinositol cascade, G-proteins and modulation of the contractile response to $Ca^{2+}$. J Biol Chem 264:5339–5342
35. Kitazawa T, Gaylinn BD, Denney GH, Somlyo AP (1991) G-protein-mediated $Ca^{2+}$-sensitization of smooth muscle contraction through myosin light chain phosphorylation. J Biol Chem 266:1708–1715
36. Kitazawa T, Masuo M, Somlyo AP (1991) G-protein-mediated inhibition of myosin light chain phosphatase in vascular smooth muscle. Proc Natl Acad Sci USA 88:9307–9310
37. Nishimura J, Khalil RA, Drenth JP, Van Breemen C (1990) Evidence for increased myofilament $Ca^{2+}$-sensitivity in norepinephrine-activated vascular smooth muscle. Am J Physiol 259:H2–H8
38. Kobayashi S, Gong MC, Somlyo AV, Somlyo AP (1991) $Ca^{2+}$ channel blockers distinguish between G-protein-coupled, pharmacomechanical $Ca^{2+}$ release and $Ca^{2+}$-sensitization in smooth muscle. Am J Physiol 260:C364–C370
39. Somlyo AV, Kitazawa T, Horiuti K, Kobayashi S, Trentham DR, Somlyo AP (1990) Heparin-sensitive inositol trisphosphate signalling and the role of G-proteins in $Ca^{2+}$-release and contractile regulation in smooth muscle. In: Sperelakis N, Wood JD (eds) Frontiers in smooth muscle research. Alan R Liss, New York, pp 167–182
40. Takai A, Troschka M, Mieskes G, Somlyo AV (1989) Protein phosphatase composition in smooth muscle of guinea-pig ileum studied with okadaic acid and inhibitor 2. Biochem J 262:617–623
41. Sellers JR, Pato MD (1984) The binding of smooth muscle myosin light chain kinase and phosphatases to actin and myosin. J Biol Chem 259:7740–7746
42. Horiuti M, Magae J, Han Y-G, Hartshorne DJ, Karaki H (1991) A novel protein phosphatase inhibitor, tautomycin: effect on smooth muscle. FEBS Lett 285:145–148
43. Ichikawa K, Ito M, Okubo S, Konishi T, Nakano T, Mino T, Nakamura F, Naka M, Tanaka T (1993) Calponin phosphatase from smooth muscle: a possible role of type 1 protein phosphatase in smooth muscle relaxation. Biochem Biophys Res Commun 193:827–833
44. Tang D-C, Kubota Y, Kamm KE, Stull JT (1993) GTPγS-induced phosphorylation of myosin light chain kinase in smooth muscle. FEBS Lett 331:272–275
45. Shirazi A, Iizuka K, Fadden P, Mosse C, Somlyo AP, Somlyo AV, Haystead T (1994) Purification and characterization of the mammalian myosin light chain phosphatase holoenzyme: the differential effects of the holoenzyme and its subunits on smooth muscle. J Biol Chem 269:31598–31606
46. Chatterjee M, Tejada M (1986) Phorbol ester-induced contraction in chemically skinned vascular smooth muscle. Am J Physiol 251:C356–C361
47. Andrea JE, Walsh MP (1992) Protein kinase C of smooth muscle. Hypertension 20:585–595
48. Ikebe M, Hartshore DJ, Elzinga M (1987) Phosphorylation of the 20,000-dalton light chain of smooth muscle myosin by the calcium-activated, phospholipid-dependent protein kinase. J Biol Chem 262:9569–9573
49. Inagaki M, Yokokura H, Itoh T, Kanmura Y, Kuriyama H, Hidaka H (1987) Purified rabbit brain protein kinase C relaxes skinned vascular smooth muscle and phosphates myosin light chain. Arch Biochem Biophys 254:136–141
50. Sutton TA, Haeberle JR (1990) Phosphorylation by protein kinase C of the 20,000-dalton light chain of myosin in intact and chemically skinned vascular smooth muscle. J Biol Chem 265:2749–2754
51. Parente JE, Walsh MP, Kerrick WGL, Hoar PE (1992) Effects of the constitutively active

proteolytic fragment of protein kinase C on the contractile properties of demembranated smooth muscle fibres. J Muscle Res Cell Motil 13:90–99

52. Itoh H, Shimomura A, Okubo S, Ichikawa K, Ito M, Konishi T, Nakano T (1993) Inhibition of myosin light chain phosphatase during $Ca^{2+}$-independent vasocontraction. Am J Physiol 265: C1319–C1324
53. Gong MC, Fuglsang A, Alessi D, Kobayashi S, Cohen P, Somlyo AV, Somlyo AP (1992) Arachidonic acid inhibits myosin light chain phosphatase and sensitizes smooth muscle to calcium. J Biol Chem 267:21492–21498
54. Gong MC, Kinter MT, Somlyo AV, Somlyo AP (1995) Arachidonic acid and diacylglyceral release associated with inhibition of myosin light chain dephosphorylation in rabbit smooth muscle. J Physiol (Lond) 486:113–122
55. Shimada T, Somlyo AP (1992) Modulation of voltage-dependent Ca channel current by arachidonic acid and other long-chain fatty acids in rabbit intestinal smooth muscle. J Gen Physiol 100:27–44

# Smooth Muscle Myosin Phosphatase

Masaaki Ito[1], Hiroyuki Shimizu[1], Masatoshi Miyahara[1], Jian-Hua Feng[1], Setsuya Okubo[1], Kazuhito Ichikawa[1], Tokuji Konishi[1], David J. Hartshorne[2], and Takeshi Nakano[1]

*Summary.* The endogenous phosphatase of chicken gizzard actomyosin contains two subunits of 58 and 38 kDa, respectively, as we showed previously. This phosphatase was active with both isolated myosin light chain and intact myosin that had been phosphorylated by myosin light chain kinase as substrates, suggesting that phosphorylated myosin might be a substrate in situ. The 38-kDa subunit was identified as a catalytic subunit of a type 1δ protein phosphatase (PP1δc), and the 58-kDa subunit was revealed to play a regulatory role in the binding of myosin. Studies with a monoclonal antibody (MoAb) to the 58-kDa subunit revealed that the 58-kDa protein was a product of proteolytic degradation of a protein of 130–133 kDa. The distribution of the 130-kDa component in chicken tissues was examined with the MoAb. An immunoreactive band of appropriate mobility was detected in analyses of all tissues except liver and skeletal muscle, and higher concentrations of the 130-kDa component were found in samples of smooth muscle. Intact myosin phosphatase was purified from chicken gizzard with the MoAb as probe. The purified holoenzyme was a heterotrimer that consisted of PP1δc (38 kDa) and regulatory subunits of 130/133 kDa and 20 kDa. cDNA clones encoding the 130/133-kDa subunits were isolated from chicken gizzard cDNA libraries. The isolation of overlapping clones suggested the presence of the two isoforms. There were open reading frames of 2889 and 3012 bases that encoded proteins of 963 and 1004 amino acids with masses of 106.7 and 111.6 kDa, respectively. The deduced sequence of the larger isoform was 123 nucleotides longer in its central coding region than the other isoform. The characteristic N-terminal region was almost entirely composed of eight tandem repeats of 33 amino acids, which have been called the *cdc10/SWI6*, or ankyrin, repeat. The 58-kDa fragment with the ability to bind to both myosin and the catalytic subunit was revealed to be located in the N-terminal half of the molecule. A cDNA clone (1338 bp) encoding chicken gizzard PP1δc was also isolated and sequenced. This clone contained an open reading frame that encoded a protein of 327 amino acids with a calculated molecular mass of 37 kDa. Its amino acid sequence was identical to the analogous isoform from rats. These results suggest that smooth muscle myosin phosphatase is a novel holoenzyme composed of a type 1δ protein phosphatase and unique regulatory subunits.

[1] First Department of Internal Medicine, Mie University School of Medicine, 2-174 Edobashi, Tsu, Mie, 514 Japan

[2] Muscle Biology Group, Shantz Building, University of Arizona, Tucson, AZ 85721, USA

*Key Words*: Smooth muscle—Myosin phosphatase—Type 1 protein phosphatase—Myosin-binding subunit

## Introduction

Phosphorylation of 20-kDa myosin light chain ($MLC_{20}$) is a dominant aspect of regulatory mechanisms in smooth muscle and nonmuscle cells [1]. It is widely accepted that this phosphorylation is catalyzed by a $Ca^{2+}$/calmodulin-dependent myosin light chain kinase (MLCK) and increases the actin-activated ATPase activity of myosin. Thus the contractile activity of smooth muscle at any given time depends on the extent of phosphorylation of myosin, which in turn reflects the balance between the activities of kinases and phosphatases.

Experiments using fluorescent dyes to measure the internal level of $Ca^{2+}$ ions have shown that the relation between phosphorylation and $Ca^{2+}$ ions is variable in intact smooth muscle and is sensitive to the nature of stimulation [2–5]. In general, agonist-induced stimulation results in a greater increase in phosphorylation of $MLC_{20}$ relative to the cytosolic concentration of $Ca^{2+}$ ions than that obtained by $K^+$ depolarization. The increased sensitivity to $Ca^{2+}$ ions implies an alteration in the balance of the activities of kinases and phosphatases. MLCK has been purified from several tissues and its characteristics, including its primary structure and the mechanism of its activation by the $Ca^{2+}$–calmodulin complex, are well understood [1; see also Matsushima et al., this volume]. The activity of MLCK is modified by its phosphorylation, which results in its reduced activity and decreased sensitivity to $Ca^{2+}$ ions [6–11]. On the other hand, it is of particular interest that guanosine 5′-O-(3-thiotriphosphate) (GTPγS) shifts the $Ca^{2+}$–force relation to the left and this G protein-coupled $Ca^{2+}$-sensitizing effect of excitatory agonists is mediated by an increase in phosphorylation of $MLC_{20}$ at a given concentration of $Ca^{2+}$ ions as the result of inhibition of $MLC_{20}$ phosphatase [12–17]. Moreover, reduced activity of phosphatase with heavy meromyosin as its substrate was found in GTPγS-treated tracheal smooth muscle [18]. In addition, the activation of protein kinase C by phorbol esters also increases the sensitivity of preparations of permeabilized smooth muscle to $Ca^{2+}$ ions [19–22]. It has been shown that this protein kinase C-mediated increase in sensitivity to $Ca^{2+}$ ions in smooth muscle can also be induced by the inhibition of myosin phosphatase [23, 24]. Thus GTPγS and phorbol esters have similar effects on smooth muscle contraction via the inhibition of myosin phosphatase.

The underlying mechanisms for these effects are not yet known. One major difficulty is that the relevant phosphatases have not been biochemically defined. The object of the present study was to purify and characterize myosin phosphatase from smooth muscle. The primary structures of catalytic and myosin-binding regulatory subunits of this phosphatase were determined.

## Myosin Phosphatase from Chicken Gizzard Crude Actomyosin

Myosin phosphatase is a persistent contaminant of preparations of myosin, and strong binding to myosin is one of the characteristic properties of myosin phosphatase. We attempted to purify and characterize the endogenous phosphatase of actomyosin from chicken gizzard [25]. The phosphatase was extracted from the crude actomyosin in a buffer that contained 50 mM $MgCl_2$, and it was purified by sequential chromatography on diethylaminoethyl (DEAE)-Cellulofine (Seikagaku Kogyo, Tokyo, Japan), heparin–Sepharose (Pharmacia, Uppsala, Sweden), and thiophosphorylated myosin light chain–Sepharose. The purified phosphatase consisted of two subunits of 58 and 38 kDa, respectively, as determined by sodium dodecyl sulfate (SDS)-polyacrylamide gel electrophoresis (SDS-PAGE). This preparation was active with both isolated $MLC_{20}$ and intact myosin as substrates, and so this phosphatase appeared to be a myosin phosphatase. The 38-kDa subunit was identified as a catalytic subunit of a type 1δ protein phosphatase (PP1δc) with Western blotting using polyclonal antibodies directed against the distinct C-terminal peptides of PP1α, PP1δ, PP1γ1, and type 2A (PP2A) catalytic subunits.

The activity of this myosin phosphatase was enhanced severalfold by $Co^{2+}$ or $Mn^{2+}$ ions, whereas other cations, including $Mg^{2+}$ and $Ca^{2+}$ ions, inhibited the phosphatase activity. Maximum activation was observed at approximately 0.3 mM $Co^{2+}$ ions. This activation required a relatively long preincubation (several minutes), and the addition of EDTA or EGTA to the activated species had no effect, suggesting that this activation is irreversible.

Okadaic acid (OA), a potent inhibitor of a type 1 protein phosphatase (PP1) and PP2A, inhibited the activity of myosin phosphatase with an inhibitory concentration

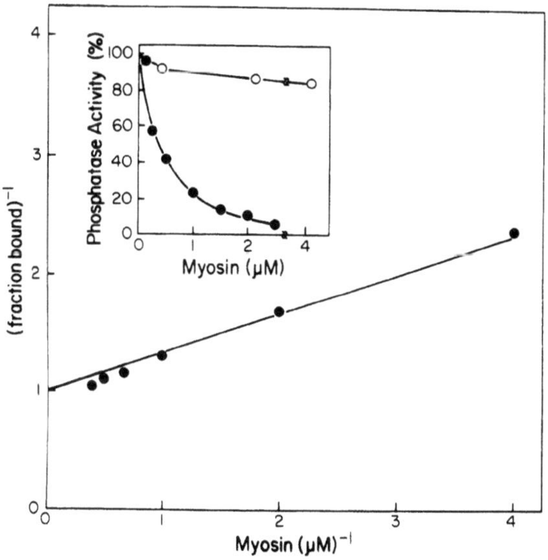


**Fig. 1.** Binding of myosin phosphatase to myosin. The binding assay was performed essentially as described previously [26]. Varying concentrations of myosin in 10 mM 3-(N-morpholino)-propanesulfonic acid (MOPS) (pH 7.0), 50 mM KCl, 1 mM $MgCl_2$, 1 mM dithiothreitol, 0.1 mM ethylene glycol bis(β-aminoethyl ether)-N,N,N',N'-tetraacetic acid (EGTA), and bovine serum albumin 0.5 mg/ml were incubated for 10 min at 25°C with myosin phosphatase 0.1–0.2 μg/ml and then centrifuged at 100 000 *g* for 15 min. The supernatant was assayed for phosphatase activity in the presence of 0.3 mM $CoCl_2$. A similar experiment was carried out to determine binding of the catalytic subunit to myosin [solvent: 10 mM Tris HCl (pH 7.5), 50 mM KCl, 5 mM $MgCl_2$, 0.25 mM dithiothreitol, 0.2 mM EGTA, 0.25 μM (*p*-amidinophenyl) methanesulfonyl fluoride hydrochloride (aPMSF), and 0.5 mM benzamidine]. **Inset.** Binding of myosin phosphatase to myosin (*closed circles*) and the binding of the catalytic subunit to myosin (*open circles*). The data for binding of myosin phosphatase to myosin are replotted as a double-reciprocal plot of the fraction bound versus the concentration of myosin. (From [25], with permission)

of 50% ($IC_{50}$) of 40 nM. Basal activity and $Co^{2+}$-activated activity had the same sensitivity to OA. These results are consistent with the identification of myosin phosphatase as PP1. The activity of PP1 is generally inhibited by nanomolar concentrations of two small heat- and acid-stable proteins, known as inhibitor-1 (I-1) and inhibitor 2 (I-2). The basal and $Co^{2+}$-activated activities of myosin phosphatase were considerably less sensitive to I-2 and were inhibited only at a relatively high concentration of I-2. However, when I-2 was added to the reaction mixture before $Co^{2+}$ ions, the activation of myosin phosphatase by $Co^{2+}$ ions was prevented. These results indicate that I-2 did not directly inhibit the enzymatic activity but blocked activation of the enzyme.

The most remarkable characteristic of this phosphatase was its ability to bind to myosin (Fig. 1). The inset in Figure 1 shows the experimental design whereby phosphatase activity was assayed in the supernatant after centrifugation with myosin at various concentrations. With increasing amounts of myosin, the myosin phosphatase activity in the supernatant decreased, suggesting that myosin phosphatase bound to myosin. The double-reciprocal plot of the bound fraction versus the concentration of myosin gave a dissociation constant of approximately 0.3 μM. In addition, the 38-kDa catalytic subunit isolated from this myosin phosphatase did not by itself bind to myosin, suggesting that the binding of myosin is a property of the 58-kDa subunit.

## Intact Smooth Muscle Myosin Phosphatase

A monoclonal antibody (MoAb) was raised against the 58-kDa subunit of myosin phosphatase that allowed us to examine the composition of the native phosphatase and its distribution in various tissues [27, 28]. The MoAb that cross-reacted with the 58-kDa subunit of the purified phosphatase detected only a 130-kDa protein in an extract of chicken gizzard; no 58-kDa component was found. In addition, a minor band of a 133-kDa protein cross-reacted with the MoAb. These 130/133-kDa components were also detected in a purified preparation of smooth muscle myosin that was shown to be contaminated by myosin phosphatase. Brief tryptic digestion of gizzard myosin converted the 130-kDa protein to the 58-kDa protein. All these results suggested that the 58-kDa subunit was a proteolytic fragment of the 130/133-kDa subunits. In addition, immunoprecipitation using the MoAb–protein A complex demonstrated that the 130/133-kDa subunits are associated with the PP1δc in the extract of chicken gizzard.

The distribution of the 130-kDa component in chicken tissues was screened with the anti-58 kDa MoAb. As shown in Figure 2, immunoreactive proteins of appropriate mass were detected in all tissues except skeletal muscle and liver. Higher concentrations of the 130-kDa component were evident in smooth muscle (i.e., gizzard, aorta, stomach, small intestine, and oviduct) than in brain, spleen, heart, kidney, and lung. The widespread distribution of the 130-kDa subunit clearly demonstrated that this phosphatase system is not confined to smooth muscle. There were some differences in their apparent molecular mass. In gizzard a major band of 130 kDa and a minor band of 133 kDa were detected. Two bands of 137 and 125 kDa were observed in the analyses of spleen, oviduct, and small intestine; and two bands of 143 and 137 kDa were found in heart tissue. A single band of about 137 kDa was observed during the analyses of brain, aorta, kidney, and lung. In the stomach, a 137-kDa protein and an additional 58-kDa component were detected; the latter may have been a fragment of the former generated by an intrinsic protease. The presence of these distinct immunoreactive

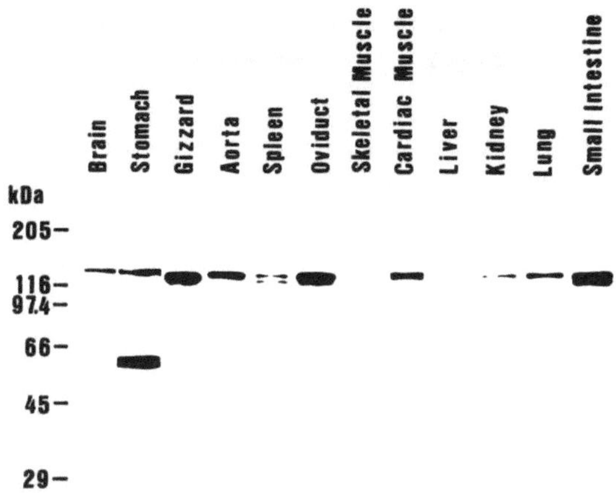


**Fig. 2.** Tissue distribution of proteins that immunoreact with a monoclonal antibody against the 58-kDa fragment. Each tissue was freshly removed from a chicken and rapidly frozen in liquid nitrogen. For assays, tissues were homogenized with a Polytron in several volumes of 500 mM NaCl, 20 mM Tris HCl (pH 7.5), 1 mM EGTA, 1 mM ethylenediaminetetraacetate (EDTA), 1 mM benzamidine, 1 mM aPMSF, leupeptin 10 μg/ml, and 0.1 mM diisopropyl fluorophosphate. After centrifugation at 20 000 g for 30 min, the supernatant (30 μg) was subjected to sodium dodecyl sulfate polyacrylamide gel electrophoresis (SDS-PAGE) with subsequent immunoblotting. (From [27], with permission)

components suggests the presence of different isoforms of this subunit in various tissues.

Because the myosin-binding subunit of smooth muscle myosin phosphatase was revealed to be a protein of 130/133 kDa, intact myosin phosphatase was purified to homogeneity from frozen and thawed chicken gizzard with the MoAb as probe. In this way, we detected the holoenzyme of myosin phosphatase instead of measuring phosphatase activity. The phosphatase was purified by the method of Alessi et al. [29] with the following modifications: After chromatography on DEAE-Sepharose Fast Flow (Pharmacia), myosin phosphatase was purified sequentially by chromatography on SP-Sepharose Fast Flow (Pharmacia), DEAE-Toyopearl 650S (Tosoh, Tokyo, Japan), Mono S HR 5/5 (Pharmacia), and Mono Q HR 5/5 (Pharmacia). The effective step in this purification was chromatography of SP-Sepharose, during which myosin phosphatase with degraded 130/133-kDa subunits (~58 kDa) was eluted at a much lower concentration of NaCl and could be separated from intact phosphatase. Chromatography on SP-Sepharose can also separate intact myosin phosphatase from MLCK, a major protein at this step, and proteases.

As shown in Figure 3, the holoenzyme purified from chicken gizzard consisted of three subunits of 130, 38, and 20 kDa, respectively. In addition to the 130-kDa protein, a minor component of 133 kDa also was observed. These three subunits appeared to be present in equimolar amounts, as judged by staining with Commassie blue, suggesting that the holoenzyme is a heterotrimer composed of one of each subunit. Both the 130- and 133-kDa subunits were recognized by the MoAb against the 58-kDa fragment. During gel filtration of Superdex 200 HR 10/30 (Pharmacia), the three subunits of 130/133, 38, and 20 kDa were eluted as a single peak that corresponded to a molecular mass of 250 kDa. The 38-kDa component was reconfirmed to be PP1δc by use of a polyclonal antibody specific to this isoform. It also became evident that the intact heterotrimeric phosphatase dephosphorylated heavy meromyosin that had been phosphorylated by MLCK and had the ability to bind to myosin.

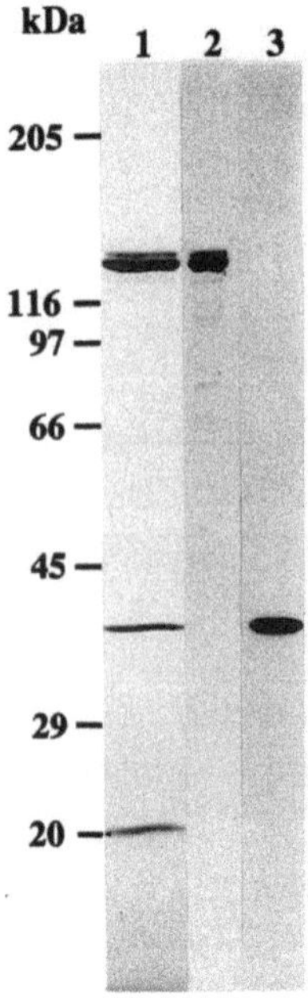


**Fig. 3.** Subunit composition of intact myosin phosphatase. The purified intact myosin phosphatase was subjected to SDS-PAGE and stained with Commassie blue (lane 1) or detected in immunoblot reactions with the monoclonal antibody against the 58-kDa fragment (lane 2) or antibody against PP1δc. Molecular mass markers are indicated on the left. (Modified from [28], with permission)

Brief digestion of intact myosin phosphatase by α-chymotrypsin converted the 130/133-kDa subunits to a protein of 58-kDa, whereas the 38-kDa catalytic subunit remained intact. A 20-kDa regulatory subunit was also digested by α-chymotrypsin. This α-chymotryptic phosphatase of 58- and 38-kDa subunits seemed to be almost identical to the myosin phosphatase from crude actomyosin, the 130-kDa subunit of which was degraded to a 58-kDa protein by an intrinsic protease during purification. During gel filtration of this α-chymotryptic myosin phosphatase, the 38-kDa catalytic subunit co-eluted with the 58-kDa fragment, suggesting that the 58-kDa fragment had the ability to bind to the catalytic subunit. Co-sedimentation with myosin revealed that the phosphatase of 58- and 38-kDa subunits retained the ability to bind to myosin. All these results indicate that the 130/133-kDa subunits have at least two functions—one being to bind the catalytic subunit and the other to bind myosin—and both domains are located mainly in the 58-kDa fragment.

## Isolation and Characterization of cDNA Clones That Encode the 130/133-kDa and Catalytic Subunits

To investigate the primary structure and the role of the 130/133-kDa subunits of myosin phosphatase we attempted molecular cloning of cDNAs that encoded these subunits. The immunoscreening of a chicken gizzard λgt11 cDNA library with the MoAb identified one positive clone (2049 bp), and nucleotide sequencing confirmed that this clone encoded the amino acid sequences of seven peptides isolated from the 58-kDa fragment of the 130/133-kDa subunits of myosin phosphatase. A chicken gizzard Uni-ZAP XR library was further screened with this clone, and several clones were obtained that covered the entire sequence of these subunits. The overlapping clones spanned a region of 4732 bp and suggested the existence of two isoforms of these subunits [28] (Fig. 4). The larger isoform (M133) had an additional region of

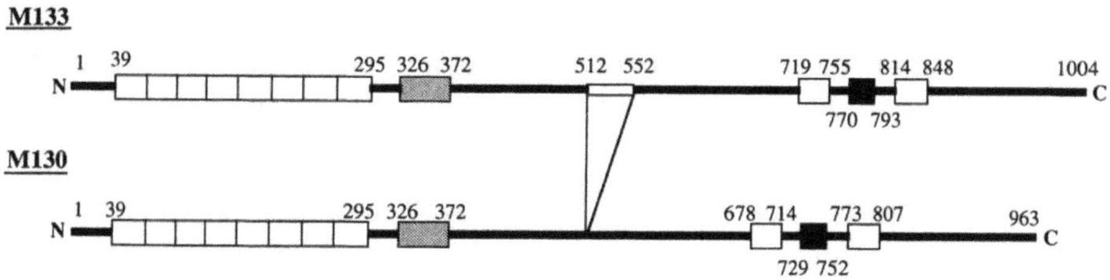


**Fig. 4.** Structures of M130 and M133. The M133 and M130 polypeptides have eight *cdc10/SWI6* repeat sequences in residues 39–295; an acidic cluster in residues 326–372; two acidic and basic clusters in residues 719–755/814–848 in M133 and 678–714/773–807 in M130; and an S- and T-rich sequence in residues 769–793 in M133 and 728–752 in M130. The insert 512–552 in M133 also is indicated. (Modified from [28], with permission)

123 bp nucleotides within the central coding region of the smaller isoform (M130). The sequencing data from all the clones together established open reading frames of 2889 and 3012 bases that encoded proteins of 963 and 1004 amino acids with molecular masses of 106.7 and 111.6 kDa, respectively. The deduced molecular masses were smaller than the values of 130 and 133 kDa estimated by SDS-PAGE.

The analysis of the deduced sequences revealed that the cDNAs encoded a novel protein that had not previously been reported. The first characteristic region was composed almost entirely of a repeated sequence motif at the N-terminus. As shown in Figure 5, the core of this domain is composed of eight successive repeats. Each repeat contains 33 amino acids except the fourth, fifth, and eighth repeats. Twenty of the thirty-three positions are conserved; that is, only one amino acid or two closely related amino acids are found at the same position in half or more of the repeats. The fifth repeat is less homologous. This type of repeat has been called the *cdc10/SWI6*, or ankyrin, repeat.

```
 39  D D G A V F L A A C S S G D T E E V L R L L E R G A D I N Y A N V     71
 72  D G L T A L H Q A C I D D N V D M V K F L V E N G A N I N Q P D N    104
105  E G W I P L H A A A S C G Y L D I A E Y L I S Q G A H V G A V N S    137
138  E G D T P L D I - A E E E A M E - - - E L L Q N E V N R Q G V D I    166
167  E - - - A A R K E E E R I M L R D A R Q W L N S G H I N D V R H A K  197
198  S G G T A L H V A A A K G Y T E V L K L L I Q A R Y D V N I K D Y    230
231  D G W T P L H A A A H W G K E E A C R I L V E N L C D M E A V N K    263
264  V G Q T A F D V A D - E D I L G Y L E E L Q K K Q N L L H S E K R    295

Consensus
 M130     D G - T A L H A A A - - G - - E - A K - L - - - G A D - N - V D -
 /M133    E       V     V               D   V R             V N   D   A N

 cdc-10   N G D T A L H W A A R I A N - - L V - A L - - - G A N P N P Q D E
 /SWI6                N I         G       I
 ANKYRIN  - G - T P L H - A A - - G H - - - V - - L L - - G A - - N - - - -
                                            A                     D
 Notch    - G - T P L H L A A R - - - - - A - - - L L - N - A D - N A - D -

 lin-12   - G R T A L H - A A - N - - - - - V - Y L - - - - A - - - - - D -
                      M
 glp-1    - G R T A L H L A A - - - - - - - V - - L - - - - - - - D - - D -
                      M
 fem-1    - G - T P L - - A A - - G H - - V V K - L L E - G - D - - - - - -
                                          I
 hr       - G - - A L Y Y A - A - - N - - - V K - L L - - G - - - - - - D -
```

**Fig. 5.** Alignment of the repeated sequences in M130/M133. Positions that are occupied by a single amino acid or two closely related amino acids in 50% or more of the repeats are boxed. A consensus sequence including these residues is shown below the sequence of the repeats for comparison with similarly derived consensus sequences from other proteins. (Modified from [28], with permission)

Hydropathic analysis of the deduced amino acid sequence indicated the presence of three major hydrophilic regions in M130/M133 (Fig. 4). The first hydrophilic region approximately covers the negatively charged cluster of glutamic and aspartic acid residues that corresponds to amino acids $Glu^{326}$–$Asp^{372}$. The second and third hydrophilic regions correspond approximately to two clusters of acidic and basic amino acids, namely, $Arg^{719}$–$Arg^{755}$ and $Lys^{814}$–$Arg^{848}$, respectively, in M133 ($Arg^{678}$–$Arg^{714}$ and $Lys^{773}$–$Arg^{807}$, respectively, in M130). One other significant sequence of amino acids, a serine- and threonine-rich sequence, is present at positions $Thr^{770}$–$Ser^{793}$ in M133 ($Thr^{729}$–$Ser^{752}$ in M130).

The sizes and tissue distributions of mRNAs that encode M130/M133 were investigated by Northern blot anaysis. mRNA of 5.7 kb was detected in gizzard (Fig. 6), brain, lung, heart, aorta, stomach, oviduct, spleen, kidney, and small intestine but not in liver or skeletal muscle. There was an additional, fainter band of an mRNA of 10.5 kb that was detected only in gizzard. It is unknown whether this minor species represented a second mRNA for myosin phosphatase or was the result of cross-hybridization with a related gene product.

To compare the PP1δc of myosin phosphatase with the analogous protein from other species, the cDNA for gizzard PP1δc was cloned and sequenced. A chicken gizzard cDNA library was screened with a labeled cDNA fragment (nucleotides 227–882) that encoded rat PP1δc [30], which had been amplified by reverse transcription and the polymerase chain reaction. A cDNA of 1.3 kb was cloned and was selected for sequence analysis. This clone consisted of 1338 bp and had an open reading frame of 981 bp with a 5′-noncoding region of 181 bp and a 3′-noncoding region of 176 bp (Fig. 7). The predicted protein contained 327 amino acid residues with a molecular mass of 37 kDa. The deduced nucleotide sequence in the open reading frame was 86% identical, and the deduced amino acid sequence was completely identical to those of rat PP1δ [30].

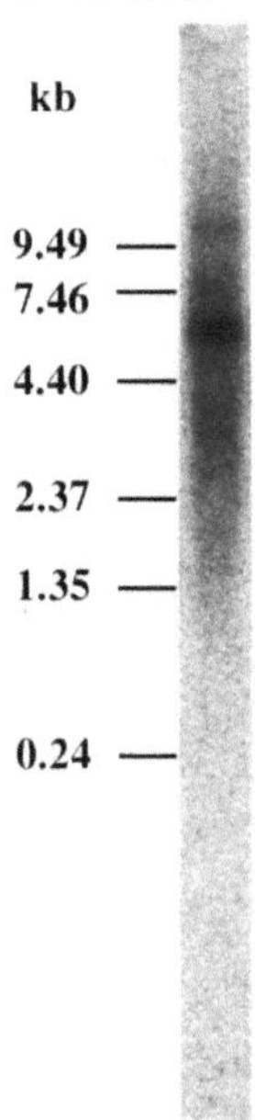


**Fig. 6.** Analysis of chicken gizzard RNA by Northern hybridization. Total RNA (30 μg/lane) prepared from chicken gizzard was subjected to electrophoresis in an agarose gel under denaturing conditions, and bands of RNA were then transferred to a nitrocellulose membrane. Numbers indicate the size of the molecular markers, in kilobases (*kb*). The membrane was hybridized overnight with the $^{32}P$-labeled probe (from clone Z1, 2320 bp) and was washed at high stringency [28]

```
                                                           G    1
CGCGTTCGGCTGGCTGCGTTGTGACGCCGGGCGGCCGGAGGGGACTGCCGGAGGGGGGCG   61
GGGGGAGCCGTCTGCGCCGCGGCCGCCGCTGCTGCTGCCCGCTCCTCCTCCTCCTGCTGC  121
TCGGCGGACGCCCTCCACCCCGCCCTCTCCCCGCCCGGGACCGAGTGTGGTGGCCACAAA  181
ATGGCGGATGGGGAGCTGAACGTGGACAGCCTCATCACCCGGCTGCTGGAAGTACGAGGA  241
 M  A  D  G  E  L  N  V  D  S  L  I  T  R  L  L  E  V  R  G   20
TGTCGTCCTGGGAAGATTGTTCAAATGACAGAAGCAGAAGTTCGGGGCCTGTGCATAAAA  301
 C  R  P  G  K  I  V  Q  M  T  E  A  E  V  R  G  L  C  I  K   40
TCACGAGAAATATTTCTTAGCCAGCCTATTCTCCTGGAACTAGAGGCACCGCTGAAAATC  361
 S  R  E  I  F  L  S  Q  P  I  L  L  E  L  E  A  P  L  K  I   60
 -------------------------
TGTGGTGATATTCATGGTCAGTATACAGATTTGCTTAGACTGTTTGAATATGGAGGATTC  421
 C  G  D  I  H  G  Q  Y  T  D  L  L  R  L  F  E  Y  G  G  F   80
CCACCAGAAGCGAATTATCTTTTCCTTGGAGATTATGTGGACAGAGGCAAACAATCTCTG  481
 P  P  E  A  N  Y  L  F  L  G  D  Y  V  D  R  G  K  Q  S  L  100
GAAACAATTTGCCTACTTCTGGCATATAAAATCAAGTACCCGGAGAACTTCTTTCTCTTA  541
 E  T  I  C  L  L  L  A  Y  K  I  K  Y  P  E  N  F  F  L  L  120
                                     ----------------------
AGAGGAAATCATGAGTGCGCTAGCATCAACCGGATCTATGGATTCTATGATGAATGTAAG  601
 R  G  N  H  E  C  A  S  I  N  R  I  Y  G  F  Y  D  E  C  K  140
 -
CGGAGGTTTAATATTAAGCTCTGGAAGACCTTCACAGACTGTTTTAATTGTCTGCCTATT  661
 R  R  F  N  I  K  L  W  K  T  F  T  D  C  F  N  C  L  P  I  160
                  ----------
GCAGCCATTGTGGATGAGAAGATCTTCTGCTGTCATGGAGGACTGTCTCCAGATCTCCAG  721
 A  A  I  V  D  E  K  I  F  C  C  H  G  G  L  S  P  D  L  Q  180
TCAATGGAGCAGATCCGGAGAATTATGAGACCTACAGATGTTCCTGACACAGGCTTACTA  781
 S  M  E  Q  I  R  R  I  M  R  P  T  D  V  P  D  T  G  L  L  200
TGTGACCTGCTGTGGTCTGACCCAGACAAAGATGTGCAAGGTTGGGGTGAAAATGATCGT  841
 C  D  L  L  W  S  D  P  D  K  D  V  Q  G  W  G  E  N  D  R  220
GGGGTCTCTTTTACTTTCGGTGCTGATGTGGTCAGTAAATTTCTGAATCGTCATGATTTG  901
 G  V  S  F  T  F  G  A  D  V  V  S  K  F  L  N  R  H  D  L  240
GATTTAATCTGTCGAGCTCACCAGGTGGTTGAAGATGGATACGAATTCTTTGCTAAACGG  961
 D  L  I  C  R  A  H  Q  V  V  E  D  G  Y  E  F  F  A  K  R  260
CAGTTGGTCACCCTATTCTCAGCTCCAAATTATTGTGGAGAGTTTGACAATGCAGGTGGC 1021
 Q  L  V  T  L  F  S  A  P  N  Y  C  G  E  F  D  N  A  G  G  280
ATGATGAGTGTGGACGAAACTCTGATGTGTTCCTTTCAGATATTGAAGCCATCTGAAAAG 1081
 M  M  S  V  D  E  T  L  M  C  S  F  Q  I  L  K  P  S  E  K  300
AAAGCCAAGTACCAGTATGGTGGACTGAATTCTGGGCGTCCAGTCACTCCACCTCGCACA 1141
 K  A  K  Y  Q  Y  G  G  L  N  S  G  R  P  V  T  P  P  R  T  320
GCTAATCCTCCGAAGAAGAGGTGAAGAAAGGAATAATGTAGAAAAACCATCAGATCTGTC 1201
 A  N  P  P  K  K  R  *                                       327
AAGGACAAAACTCCATATTACAGTATATAATAAGTGTGCACTGTAAAACCATCCAGCCAT 1261
TTGACACCCTTTATGATGTCACACGTTTAACTTAAAGAGACGGTAAAGGATCTTTATTAA 1321
TTTTTCTAGTAGAAAGA                                            1338
```

**Fig. 7.** Nucleotide and deduced amino acid sequences of chicken gizzard PP1δc. The predicted amino acid sequence is displayed below the nucleotide sequence. Underlining indicates peptide sequences derived from the 38 kDa catalytic subunit. The stop codon is indicated by an asterisk

## Discussion

In the present study we showed that smooth muscle myosin phosphatase is a new holoenzyme of type 1 protein phosphatase. It is composed of three subunits, a 38-kDa PP1δ catalytic subunit and two regulatory subunits of 130/133 and 20 kDa; the 130/133-kDa subunits anchor the catalytic subunit to the target locus (i.e., myosin); the functions of the 20-kDa subunit are not known. These results are consistent with those reported by Alessi et al. [29]. There are many reports describing the purification of phosphatases from smooth muscle, some of which were able to dephosphorylate not only isolated $MLC_{20}$ but also intact myosin or heavy meromyosin [reviews: 25, 31]. Pato and Kerc isolated a myosin phosphatase, designated SMP IV, from turkey gizzard [32]. The 58/38-kDa myosin phosphatase purified from crude actomyosin in the present study resembled SMP-IV, although several properties of our myosin phosphatase were different from those of SMP-IV, namely, the conditions of extraction and metal ion requirements. However, the two phosphatases have identical subunit compositions: Both dephosphorylate myosin or heavy meromyosin in addition to isolated $MLC_{20}$; both bind to myosin; and both are relatively insensitive to I-2. Recently, other myosin phosphatases composed of a single subunit with the ability to bind unphosphorylated myosin were reported [33, 34]. From our results we can conclude that the trimeric phosphatase composed of subunits of 130/133, 38, and 20 kDa was an intact myosin phosphatase associated with actomyosin in smooth muscle, though the possibility of the existence of other types of myosin phosphatase in smooth muscle cannot be ruled out.

From the molecular cloning of the catalytic subunit, it was evident that the amino acid sequence of chicken gizzard PP1δc is identical to that of the rat [30], rabbit [35], and human [36]. This result suggests that these isoenzymes comprise one of the most highly conserved proteins in the animal kingdom. There are at least three major isoforms of the mammalian catalytic subunit of PP1, designated PP1α, PP1γ, and PP1δ (termed PP1β by Dombrádi et al. [35]); the differences among them are found mainly at the C-terminus [30]. These isoforms might form complexes with distinct regulatory subunits to yield different types of holoenzyme, which may explain why the catalytic subunit of myosin phosphatase is the δ isoform. However, all the various recombinant isoforms, after expression in *Escherichia coli*, were reported to be able to associate with both the G subunit and M complexes of PP1 in vitro [37]. The relations between the isoforms of the PP1 catalytic subunit and the different types of regulatory subunit are still obscure, and additional experiments are necessary to clarify these relations.

One of the most striking features of the present myosin phosphatase is its activation by $Co^{2+}$ and $Mn^{2+}$ ions. However, the mechanisms of activation by $Co^{2+}$ ions is not understood, and such a phenomenon is the focus of some controversy [review: 25], the activation seems to be not due to protein modification (e.g., dephosphorylation or degradation), and it might involve a change in the conformation of the catalytic subunit. Another atypical feature of myosin phosphatase as a member of the PP1 family is its relative resistance to I-2, as inhibition by I-2 is one of the criteria for PP1 [38]. Similar characteristics with regard to insensitivity to I-2 were reported previously for several other myosin phosphatases [29, 31, 32]. Because the amino acid sequence of the PP1 catalytic subunit is highly conserved and different recombinant isoforms of the PP1 catalytic subunit have been reported to be inhibited by I-2 to the same extent [37, 39], the resistance of certain myosin phosphatases to I-2 appears to be due to the presence of their respective unique regulatory subunits.

From the results of immunoblotting and molecular cloning it became evident that two isoforms of the 130-kDa regulatory subunit of myosin phosphatase are expressed in gizzard smooth muscle. The larger isoform has an insert of 41 amino acids in the center of the molecule. The sequences of the two isoforms exhibited no identity to known sequences, but some similarities to other proteins were evident. The amino acid sequences of the repeats are aligned for comparison with consensus repeat sequences from other proteins in Figure 5. The characteristic repeats are also present in slightly altered form in human erythroid ankyrin [40]; in products of four invertebrate genes—*Notch* [41], *lin-12* [42], *glp-1* [43], *fem-1* [44]—that regulate tissue differentiation; in three yeast genes—*cdc10* [45], *SWI6* [46], *SWI4* [47]—that have functions related to the cell cycle; and in a host range gene (*hr*) of vaccinia virus [48].

Furthermore, a long region in M130/M133 (from $Asp^{83}$ to $Ser^{881}$ in M133) that includes the *cdc10/SWI6* repeat exhibits homology to region IV of *Notch* protein of *Drosophila* at the 17% level [40]. The definitive function of the repeat is unknown, but this observation strengthens the hypothesis that the repeats are involved in control of cell growth and differentiation. Ankyrin repeats are thought to serve as domains at which integral membrane proteins and cytoskeletal proteins interact. It is suggested that M130/M133 acts to target the catalytic subunit not only to myosin but also to other cytoskeletal proteins. In addition, some regions in M130/M133 exhibit homology to the amino acid sequences of nucleoplasmin [49] and malaria antigen [50]. However, M130/M133 exhibits no significant homology to other PP1-binding proteins, namely, the glycogen-associated protein phosphatase [51], inhibitor-1 [52], inhibitor-2 [53], and the product of the fission yeast *sds22*$^{+}$ gene [54], or to the retinoblastoma protein ($p110^{RB}$) [55].

The 58-kDa fragment of M130/M133 has important functions, as it is involved in binding to both myosin and the catalytic subunit. From the following evidence, it is concluded that the 58-kDa fragment is located at the N-terminal half of the M130/M133 molecules. First, all seven peptides isolated from the 58-kDa fragment purified from chicken gizzard actomyosin were concentrated in the region between residues $K^{286}$ and $S^{467}$. Second, direct sequence analysis of the 58-kDa fragment, digested by either intrinsic gizzard protease or α-chymotrypsin, was not possible because of a blocked amino-terminus. Third, the 58-kDa fragment generated by α-chymotrypsin did not yield a doublet pattern like that of M130/M133 during SDS-PAGE, a result suggesting that the 58-kDa fragment does not include the inserted region $Arg^{512}$–$Arg^{552}$ of M133.

Although the activity of myosin phosphatase is believed to be highly regulated in vivo, the mechanism for such regulation remains to be characterized. One attractive mechanism for this regulation was proposed by Somlyo and his colleagues. They suggested that arachidonic acid, which sensitized smooth muscle to $Ca^{2+}$ ions by inhibiting the dephosphorylation of myosin light chain in situ, dissociated the trimeric myosin phosphatase and inhibited its activity toward heavy meromyosin [56]. Another possible regulatory mechanism might involve the phosphorylation of myosin phosphatase itself. The M130/M133 subunits of chicken gizzard myosin phosphatase can potentially undergo phosphorylation at several sites by cAMP-dependent protein kinase, protein kinase C, $p34^{cdc2}$, or glycogen synthase kinase-3. It is suggested that the activity of myosin phosphatase is regulated via the phosphorylation of M130/M133 by these kinases. Our preliminary evidence indicates that the M130/M133 and the 20-kDa subunit can be phosphorylated in vitro by cAMP-dependent protein kinase, but the effects of such phosphorylation, if any, are not known.

The phosphates related to M130/M133 were observed to be widely distributed in many tissues, excluding skeletal muscle and liver, by both immunoblot and Northern blot analyses. Recently, the myosin-targeting subunit of skeletal muscle myosin phosphatase was reported to be distinct from that of the smooth muscle enzyme [57]. The wide distribution of M130/M133 suggests that this group of phosphatases might have other functions and that their activities are not confined to the regulation of smooth muscle contraction.

*Acknowledgments.* This work was supported in part by Grants-in-Aid for Scientific Research and for Cancer Research from the Ministry of Education, Science and Culture of Japan, by grants from the Japanese Foundation for Cardiovascular Research, the Kato Memorial Bioscience Foundation, the Kanae Foundation of Research for New Medicine, the Study Group of Molecular Cardiology and Cell Signaling Project in Mie University (to M.I. and T.N.), and by grants HL 23615 and HL 20984 from the US National Institutes of Health (to D.J.H.).

## References

1. Hartshorne DJ (1987) Biochemistry of the contractile process in smooth muscle. In: Johnson LR (ed) Physiology of the gastrointestinal tract, 2nd edn, vol I. Raven Press, New York, pp 423–482
2. Morgan JP, Morgan KG (1984) Stimulus-specific patterns of intracellular calcium levels in smooth muscle of ferret portal vein. J Physiol (Lond) 351:155–167
3. Somlyo AP, Himpens B (1989) Cell calcium and its regulation in smooth muscle. FASEB J 3:2266–2276
4. Rembold CM, Murphy RA (1988) Myoplasmic [$Ca^{2+}$] determines myosin phosphorylation in agonist-stimulated swine arterial smooth muscle. Circ Res 63:593–603
5. Karaki H (1989) $Ca^{2-}$ localization and sensitivity in vascular smooth muscle. Trends Pharmacol Sci 10:320–325
6. Adelstein RS, Conti MA, Hathaway DR, Klee CB (1978) Phosphorylation of smooth muscle myosin light chain kinase by the catalytic subunit of adenosine 3′:5′-monophosphate-dependent protein kinase. J Biol Chem 253:8347–8350
7. Ikebe M, Inagaki M, Kanamaru K, Hidaka H (1985) Phosphorylation of smooth muscle myosin light chain kinase by $Ca^{2+}$-activated, phospholipid-dependent protein kinase. J Biol Chem 260:4547–4550
8. Nishikawa M, Shirakawa S, Adelstein RS (1985) Phosphorylation of smooth muscle myosin light chain kinase by protein kinase C: comparative study of the phosphorylated sites. J Biol Chem 260:8978–8983
9. Stull JT, Hsu LC, Tansey MG, Kamm KE (1990) Myosin light chain kinase phosphorylation in tracheal smooth muscle. J Biol Chem 265:16683–16690
10. Tansey MG, Word RA, Hidaka H, Singer HA, Schworer CM, Kamm KE, Stull JT (1992) Phosphorylation of myosin light chain kinase by the multifunctional calmodulin-dependent protein kinase II in smooth muscle cells. J Biol Chem 267:12511–12516
11. Tansey MG, Luby-Phelps K, Kamm KE, Stull JT (1994) $Ca^{2+}$-dependent phosphorylation of myosin light chain kinase decreases the $Ca^{2+}$ sensitivity of light chain phosphorylation within smooth muscle cells. J Biol Chem 269:9912–9920
12. Kitazawa T, Kobayashi S, Horiuti K, Somlyo AV, Somlyo AP (1989) Receptor-coupled, permeabilized smooth muscle: role of the phosphatidylinositol cascade, G-proteins and modulation of the contractile response to $Ca^{2+}$. J Biol Chem 264:5339–5342
13. Fujiwara T, Itoh T, Kubota Y, Kuriyama H (1989) Effects of guanosine nucleotides on skinned smooth muscle tissues of the rabbit mesenteric artery. J Physiol (Lond) 408:535–547
14. Kitazawa T, Somlyo AP (1990) Desensitization and muscarinic re-sensitization of force and myosin light chain phosphorylation to cytoplasmic $Ca^{2+}$ in smooth muscle. Biochem Biphys Res Commun 172:1291–1297
15. Nishimura J, Khalil RA, Drenth JP, van Breemen C (1990) Evidence for increased myofilament $Ca^{2+}$ sensitivity in norepinephrine-activated vascular smooth muscle. Am J Physiol 259:H2–H8
16. Kitazawa T, Masuo M, Somlyo AP (1991) G protein-mediated inhibition of myosin light-chain phosphatase in vascular smooth muscle. Proc Natl Acad Sci USA 88:9307–9310
17. Kitazawa T, Gaylinn BD, Denney GH, Somlyo AP (1991) G-protein-mediated $Ca^{2+}$ sensitization of smooth muscle contraction through myosin light chain phosphorylation. J Biol Chem 266:1708–1715
18. Kubota Y, Nomura M, Kamm KE, Mumby MC, Stull JT (1992) GTPγS-dependent regulation of smooth muscle contractile elements. Am J Physiol 262:C405–C410
19. Itoh T, Kanmura Y, Kuriyama H, Simimoto K (1986) A phorbol ester has dual actions on the mechanical response in the rabbit mesenteric and porcine coronary arteries. J Physiol (Lond) 375:515–534

20. Chatterjee M, Tejada M (1986) Phorbor ester-induced contraction in chemically skinned vascular smooth muscle. Am J Physiol 251:C356–C361
21. Fujiwara T, Itoh T, Kubota Y, Kuriyama H (1988) Actions of a phorbor ester on factors regulating contraction in rabbit mesenteric artery. Circ Res 63:893–902
22. Sato K, Hori M, Ozaki H, Takano-Ohmuro H, Tsuchiya T, Sugi H, Karaki H (1992) Myosin phosphorylation-independent contraction induced by phorbol ester in vascular smooth muscle. J Pharmacol Exp Ther 261:497–505
23. Itoh H, Shimomura A, Okubo S, Ichikawa K, Ito M, Konishi T, Nakano T (1993) Inhibition of myosin light chain phosphatase during $Ca^{2+}$-independent vasocontraction. Am J Physiol 265:C1319–C1324
24. Masuo M, Reardon S, Ikebe M, Kitazawa T (1994) A novel mechanism for the $Ca^{2+}$-sensitizing effect of protein kinase C on vascular smooth muscle: inhibition of myosin light chain phosphatase. J Gen Physiol 104:265–286
25. Okubo S, Erdödi F, Ito M, Ichikawa K, Konishi T, Nakano T, Kawamura T, Brautigan DL, Hartshorne DJ (1993) Characterization of a myosin-bound phosphatase from smooth muscle. Adv Protein Phosphatases 7:295–314
26. Sellers JR, Pato MD (1984) The binding of smooth muscle myosin light chain kinase and phosphatases to actin and myosin. J Biol Chem 259:7740–7746
27. Okubo S, Ito M, Takashiba Y, Ichikawa K, Miyahara M, Shimizu H, Konishi T, Shima H, Nagao M, Hartshorne DJ, Nakano T (1994) A regulatory subunit of smooth muscle myosin bound phosphatase. Biochem Biophys Res Commun 200:429–434
28. Shimizu H, Ito M, Miyahara M, Ichikawa K, Okubo S, Konishi T, Naka M, Tanaka T, Hirano K, Hartshorne DJ, Nakano T (1994) Characterization of the myosin-binding subunit of smooth muscle myosin phosphatase. J Biol Chem 269:30407–30411
29. Alessi D, MacDougall LK, Sola MM, Ikebe M, Cohen P (1992) The control of protein phosphatase-1 by targetting subunits: the major myosin phosphatase in avian smooth muscle is a novel form of protein phosphatase-1. Eur J Biochem 210:1023–1035
30. Sasaki K, Shima H, Kitagawa Y, Irino S, Sugimura T, Nagao M (1990) Identification of members of the protein phosphatase 1 gene family in the rat and enhanced expression of protein phosphatase 1α gene in rat hepatocellular carcinomas. Jpn J Cancer Res 81:1272–1280
31. Mitsui T, Inagaki M, Ikebe M (1992) Purification and characterization of smooth muscle myosin-associated phosphatase from chicken gizzards. J Biol Chem 267:16727–16735
32. Pato MD, Kerc E (1985) Purification and characterization of a smooth muscle myosin phosphatase from turkey gizzards. J Biol Chem 260:12359–12366
33. Graham Tulloch A, Pato MD (1991) Turkey gizzard smooth muscle myosin phosphatase-III is a novel protein phosphatase. J Biol Chem 266:20168–20174
34. Nowak G, Rainer F, Sobieszek A (1993) Purification and characterization of the myofibrillar form of myosin light-chain phosphatase from turkey gizzard smooth muscle. Biochim Biophys Acta 1203:230–235
35. Dombrádi V, Axton JM, Brewis ND, da Cruz e Silva EF, Alphey L, Cohen PTW (1990) *Drosophila* contains three genes that encode distinct isoforms of protein phosphatase 1. Eur J Biochem 194:739–745
36. Barker HM, Brewis ND, Street AJ, Spurr NK, Cohen PTW (1994) Three genes for protein phosphatase 1 map to different human chromosomes: sequence, expression and gene localization of protein serine/threonine phosphatase 1 beta (PPP1CB). Biochim Biophys Acta 1220:212–218
37. Alessi DR, Street AJ, Cohen P, Cohen PTW (1993) Inhibitor-2 functions like a chaperone to fold three expressed isoforms of mammalian protein phosphatase-1 into a conformation with the specificity and regulatory properties of the native enzyme. Eur J Biochem 213:1055–1066
38. Cohen P (1989) The structure and regulation of protein phosphatases. Annu Rev Biochem 58:453–508
39. Zhang Z, Bai G, Shima H, Zhao S, Nagao M, Lee EYC (1993) Expression and characterization of rat protein phosphatase-1α, -1γ1, -1γ2, and -1δ. Arch Biochem Biophys 303:402–406
40. Lux SE, John KM, Bennett V (1990) Analysis of cDNA for human erythocyte ankyrin indicates a repeated structure with homology to tissue-differentiation and cell-cycle control proteins. Nature 344:36–42

41. Wharton KA, Johansen KM, Xu T, Artavanis-Tsakonas S (1985) Nucleotide sequence from the neurogenic locus Notch implies a gene product that shares homology with proteins containing EGF-like repeats. Cell 43:567–581
42. Yochem J, Weston K, Greenwald I (1988) The *Caenorhabditis elegans lin-12* gene encodes a transmembrane protein with overall similarity to *Drosophila Notch.* Nature 335:547–550.
43. Yochem J, Greenwald I (1989) *glp-1* and *lin-12*, genes implicated in distinct cell-cell interactions in C. elegans, encode similar transmembrane proteins. Cell 58:553–563
44. Spence AM, Coulson A, Hodgkin J (1990) The product of *fem-1*, a nematode sex-determining gene, contains a motif found in cell cycle control proteins and receptors for cell-cell interactions. Cell 60:981–990
45. Aves SJ, Durkacz BW, Carr A, Nurse P (1985) Cloning, sequencing and transcriptional control of the *Schizosaccharomyces pombe cdc10* 'start' gene. EMBO J 4:457–463
46. Breeden L, Nasmyth K (1987) Similarity between cell-cycle genes of budding yeast and fission yeast and the *Notch* gene of *Drosophila.* Nature 329:651–654
47. Andrews BJ, Herskowitz I (1989) Identification of a DNA binding factor involved in cell-cell control of the yeast *HO* gene. Cell 57:21–29
48. Gillard S, Spehner D, Drillien R, Kirn A (1986) Localization and sequence of a vaccinia virus gene required for multiplication in human cells. Proc Natl Acad Sci USA 83:5573–5577
49. Schmidt-Zachmann MS, Hügle-Dörr B, Franke WW (1987) A constitutive nucleolar protein identified as a member of the nucleoplasmin family. EMBO J 6:1881–1890
50. Weber JL, Lyon JA, Wolff RH, Hall T, Lowell GH, Chulay JD (1988) Primary structure of a *Plasmodium falciparum* malaria antigen located at the merozoite surface and within the parasitophorous vacuole. J Biol Chem 263:11421–11425
51. Tang PM, Bondor JA, Swiderek KM, DePaoli-Roach AA (1991) Molecular cloning and expression of the regulatory (RG1) subunit of the glycogen-associated protein phosphatase. J Biol Chem 266:15782–15789
52. Elbrecht A, DiRenzo J, Smith RG, Shenolikar S (1990) Molecular cloning of protein phosphatase inhibitor-1 and its expression in rat and rabbit tissues. J Biol Chem 265:13415–13418
53. Holmes CFB, Campbell DG, Caudwell FB, Aitken A, Cohen P (1986) The protein phosphatases involved in cellular regulation: primary structure of inhibitor-2 from rabbit skeletal muscle. Eur J Biochem 155:173–182
54. Ohkura H, Yanagida M (1991) S. pombe gene *sds22*$^{+}$ essential for a midmitotic transition encodes a leucine-rich repeat protein that positively modulates protein phosphatase-1. Cell 64:149–157
55. Durfee T, Becherer K, Chen PL, Yeh SH, Yang Y, Kilburn AE, Lee WH, Elledge SJ (1993) The retinoblastoma protein associates with the protein phosphatase type 1 catalytic subunit. Genes Dev 7:555–569
56. Gong MC, Fuglsang A, Alessi D, Kobayashi S, Cohen P, Somlyo AV, Somlyo AP (1992) Arachidonic acid inhibits myosin light chain phosphatase and sensitizes smooth muscle to calcium. J Biol Chem 267:21492–21498
57. Dent P, MacDougall LK, Mackintosh C, Campbell DG, Cohen P (1992) A myofibrillar protein phosphatase from rabbit skeletal muscle contains the β isoform of protein phosphatase-1 complexed to a regulatory subunit which greatly enhances the dephosphorylation of myosin. Eur J Biochem 210:1037–1044

# Structure and Expression of Vascular Smooth Muscle Calponin

Katsuhito Takahashi, Hiroaki Masuda, Michiko Takagi, Yoshimi Eshita, and Nobuhiko Shibata

*Summary.* cDNA clones encoding vascular smooth muscle calponin have been isolated from a cDNA library constructed from human and rat aortic smooth muscle cells (SMCs). The $NH_2$-terminal region of calponin shares homology with a putative negative regulatory domain of the GDP-GTP exchanger for Rho-like GTP-binding proteins. Expression of calponin is down-modulated in proliferating SMCs of atherosclerotic neointima and injured arterial media. Direct transfer of the recombinant human calponin gene into rabbit balloon-injured arteries in vivo increased calponin expression by SMCs and suppressed arterial myointimal hyperplasia. The cellular distribution of calponin has been shown to be dramatically changed upon stimulation with an agonist that is known to activate protein kinase C in vascular SMCs. In addition to regulation of the contraction–relaxation cycle of smooth muscle, the findings reviewed here suggest that calponin might also be involved in the control of noncontractile function of SMCs, such as adhesion, migration, and proliferation.

*Key Words.* Vascular smooth muscle—Calponin—SM22

## Introduction

Calponin is an actin-, tropomyosin-, and calmodulin-binding protein [1–14] that is implicated in the auxiliary regulatory role of smooth muscle contraction. The association of calponin with the actin filament results in inhibition of actin-activated myosin MgATPase activity [6, 15–23], which is mainly due to a decrease in the cycling rate of myosin ATPase [6, 17, 19]. Suppression of the actin–myosin interaction by calponin has also been demonstrated in the sliding actin in vitro motility assay system [21, 23], in which calponin binding to actin results in inhibition of the movement of actin by either phosphorylated smooth muscle myosin or rabbit skeletal muscle heavy meromyosin. Studies using the in vitro motility assay have shown that calponin inhibits the rate of dissociation of the high-affinity actomyosin complex and consequently increases maximum force production by smooth muscle myosin [23]. These observations suggest that calponin may be an integral component of the latch-state in mammalian smooth muscle [23].

---

Department of Medicine, Center for Adult Diseases, 1-3-3 Nakamichi, Higashinari-ku, Osaka, 537 Japan

In this review, we describe recent findings on the structural and expressional characterization of vascular smooth muscle calponin.

## Calponin Structure

The human and rat calponin α cDNA, encoding a polypeptide of 297 amino acids with 97% similarity ($M_r$ 33161 daltons for humans and 33326 daltons for rats), were isolated from aortic smooth muscle, cloned, and sequenced [24–26]. The alignment of the predicted amino acid sequences indicates that although substitutions occur throughout the structure there are highly homologous regions among the human, rat, and chicken calponins [5]. They are between $Phe^{6}$ and $Asp^{27}$, $Trp^{37}$ and $Val^{72}$, $Tyr^{98}$ and $Ser^{129}$, $Lys^{156}$ and $Glu^{236}$, and $Leu^{245}$ and $Cys^{273}$ of the rat and human calponin sequence. Within residues 162–236 and 245–273, there are three internal repeats (see Fig. 1, below). Each repeat contains 29–31 amino acids with 62%–68% identity, which are well conserved with the chicken calponin sequence [5] (89 identical residues and 4 conservative substitutions in a 93-amino-acid overlap). The corresponding DNA sequences are also highly similar between each internal repeat (71%–77% nucleotide identity) as well as among human, rat, and chicken calponins (88%–90% nucleotide identity), indicating that the calponin repeats are presumably formed by successive gene duplications and may play an essential role in its function. A part of the putative actin-binding sequence identified in the chicken gizzard calponin sequence, sharing an actin-binding consensus with various actin-binding proteins [27], is conserved in the rat and human calponin sequence between residues $Val^{142}$ and $Tyr^{182}$.

The cDNA clones encoding a novel acidic calponin isoform have been isolated from rat aortic vascular smooth muscle cells [28]. The amino-terminal 273 residues of the deduced 330-amino-acid polypeptide ($M_r$ 36377 daltons) of the acidic calponin isoform are highly homologous to basic smooth muscle calponin isoforms, but the additional 57 residues at the carboxyl-terminus comprise a strongly acidic domain [28]. The acidic calponin has been shown to be expressed in both cultured rat aortic vascular smooth muscle cells and nonmuscle tissues of adult rat [28, 29]. Furthermore, based on sequences of mouse and pig cDNA clones isolated from uterus smooth muscle cells, at least two isoforms of basic calponin, termed h1 and h2, have been identified [30]. The calponin h1 isoform is the equivalent of the calponin α. The divergence between sequences of the acidic calponin and either the h1(α) and h2 calponin indicates that each of the three forms is encoded by three distinct genes.

As noted in the sequence of chicken gizzard calponin [5], the amino acid sequences of human and rat aortic calponin share homology with SM22 proteins [31] (44% identical and 69% conservative residues in a 186-amino-acid overlap between human aortic calponin α and human SM22 [24]). SM22 is the equivalent of a product of the human gene WS3-10, which is overexpressed in senescent human fibroblasts derived from patients with Werner syndrome [32]. SM22 is homologous to NP25, a novel protein that is differentially expressed by neuronal subpopulations of the rat central nervous system [33]. Inspection of the aligned sequences of calponin and SM22 indicates that, although incomplete, both the predicted human and chicken SM22 sequences [31] also contain a characteristic motif of the calponin repeat with 27 (human SM22) or 24 (chicken SM22) amino acids in length at the carboxyl-terminus. Twenty of the twenty-five positions in an overlap between the sequences of human and chicken SM22 and those of the each calponin repeat are well conserved (see

Fig. 1, below). They form a signature consensus sequence, I(G/S)LQMG(T/S) NKGASQ-GMT-YG-RQ- [single-letter amino acid code]; residues that contain more than three amino acids at the same position are marked by hyphens. Interestingly, as noted by Vancompernolle et al. [34], this consensus motif is strikingly similar to a consensus sequence in the repeats of the *unc*-87 gene product of *Caenorhabditis elegans* (a 357-amino acid peptide with $M_r$ 39 759 daltons [35, 26]) that is located in the I-bands of the body wall skeletal muscle and composed of seven repeats of the 41–53 amino acid segment [35, 36]. The null mutants of the *unc*-87 gene (unc-87, st1005) show paralysis and disorganization of the thin filaments [37], suggesting that those proteins containing the characteristic motif of calponin's repeating structure may be members of an emerging new family of muscle and nonmuscle proteins involved in the regulation and organization of the actin filaments.

The $NH_2$-terminal half of the cloned aortic calponin molecules show extensive similarities with the $NH_2$-terminal region of the *vav* proto-oncogene product [24, 38] (Figs. 1 and 2). Over the $NH_2$-terminal 89 amino acid residues of the *vav* proto-oncogene product (residues 32–121 of the human Vav sequence), there are 36% identity and 62% similarity to the human and rat aortic calponin α sequence. The *vav* gene was originally identified as a locus that induced transformation of an NIH 3T3

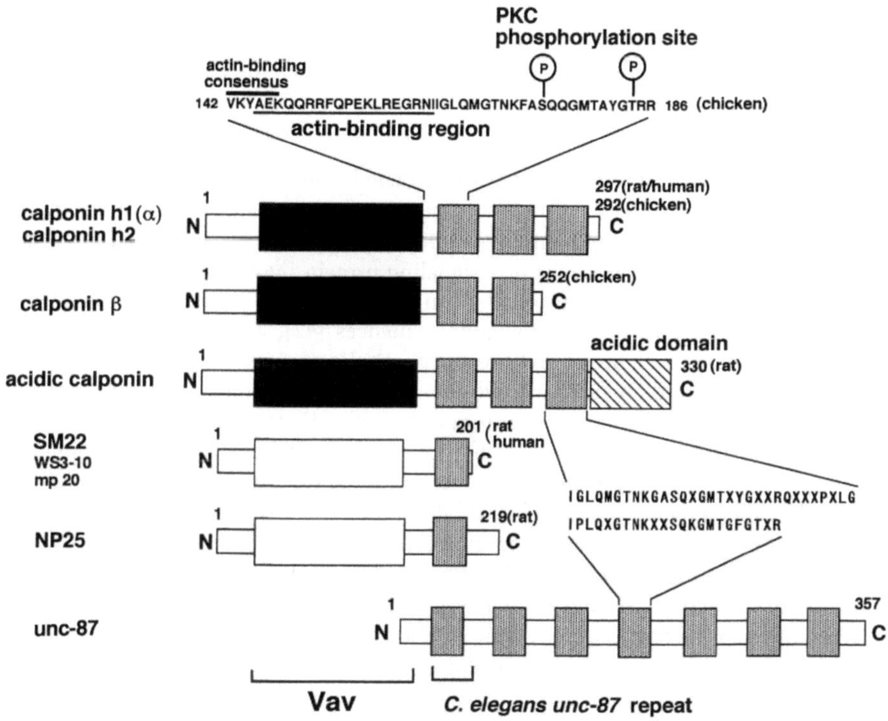


**Fig. 1.** Domain structure of calponin, SM22, and homologous proteins. The amino acid sequence in the actin-binding region described [66] is from that of chicken gizzard calponin (residues 142–186). SM22 also shares homology to the $NH_2$-terminal region of the Vav and CDC24 proteins [38]. A signature consensus sequence among calponin and *C. elegans unc-87* gene product [35, 36] is indicated. Possible phosphorylation sites by calcium/calmodulin-dependent protein kinase II or protein kinase C [39–41] are concentrated in this repeating structure as indicated in the sequence (*P*)

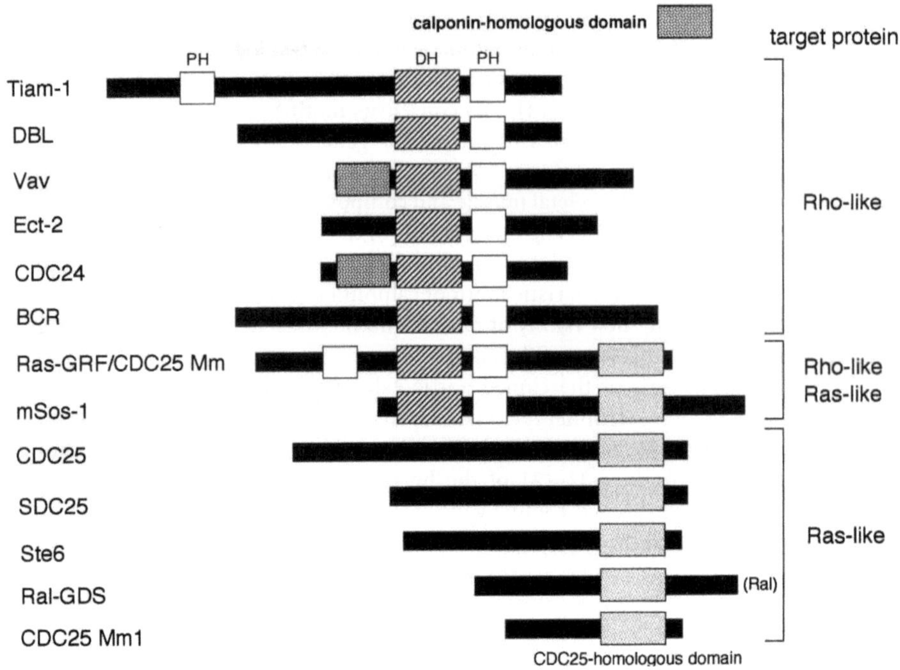


**Fig. 2.** Domain structure of the Rho GDP-GTP exchange factor [42]. *DH*, Dbl-homologous domain (*hatched box*); *PH*, pleckstrin-homologous domain (*open box*)

mouse fibroblast by transfection of DNA extracted from a human esophageal carcinoma [43–45]. The *vav* proto-oncogene product, specifically expressed in hematopoietic cells, is an 845-amino-acid peptide (apparent $M_r$ 95000 daltons) with characteristic motifs reminiscent of a signaling protein [46–50] participating in the catalysis of guanine nucleotide exchange on the Rho-like GTP-binding protein: Dbl-homologous and Pleckstrin-homologous sequences [42], a cysteine-rich sequence homologous to the zinc finger motif or the phorbol ester-binding domain of protein kinase C [44, 45], and one SH2 and two SH3 domains (*src* homology region) [51]. Of particular interest is the finding that removal of the $NH_2$-terminal 67 residues of the Vav protein, which contains a region with the highest homology to calponin, is sufficient to activate the transforming potential of the *vav* proto-oncogene [44, 45]. It is possible that a region homologous to calponin on the Vav sequence functions as a negative regulatory domain on the guanine nucleotide exchange activity by the Db1- and Pleckstrin-homologous domain [42, 52]. Furthermore, the $NH_2$-terminal portion of calponin is similar to that of CDC24 of budding yeast, an essential cell division cycle gene that is required for cytoskeletal organization at the site at which budding occurs [53, 54]. Functional complementation studies as well as in vitro biochemical analysis indicate that CDC24 catalyzes the dissociation of GDP from CDC42, a yeast Rho-like protein, and thereby qualifies as a highly selective guanine nucleotide exchange factor for the Rho subfamily of Ras-like small GTP-binding proteins [54]. Although cellular functions of the GDP-GTP exchange factors for Rho-like proteins in smooth muscle regulation remain to be elucidated, the extensive sequence similarities between calponin and a putative negative regulatory domain of Vav or CDC24 proteins suggest that calponin might also be coupled to a cellular signaling pathway that inhibits the activities of the Rho-like GTP-binding proteins. Important in this connection is the finding that the GTP-bound form of Rho augments calcium sensitivity of the smooth muscle contractile apparatus [55]. It is also reported that Tiam-1, a member of the GDP-GTP exchange protein family for Rho-like proteins, is involved in inducing invasive properties of tumor cells [42]. Because invasion and proliferation of smooth muscle cells from the arterial media to the intima is an underlying mechanism for atherogenesis or restenosis after balloon injury [56], it is interesting to study the expression of vascular smooth muscle calponin in normal and diseased arteries.

## Expression of Vascular Smooth Muscle Calponin in Normal and Diseased Arteries

Calponin is normally expressed by differentiated smooth muscle cells (SMCs). When cultured, quiescent SMCs are stimulated by mitogen, calponin expression is rapidly down-regulated at both mRNA and protein levels [7, 11, 26, 57, 58], and the extent of down-regulation has been correlated with a proliferating SMC phenotype in vitro. Analysis of the spatiotemporal distribution of calponin expression during avian embryonic development indicates that the calponin gene belongs to a group of late differentiation determinants in smooth muscles [58]. These findings imply that down-modulation of the calponin gene could be a sensitive indicator for detecting phenotypic modulation of smooth muscle cells in the diseased arteries. Using an anti-calponin polyclonal antibody [59, 60] and a calponin-specific cRNA as probes [61], expression of the calponin gene in atherosclerotic human arteries and balloon-injured rabbit carotid arteries was examined. In the early stage of the atherosclerotic lesions with little or no intimal hyperplasia, most of the medial SMCs that express the smooth muscle-specific $\alpha$-actin isoform at high levels also express immunoreactive calponin. SMCs that undergo proliferation and migration into the intima, however, express $\alpha$-actin but not immunoreactive calponin. In human atherectomized lesions freshly excised from coronary and femoral arteries, calponin expression was down-regulated in most intimal $\alpha$-actin- and SM22-positive SMCs that underwent migration and proliferation [62, 63].

Using a rabbit model of vascular injury, we found that calponin expression in the medial SMCs was down-regulated early (days 3–14) after balloon injury, concomitant with an increase in the number of proliferating SMCs [64; unpublished data]. It results in significantly increased intimal thickening within 2 weeks. At 6 weeks after balloon angioplasty, SMC proliferation and calponin expression in both intimal and medial smooth muscle cells had returned to control levels.

By direct transfer of the human calponin gene into the cells and arteries of live animals, we have shown that a constitutive calponin expression decreased the rate of migration and proliferation of stimulated SMCs. Liposome-mediated delivery of the recombinant human calponin gene into rabbit balloon-injured arteries in vivo increased calponin expression by the SMCs and suppressed arterial myointimal hyperplasia [65]. The results indicate that calponin plays a role in the control of SMC proliferation and migration and suggest that a reduction of calponin in the arterial media might be responsible for the initiation of neointimal SMC hyperplasia. Genetic restoration of calponin in the artery wall may thus be of potential therapeutic value for cardiovascular disease.

## Intracellular Expression of Calponin in Contractile Vascular Smooth Muscle Cells

Morgan and her colleagues have used immunofluorescence and digital imaging microscopy to determine the cellular distribution of calponin in single vascular SMCs [14]. In resting cells calponin is distributed throughout the cytosol, associated with filamentous actin structures. Upon stimulation with an agonist that is known to activate protein kinase C in these cells, however, these authors found that the calponin distribution changed from a primarily cytosolic one to one that is primarily surface cortex-associated. Interestingly, this agonist-induced redistribution of calponin was not observed when the cell was stimulated with depolarization, and was partially inhibited by the protein kinase C inhibitor calphostin. Furthermore, the redistribution of calponin occurred simultaneously with protein kinase C translocation and preceded contraction of the cells. Although the sites of relocalization for calponin are currently uncertain, these findings suggest two interesting possibilities: (1) The physiological function of calponin may be to mediate agonist-activated contraction via a protein kinase C-dependent pathway [39–41]. (2) In addition to the inhibition of actomyosin ATPase activity, calponin might be involved in the regulation of cortical actin cytoskeleton, such as modulation of the cell surface signal transduction or cell–matrix or cell–cell adhesion properties.

## References


1. Takahashi K, Hiwada K, Kokubu T (1986) Isolation and characterization of a 34000-dalton calmodulin- and F-actin-binding protein from chicken gizzard smooth muscle. Biochem Biophys Res Commun 141:20–26
2. Takahashi K, Hiwada K, Kokubu T (1987) Occurrence of anti-gizzard p34K antibody cross-reactive components in bovine smooth muscles and non-smooth muscle tissues. Life Sci 41:291–296
3. Takahashi K, Hiwada K, Kokubu T (1988) Vascular smooth muscle calponin: a novel troponin T-like protein. Hypertension 11:620–626
4. Takahashi K, Abe M, Hiwada K, Kokubu T (1988) A novel troponin T-like protein (calponin) in vascular smooth muscle: interaction with tropomyosin paracrystals. J Hypertens 6:S40–S43
5. Takahashi K, Nadal-Ginard B (1991) Molecular cloning and sequence analysis of smooth muscle calponin. J Biol Chem 266:13284–13288
6. Nishida W, Abe M, Takahashi K, Hiwada K (1990) Do thin filaments of smooth muscle contain calponin? A new method for the preparation. FEBS Lett 268:165–168
7. Gimona M, Herzog M, Vandekerckhove J, Small JV (1990) Smooth muscle specific expression of calponin. FEBS Lett 274:159–162
8. Takeuchi K, Takahashi K, Abe M, Nishida W, Hiwada K, Nabeya T, Maruyama K (1991) Co-localization of immunoreactive forms of calponin with actin cytoskeleton in platelets, fibroblasts, and vascular smooth muscle. J Biochem (Tokyo) 109:311–316
9. Lehman W (1991) Calponin and the composition of smooth muscle thin filaments. J Muscle Res Cell Motil 12:221–224
10. Marston SB (1991) Properties of calponin isolated from sheep aorta thin filaments. FEBS Lett 292:179–182
11. Birukov KG, Stepanova OV, Nanaev AK, Shrinsky VP (1991) Expression of calponin in rabbit and human aortic smooth muscle cells. Cell Tissue Res 31:16–21
12. Walsh MP, Camichael JD, Kargacin GJ (1993) Characterization and confocal imaging of calponin in gastrointestinal smooth muscle. Am J Physiol 265:C1371–C1378
13. North AJ, Gimona M, Cross RA, Small JV (1994) Calponin is localized in both the contractile apparatus and the cytoskeleton of smooth muscle cells. J Cell Sci 107:437–444

14. Parker CA, Takahashi K, Tao T, Morgan KG (1994) Agonist-induced redistribution of calponin in contractile vascular smooth muscle cells. Am J Physiol 267:C1262–C1270
15. Winder SJ, Walsh MP (1990) Smooth muscle calponin. J Biol Chem 265:10148–10155
16. Abe M, Takahashi K, Hiwada K (1990) Effect of calponin on actin-activated myosin ATPase activity. J Biochem (Tokyo) 108:835–838
17. Horiuchi KY, Chacko S (1991) The mechanism for the inhibition of actin-activated ATPase of smooth muscle heavy meromyosin by calponin. Biochem Biophys Res Commun 176:1487–1493
18. Makuch R, Birukov K, Shrinsky V, Dabrowska R (1991) Functional interrelationship between calponin and caldesmon. Biochem J 280:33–38
19. Miki M, Walsh MP, Hartshorne DJ (1992) The mechanism of inhibition of the actin-activated myosin MgATPase by calponin. Biochem Biophys Res Commun 187:867–871
20. Winder SJ, Sutherland C, Walsh MP (1992) A comparison of the effects of calponin on smooth and skeletal muscle actomyosin systems in the presence and absence of caldesmon. Biochem J 288:733–739
21. Shrinsky V, Birukov KG, Hettasch JM, Sellers JR (1992) Inhibition of the relative movement of actin and myosin by caldesmon and calponin. J Biol Chem 267:15886–15892
22. Gong B-J, Mabuchi K, Takahashi K, Nadal-Ginard B, Tao T (1993) Characterization of wild type and mutant chicken gizzard alpha calponin expressed in E. coli. J Biochem 114:453–456
23. Haeberle JR (1994) Calponin decreases the rate of cross-bridge cycling and increases maximum force production by smooth muscle myosin in an in vitro motility assay. J Biol Chem 269:12424–12431
24. Takahashi K, Nadal-Ginard B (1991) Human vascular smooth muscle calponin and SM22 share homolgy with myc and vav proto-oncogene products: structural characterization and functional implications. Circulation 84:II–532
25. Takahashi K, Takagi M, Kojima A, Eshita Y, Nakai M, Ohgami K, Nara Y, Yamori Y, Shibata N (1993) Cloning and expression of rat vascular smooth muscle calponin: homology to the vav proto-oncogene product and differential expression in the hypertrophic versus hyperplastic smooth muscle cells of the spontaneously hypertensive rat. J Mol Cell Cardiol 25:S32
26. Shanahan CM, Weissberg PL, Metcalf JC (1993) Isolation of gene markers of differentiated and proliferating vascular smooth muscle cells. Circ Res 73:193–204
27. Mezgueldi M, Fattoum A, Derancourt J, Kassab R (1992) Mapping of the functional domains in the amino-terminal region of calponin. J Biol Chem 267:15943–15951
28. Applegate D, Feng W, Green RS, Taubman MB (1994) Cloning and expression of a novel acidic calponin isoform from rat aortic vascular smooth muscle. J Biol Chem 269: 10683–10690
29. Trabelsi-Terzidis H, Fattoum A, Represa A, Dessi F, Ben-Ari Y, Der Terrossian E (1995) Expression of an acidic isoform of calponin in rat brain: western blots on one- or two-dimensional gels and immuno-localization in cultured cells. Biochem J 306:211–215
30. Strasser P, Gimona M, Moessler H, Herzog M, Small JV (1993) Mammalian calponin: identification and expression of genetic variants. FEBS Lett 330:13–18
31. Pearlstone JR, Weber M, Lees-Miller JP, Carpenter MR, Smillie LB (1987) Amino acid sequence of SM22alpha from the chicken gizzard smooth muscle. J Biol Chem 262:5985
32. Thweatt R, Lumpkin CK, Goldstein S (1992) A novel gene encoding a smooth muscle protein is overexpressed in senescent human fibroblasts. Biochem Biophys Res Commun 187:1–7
33. Ren W-Z, Ng GYK, Wang R-X, Wu PH, O'Dowd BF, Osmond DH, George SR, Liew C-C (1994) The identification of NP25: a novel protein that is differentially expressed by neuronal subpopulations. Mol Brain Res 22:173–185
34. Vancompernolle K, Gimona M, Herzog M, Van Damme J, Vandekerckhove J, Small JV (1990) Isolation and sequence of a tropomyosin-binding fragment of turkey gizzard calponin. FEBS Lett 274:146–150
35. Goetink S, Waterston RH (1994) The caenorhabditis elegans UNC-87 protein is essential for maintenance, but not assembly, of bodywall muscle. J Cell Biol 127:71–78
36. Goetink S, Waterston RH (1994) The Caenorhabditis elegans mucle-affecting gene unc-87 encodes a novel thin filament associated protein. J Cell Biol 127:79–93
37. Waterston RH, Thomson JN, Brenner S (1980) Mutants with altered muscle structure in Caenorhabditis elegans. Dev Biol 77:271–302

38. Adams JM, Houston H, Allen J, Lints T, Harvey R (1992) The hematopoietically expressed vav proto-oncogene shares homology with the dbl GDP-GTP exchange factor, the bcr gene and a yeast gene (CDC24) involved in cytoskeletal organization. Oncogene 7:611–618
39. Winder SJ, Walsh MP (1990) Structural and functional characterization of calponin fragments. Biochem Int 22:335–341
40. Naka M, Kureishi Y, Muroga Y, Takahashi K, Ito M, Tanaka T (1990) Modulation of smooth muscle calponin by protein kinase C and calmodulin. Biochem Biophys Res Commun 171:933–937
41. Nakamura F, Mino T, Yamamoto J, Naka M, Tanaka T (1993) Identification of the regulatory site in smooth muscle calponin that is phosphorylated by protein kinase C. J Biol Chem 268:6194–6201
42. Habets GGM, Scholtes EHM, Zuydgeest D, van der Kammen RA, Stam JC, Berns A, Collard JG (1994) Identification of an invasion-inducing gene, Tiam-1, that encodes a protein with homology to GDP-GTP exchange for rho-like proteins. Cell 77:537–549
43. Katzav S, Martin-Zanca D, Barbacid M (1989) Vav, a novel human oncogene derived from a locus ubiquitously expressed in hematopoietic cells. EMBO J 8:2283–2290
44. Coppola J, Bryant S, Koda T, Conway D, Barbaid M (1991) Mechanism of activation of the vav protooncogene. Cell Growth Differ 2:95–105
45. Katazav S, Cleveland JL, Heslop HE, Pulido D (1991) Loss of the amino-terminal helix-loop-helix domain of the vav proto-oncogene activates its transforming potential. Mol Cell Biol 11:1912–1920
46. Bustelo XR, Ledbetter JA, Barbacid M (1992) Product of vav proto-oncogene defines a new class of tyrosine protein kinase substrates. Nature 356:68–70
47. Margolis B, Hu P, Katzav S, Li W, Oliver JM, Ullrich A, Sclessinger J (1992) Tyrosine phosphorylation of vav proto-oncogene product containing SH2 domain and transcription factor motifs. Nature 356:71–74
48. Bustelo XR, Barbacid M (1992) Tyrosine phosphorylation of the vav proto-oncogene product in human hematopoietic cells. Science 256:1196–1199
49. Alai M, Mui ALF, Cutler RL, Bustelo XR, Barbacid M, Krystal G (1992) Steel factor stimulates the tyrosine phosphorylation of the proto-oncogene product, $p95^{vav}$, in human hematopoietic cells. J Biol Chem 267:18021–18025
50. Matsuguchi T, Inhorn RC, Carlesso N, Xu G, Druker B, Griffin JD (1995) Tyrosine phosphorylation of $p95^{vav}$ in myeloid cells is regulated by GM-CSF, IL-3 and Steel factor and is constitutively increased by $p210^{BCR/ABL}$. EMBO J 14:257–265
51. Koch CA, Anderson D, Moran MF, Ellis C, Pawson T (1991) SH2 and SH3 domains: elements that control interactions of cytoplasmic signaling proteins. Science 252:668–674
52. Gulbins E, Coggeshall KM, Baier G, Katzav S, Burn P, Altman A (1993) Tyrosine kinase-stimulated guanine nucleotide exchange activity of Vav in T cell activation. Science 260:822–825
53. Ron D, Zannini M, Lewis M, Wickner RB, Hunt LT, Graziani G, Tronick SR, Aaronson SA, Eva A (1991) A region of proto-dbl essential for its transforming activity shows sequence similarity to a yeast cell cycle gene, CDC24, and the human breakpoint cluster gene, bcr. New Biol 3:372–379
54. Hart MJ, Eva A, Evans T, Aaronson SA, Cerione RA (1991) Catalysis of guanine nucleotide exchange on the CDC42Hs protein by the dbl oncogene product. Nature 354:311–314
55. Hirata K, Kikuchi A, Sasaki T, Kuroda S, Kaibuchi K, Matsuura Y, Seki H, Saida K, Takai Y (1992) Involvement of rho p21 in the GTP-enhanced calcium ion sensitivity of smooth muscle contraction. J Biol Chem 267:8719–8722
56. Ross R (1993) The pathogenesis of atherosclerosis: a perspective for the 1990s. Nature 363:801–809
57. Gimona M, Sparrow MP, Strasser P, Herzog M, Small JV (1992) Calponin and SM22 isoforms in avian and mammalian smooth muscle: absence of phosphorylation in vivo. Eur J Biochem 205:1067–1075
58. Duband J-L, Gimona M, Scatena M, Sartore S, Small JV (1993) Calponin and SM22 as diffentiation markers of smooth muscle: spactiotemporal distribution during avian embryonic development. Differentiation 55:1–11
59. Takahashi K, Eshita Y, Takagi M, Nakai M, Kato O, Kobayashi T, Seto M, Sasaki Y, Shibata N (1992) Suppression of calponin and SM22 gene expression in abnormally proliferating smooth

muscle cells in human coronary atherosclerosis and injured rabbit arteries: implications for molecular and genetic basis of vascular smooth muscle modulation. Circulation 86:I–86
60. Frid MG, Shekhonin BV, Koteliansky VE, Glukhova MA (1992) Phenotypic changes of human smooth muscle cells during development: late expression of heavy caldesmon and calponin. Dev Biol 153:183–193
61. Shanahan C, Cary NRB, Metcalf JC, Weissberg PL (1994) High expression of genes for calcification-regulating proteins in human atherosclerotic plaques. J Clin Invest 93:2393–2402
62. Negoro N, Fukui R, Tsuchikane E, Otsuji S, Kato O, Kobayasi T, Takagi M, Eshita Y, Nishida M, Kojima A, Takahashi K, Shibata N (1994) Down expression of calponin in smooth muscle of coronary artery lesions identifies a group of lesions at high risk for restenosis after athelectomy: implications for human gene therapy by direct calponin gene transfer. Circulation 90:I-142
63. Tsuchikane E, Funamoto M, Otsuji S, Tateyama H, Kato O, Kobayashi T, Fukui R, Negoro N, Takahashi K, Shibata N (1994) The relationship between clinical features and expression of calponin in tissue excised by DCA in restenotic lesions after balloon angioplasty. Circulation 90: I-393
64. Takahashi K, Takagi M, Ohgami K, Nakai M, Kojima A, Nadal-Ginard B, Shibata N (1993) Inhibition of smooth muscle cell migration and proliferation caused by transfection of the human calponin gene is associated with enhanced cell matrix adhesion and reduced PDGF responsiveness. Circulation 88:I–174
65. Takahashi K, Fukui R, Kato O, Nakazawa S, Takagi M, Kinoshita I, Hiraishi T, Kojima A, Sato S, Eshita Y, Fujio Y, Kobayashi T, Shibata N (1993) Percutaneous transluminal transfer of the human calponin gene for suppression of intimal hyperplasia following arterial balloon injury: a model for successful gene therapy for restenosis. Circulation 88:I-657
66. Mezgueldi M, Mendre C, Calas B, Kassab R, Fattoum A (1995) Characterization of the regulatory domain of gizzard calponin. J Biol Chem 270:8867–8876

# Modulation of Vascular Smooth Muscle Contraction by Calponin Phosphorylation

Toshio Tanaka, Michiko Naka, Terumasa Mino, Uhito Yuasa, Fumiaki Nakamura, Junichi Yamamoto, and Satoshi Matsushima

*Summary.* Calponin is an abundant thin filament-associated smooth muscle protein that has been implicated in the modulation of smooth muscle contraction. Recently, we found that smooth muscle calponin is an excellent substrate for protein kinase C and that phosphorylation reduced the binding of calponin to F-actin and tropomyosin. We found that $Thr^{184}$ is the preferred site of phosphorylation and is functionally the most important of the sites phosphorylated by protein kinase C in smooth muscle calponin. We have demonstrated that calponin can be rapidly dephosphorylated by type 1δ protein phosphatase, just as the 20 kDa light chain in intact myosin can be dephosphorylated in vitro. We investigated calponin phosphorylation during endothelin-1 or phorbol 12,13-dibutylate (PDBu) stimulation of intact strips of porcine coronary artery. Stimulation by endothelin-1 or PDBu resulted in a significant increase of $^{32}P$ incorporation into the calponin in association with the development of force. These results suggest that calponin phosphorylation by protein kinase C plays a potential role in the modulation of smooth muscle contraction by agonists.

*Key Words.* Calponin—Protein kinase C—Smooth muscle—Phosphorylation—Dephosphorylation

## Introduction

In response to most agonists, inositol (1,4,5)-triphosphate ($InsP_3$) and diacylglycerol (DAG) are formed by the hydrolysis of an inositol lipid precursor stored in the plasma membrane of smooth muscle cells [1]. The $InsP_3$ released into the cytoplasm mobilizes calcium from intracellular stores, whereas DAG stimulates protein kinase C (PKC) [2]. The biochemical signaling system that induces and maintains agonist-stimulated contraction in smooth muscle should be elucidated. A rapid release of calcium from non-mitochondrial organelles ensues, and it has been demonstrated that it is sufficient to provoke contraction [3]. However, the tonic phase of contraction cannot be explained solely by this second messenger. Moreover, it has been shown that a contractile response similar to that evoked by an agonist requires a combination of calcium ionophores and phorbol esters [4]. This response implies that the

Department of Molecular and Cellular Pharmacology, Mie University School of Medicine, 2-174 Edobashi, Tsu, Mie, 514 Japan

second product of phosphatidylinositol lipid hydrolysis, namely DAG, could be important in smooth muscle contraction [2].

The contractile response in muscle cells is regulated by changes in intracellular calcium ion concentrations and depends on myosin and actin interaction. The dominant molecular mechanisms for calcium activation of contractile proteins in vertebrate skeletal and cardiac muscle are due to the thin-filament calcium-regulatory system troponin–tropomyosin [5]. In this system, troponin C acts as a calcium-receptor protein and troponin T as a tropomyosin-binding subunit. In vertebrate smooth muscle where no troponin complex exists, the contractile activity is thought to be regulated by $Ca^{2+}$/calmodulin-dependent myosin ligh chain phosphorylation [6–8]. $Ca^{2+}$/calmodulin-dependent phosphorylation of the 20 kDa myosin light chain subunits of myosin has been reported to increase actin-activated MgATPase activity and initiate smooth muscle contraction [6]. Although increases in intracellular calcium ion concentration are a primary determinant from myosin light chain phosphorylation [9], these elements are not the only cellular processes regulating contraction [8]. Many examples of dissociation between intracellular free $Ca^{2+}$ concentration levels, myosin light chain phosphorylation levels, cross-bridge cycling rates, and force levels have been reported [10]; and the possible existence of an additional regulatory mechanism for smooth muscle contraction is difficult to negate. Studies suggest the existence of (1) an agonist-induced increase in calcium sensitivity of the contractile proteins or (2) a calcium-independent process.

A new troponin T-like protein, calponin has been identified in smooth muscle and nonmuscle cells [11]. Calponin isolated from smooth muscle was reported to bind F-actin, tropomyosin, and calmodulin, suggesting its involvement in the regulation of smooth muscle contraction. We found that smooth muscle calponin is an excellent substrate for protein kinase C (PKC) [12, 13].

The main goal of the present study was to investigate a possible physiological role for calponin phosphorylation in the regulation of smooth muscle contraction. In this study, we found that F-actin and tropomyosin decreased the rate of phosphorylation of calponin by PKC in different fashions, and that the phosphorylation of calponin reduces its ability to bind to these proteins. The phosphorylation of calponin at multiple sites by PKC was analyzed, and the functionally most important and preferred site of phosphorylation was identified. We have demonstrated the dephosphorylation of calponin by protein phosphatase types 1 and 2A. Furthermore, we have demonstrated the calponin phosphorylation response following stimulation by endothelin-1 or phorbol 12, 13-dibutyrate (PDBu) in $^{32}P$-labeled porcine coronary artery. In the light of our findings, we advance the argument that calponin phosphorylation plays a potential role in the regulation of smooth muscle contraction by endothelin-1 or PDBu.

## Materials and Methods

### Materials

[$\gamma$-$^{32}P$]ATP was purchased from Amersham Corporation. Cyanogen bromide-activated Sepharose 4B, CM-Sephadex, and TPCK-treated trypsin were purchased from Sigma (St. Louis, MO, USA). Phosphatidylaserine (beef brain) was purchased from Serdary Research Laboratories. All other chemicals were of the highest grade commercially available.

## Purification of Proteins

Calponin was purified from chicken gizzard as previously reported [12]. PKC was purified from rat brain by a method described elsewhere [12]. Myosin-bound phosphatase (PP-1δ) was purified to homogeneity from chicken gizzard essentially by the method of Okubo et al. [14]. Type 2A protein phosphatase (PP-2A) from chicken gizzard was purified to apparent homogeneity by the method of Pato and Adelstein [15]. Tropomyosin was purified from chicken gizzard by a modification of the procedure described by Bretscher [16]. The following proteins were purified by previously described methods: rabbit skeletal actin [17], bovine brain calmodulin [18], chicken gizzard myosin [19], and myosin light chain kinase [20].

## Phosphorylation of Calponin

Phosphorylation of purified chicken gizzard calponin was carried out at 30°C in a reaction mixture that contained 20 mM Tris HCl (pH 7.5), 5.8 μM calponin, PKC 1 μg/ml, 100 μM [γ-$^{32}$P]ATP (25 000–30 000 cpm/nM) or 100 μM ATP, 5 mM $MgCl_2$, 0.8 mM $CaCl_2$, and phosphatidylaserine 50 μg/ml. The reaction was initiated by the addition of [γ-$^{32}$P]ATP. Incorporation of radiolabeled phosphate was monitored by adding aliquots of the reaction mixture to 20% trichloracetic acid (TCA) at appropriate times. The precipitates, after centrifugation, were washed twice with 5% TCA, and then radioactivity was determined by scintillation counting. For the preparation of phosphorylated calponin we took advantage of the heat stability of calponin [12]. The reactions were stopped by addition of 5 mM EGTA, which was followed by boiling at 95°C for 5 min. The mixtures were then centrifuged at 100 000 *g* for 30 min. This procedure effectively inactivates and removes the kinase without affecting the function of the calponin. The resultant supernatants were dialyzed exhaustively against buffer that contained 20 mM Tris HCl (pH 7.5), 50 mM NaCl, 0.1 mM EGTA, and 0.5 mM dithiothreitol (DTT), and they were used for subsequent binding or fluorescence assay.

## Purification of Tryptic Phosphopeptides by HPLC

$^{32}$P-labeled calponin was purified by high-performance liquid chromatography (HPLC). In brief, 1 ml of calponin (0.3 mg/ml) was phosphorylated by incubation with PKC at 30°C for various times. The reactions were stopped by addition of 5 mM EGTA, boiling, and subsequent clarification; and 0.1% trifluoroacetic acid was added to the supernatants. Each supernatant was then applied to a Bondasphere 5 C-8 (Nihon Waters Ltd., Tokyo, Japan) reverse-phase column attached to a Shimaduz LC 6A HPLC system. HPLC-purified $^{32}$P-labeled calponin was lyophilized, and the residue was dissolved in a solution of 100 mM Tris HCl (pH 8.3), 1 mM DTT, and 20 mM KCl that contained TPCK-treated trypsin (1:100, w/w) and incubated at 25°C. After 6 h, additional trypsin was added to bring the ratio of trypsin to calponin up to 1:50 (w/w), and the incupbation was continued for another 12 h. The samples were adjusted to

0.1% trifluoracetic acid, and then tryptic fragments were separated by HPLC on a C-8 reverse-phase column with a linear gradient from 0.1% trifluoracetic acid in water to 0.1% trifluoroacetic acid in 50% acetonitrile in 100 min. The flow rate was 1 ml/min, and fractions were collected at 1-min intervals. The radioactivity of each fraction was determined by liquid scintillation counting. For further purification, fractions corresponding to the major peaks of radioactivity were pooled separately, lyophilized, and rechromatographed with more gradual gradient of acetonitrile.

## Determination of Amino Acid Sequence

The amino acid sequences of $^{32}$P-labeled phosphopeptides were determined with a gas-phase protein sequencer (model 473A; Applied Biosystems, The Perkin-Elmer Corporation, Chiba, Japan).

## Analysis of Phosphoamino Acid

Phosphoamino acids from $^{32}$P-labeled tryptic peptide were analyzed by one-dimensional thin-layer electrophoresis as described previously [21]. Purified $^{32}$P-labeled peptides were dried in a centrifugal concentrator (model CC-181; Tomy, TOMY TECH INC., Tokyo, Japan), and each sample was dissolved on 6 N HCl and hydrolyzed for 3 h at 110°C. The samples were dried and then dissolved in buffer (pH 3.5) that contained 1 μg each of phenol red, phosphoserine, phosphothreonine, and phosphotyrosine. The samples were spotted on a 20 × 20 cm cellulose plate and subjected to electrophoresis (origin closer to the negative electrode) at 1000 V for 60 min. Under these conditions, all of the applied radioactivity remained on the plate.

## Assay for Binding of Proteins to F-Actin

Assessment of the binding of unphosphorylated or phosphorylated chymotryptic fragments of calponin to F-actin were achieved by the co-sedimentation method. The samples were incubated for 30 min at 25°C in a solution that contained 20 mM Tris HCl (pH 7.5), 30 mM KCl, 2 mM $MgCl_2$, 1 mM ATP, 1 mM DTT, and 0.2 mM $CaCl_2$. The samples were then centrifuged for 1 h at 100 000 *g* (TL 100; Beckman Instruments (Japan) Ltd., Tokyo, Japan) to sediment actin and the fragements of calponin that had bound to actin. After centrifugation, the pellets and supernatants were analyzed by sodium dodecyl sulfate–polyacrylamide gel electrophoresis (SDS-PAGE). Gels were stained with Coomassie Brilliant Blue R-250, and the fragments distributed in the pellets and supernatants were quantified by densitometric scanning. Determination of stoichiometry and dissociation constants (Kd) were also carried out by the co-sedimentation method in the same solution that contained 10 mM actin and 0.75–4.00 mM calponin and in the presence and absence of 1.7 mM tropomyosin.

## Fluorescence Measurements

Dansyl-calmondulin was prepared by the method described by Kincaid et al. [22]. Typically, 30 μl of acetone containing 6 mM dansyl chloride (1.5 mol/mol of calmodulin) were added slowly with stirring at 0°C to 2 ml of a solution of calmodulin (1 mg/ml) that had been dialyzed against 10 mM $NaHCO_3$ (pH 10.0).

After incubation for 90 min at 30°C, the mixture was dialyzed exhaustively against 20 mM Tris HCl (pH 8.0), 250 mM NaCl, 5 mM $MgCl_2$, 0.1 mM EGTA, and 1 mM DTT at 4°C.

Dansyl-tropomyosin was prepared by the method described by Burtnick and Bhangu [23]. Tropomyosin was labeled with dansyl chloride in a reaction mixture composed of 40 mM sodium borate (pH 8.0), 0.5 M KCl, and 1 mM DTT for 3 h at 30°C. After labeling, the protein was dialyzed exhaustively against the same buffer as used for dialysis of dansyl-calmodulin. Fluorescence measurements were made at 25°C in a spectrofluorometer equipped with a thermostatted cell holder (RF-5000; Shimadzu Corporation, Kyoto, Japan). Emission spectra were recorded with an excitation wavelength of 340 nm for dansyl-calmodulin and 365 nm for dansyl-tropomyosin, with slit widths of 5 nm. Binding data were analyzed by double-reciprocal plots and lines were fitted by least-squares analysis.

### Assay of Phosphatase Activity

Dephosphorylation of each substrate was performed at 30°C in a reaction mixture that contained 50 mM Tris HCl (pH 7.5), substrate (1.7 μM), and PP-1δ (1.2 μg/ml) or PP-2A (0.4 μg/ml).

[$^{32}P$] radioactivity was determined by the method of Pato and Adelstein [15].

### Calponin Phosphorylation Response in $^{32}P$-Labeled Smooth Muscle

We investigated the calponin phosphorylation during endothelin-1 or PDBu stimulation of intact strips of procine coronary artery. Strips were incubated in $^{32}P$ following exposure to the agonist, rapidly frozen, and homogenized. Calponin was immunoprecipitated from homogenates followed by SDS-PAGE. The radioactive bands corresponding to the autoradiogram were excised from the gel, and $^{32}P$ incorporation of calponin was determined by liquid scintillation counting.

### Other Techniques

Measurement of chemical phosphate in protein was determined by the method described by Buss and Stull [24]. SDS-PAGE was carried out in the buffer system established by Laemmli [25]. Gels were autoradiographed, if necessary, on Kodak X-Omat RP film. Protein concentrations were determined by the method of Bradford [26]. Statistical significance was determined using paired or unpaired Student's *t*-test.

## Results and Discussion

### Effects of Actin and Tropomyosin on Calponin Phosphorylation by PKC

We found that actin and tropomyosin can inhibit the phosphorylation of calponin. Kinetic analysis of actin-induced inhibition of the phosphorylation of calponin by PKC revealed that actin inhibited the phosphorylation in a competitive fashion with respect to calponin, and the inhibition constant (Ki) value was 2.2 μM. In contrast to actin, tropomyosin inhibited the phosphorylation in a noncompetitive fasion. These results indicate that the sites of binding of actin and tropomyosin to the calponin molecule may be different from one another.

## Effects of Phosphorylation on Calponin Binding of F-Actin and Tropomyosin

The binding of phosphorylated and unphosphorylated calponin to F-actin and tropomyosin were compared by the method of co-sedimentation with F-actin and fluoresence of dansyl-tropomyosin, respectively. The apparent dissociation constant (Kd) value of calponin for F-actin in the absence of tropomyosin increased from 0.19 $\pm$ 0.06 μM for the unphosphorylated protein to 1.60 $\pm$ 0.54 μM for the phosphorylated protein. The corresponding values of calponin for F-actin in the presence of tropomyosin also increased from 0.18 $\pm$ 0.04 to 1.20 $\pm$ 0.16 μM. Tropomyosin did not significantly affect the Kd value of unphosphorylated and phosphorylated calponin for F-actin. To determine whether the phosphorylation of calponin affects its binding to tropomyosin, we examined the effect of the phosphorylation of calponin on its ability to bind dansyl-tropomyosin. The apparent Kd value of tropomyosin for calponin increased significantly, from 1.80 $\pm$ 0.19 μM for the unphosphorylated protein to 4.10 $\pm$ 0.94 μM for the phosphorylated protein ($P < 0.05$). These results suggest that phosphorylation of calponin by PKC reduces the binding affinity of calponin not only for F-actin but also for tropomyosin.

## Phosphorylation Sites and the F-Actin-, Tropomyosin-, and Calmodulin-Binding Regions in Calponin

In an effort to obtain the functional fragments of calponin, samples of unphosphorylated and phosphorylated calponin were digested with α-chymotrypsin, Limited chymotryptic digestion of both unphosphorylated and phosphorylated calponin gave rise to the same fragments. During the course of a 60-min reaction the first relatively stable products are the 22-kDa and 12-kDa fragments. Radioactivity was associated with the 22- and 12-kDa fragments as well as with the intact calponin. These results indicate that the sites of phosphorylation are present in both the 22-kDa and 12-kDa fragments generated by the digestion with α-chymotrypsin.

Using fluoresence, co-sedimentation, and affinity chromatography techniques, we determined that the 22-kDa fragment and intact calponin interacts with F-actin tropomyosin in a $Ca^{2+}$-independent manner and with calmodulin in a $Ca^{2+}$-dependent manner; but the 12-kDa fragment has no ability to bind to these proteins.

## Sites of Phosphorylation of Calponin by PKC

Fractionation of tryptic phosphorylated peptides by HPLC allowed isolation of three phosphopeptides (designated T1, T2, and T3), each of which was located in three repeating amino acid motifs of calponin. In these three phosphopeptides, the preferred site of phosphorylation is located in T2, which accounts for about 67% of the total phosphorylation by PKC within the first 5 min, during which time the initial rate of phosphorylation was constant. Both the relative initial rates and extents of

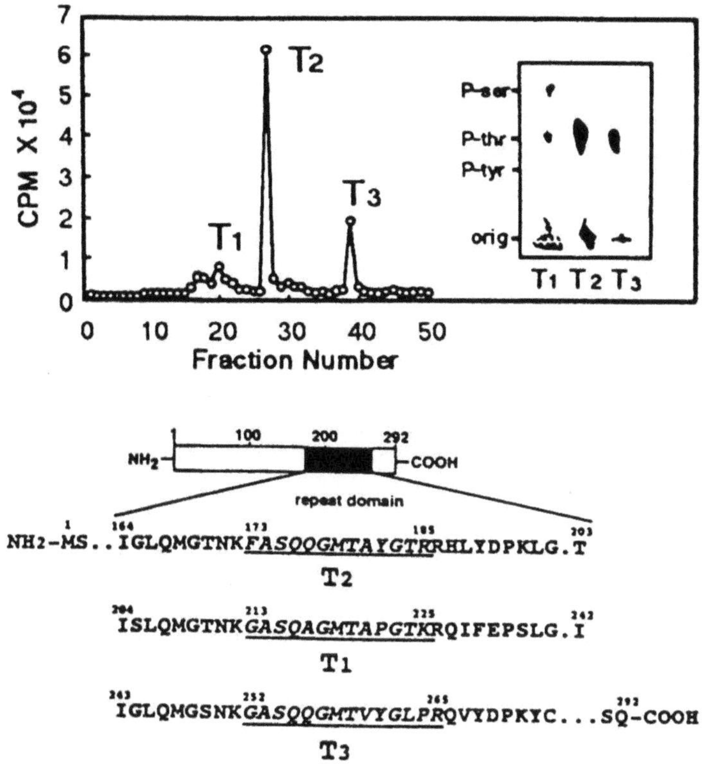


**Fig. 1.** Sequence analysis of phosphopeptides generated by tryptic proteoloysis of calponin that had been phosphorylated by protein kinase C (PKC). Calponin was exhaustively phosphorylated by PKC and was digested with trypsin; the phosphopeptides (designated *T1*, *T2*, and *T3* in order of elution) were separated by reverse-phase high-performance liquid chromatography. Purified phosphopeptides were subjected to amino acid sequencing. Their amino acid sequences are underlined. The number is based on the published sequence [27]

phosphorylation decreased in the order T2 > T3 > T1. Both serine and threonine residues were phosphorylated in T1 (213GASQAGMTAPGTK225); only the threonine residue was phosphorylated in T2 (173FASQQGMTAYGTR185) and T3 (252GASQQGMTVYGLPR265). To eliminate the possibility that either of these residues ($\text{Thr}^{180}$ and $\text{Thr}^{184}$) was the phosphorylation site, we used the synthetic peptide corresponding to T2. $\text{Thr}^{180}$ in the trypic phosphopeptide T2 and both $\text{Thr}^{180}$ and $\text{Thr}^{184}$ in the corresponding synthetic peptide were detected. The yield of $\text{Thr}^{184}$ in the tryptic phosphopeptide T2 was found to be markedly reduced. This evidence suggests that $\text{Thr}^{184}$ is the phosphorylated amino acid in T2, which accounts for about 67% of the total phosphorylation within the first 5 min. As the 22-kDa fragment contained only T2, the phosphorylation site in T2 appeared to regulate the binding of calponin to F-actin and tropomyosin. The study of amino acid sequencing of the synthetic peptide corresponding to the T2 indicates that PKC can phosphorylate $\text{Thr}^{184}$ in T2. These findings strongly suggest that $\text{Thr}^{184}$ is the preferred site of phosphorylation and is functionally the most important of the sites phosphorylated by PKC in smooth muscle calponin (Fig. 1).

## Dephosphorylation of Calponin by Type 1δ Protein Phosphatase

Dephosphorylation of each substrate was performed at 30°C in a reaction mixture that contained 50 mM Tris HCl (pH 7.5), substrate (1.7 μM), and type 1δ protein phosphatase (PP-1δ) (1.2 μg/ml) or PP-2A (0.4 μg/ml). [$^{32}$P]radioactivity was determined by the method of Pato and Adelstein [15]. Double-reciprocal plots of reaction velocity and substrate concentration revealed that the Michaelis-Menten constant (Km) values of protein phosphatase (type 1δ) for calponin and 20-kDa myosin light chain were 0.60 and 0.70 μM, respectively, and the maximal velocity (Vmax) values were 360 and 191 nmol/min/mg, respectively. The Km values of PP-2A for calponin and 20-kDa myosin light chain were 1.1 and 2.5 μM, respectively, and the Vmax values were 930 and 770 nmol/min per mg, respectively. These results indicate that calponin is a good substrate for PP-1δ as is PP-2A.

Phosphopeptide analysis of calponin that had been phosphorylated by PK-C revealed three major peaks of material with high radioactivity as previously described [28]. These peaks were designated T1, T2, and T3, according to the order of their elution from the column, as described by Nakamura et al. [28]. After phosphorylated calponin had been incubated with PP-1δ for 20 s, during which time phosphorylated calponin (1.2 mol Pi/mol calponin) was dephosphorylated to approximately 0.4 mol Pi/mol calponin, phosphopeptide analysis demonstrated that the T2 fragment, which showed the highest level of phosphorylation by PK-C, and the T3 fragment were more rapidly dephosphorylated by PP-1δ than the T1 fragment. Analysis after a 30-min incubation revealed no radioactivity peaks of phosphopeptide. These results indicate that PP-1δ preferentially dephosphorylates the T2 fragment, which is functionally the most important fragment because it is involved in regulation of binding to F-actin and tropomyosin [28]. Finally, however, all three phosphopeptide fragments were completely dephosphorylated.

The effect of the dephosphorylation of calponin by PP-1δ on the actin-activated $Mg^{2+}$-ATPase activity of smooth muscle myosin was examined. Calponin caused 84% inhibition of the actin-activated myosin ATPase activity, the activity fell from 75.1 to 12.0 nmol/min per mg. Phosphorylation of calponin by PK-C resulted in loss of inhibition of the ATPase activity. At a concentration of 2 μM, calponin that had been dephosphorylated by PP-1δ inhibited the ATPase activity to the same extent as native, unphosphorylated calponin (Fig. 2). Winder et al. [29] have reported that calponin phosphatase is identical to PP-2A, and most calponin phosphatase (>90%) exists in the soluble fraction. We demonstrated that purified PP-1δ dephosphorylated calponin as well as PP-2A. PP-1δ has the ability to bind to myosin via the 130-kDa myosin-binding regulatory subunit [30] and is present in the actomyosin fraction of chicken gizzard [14]. Furthermore, PP-1δ tightly binds to myosin and is not dissociated from myosin even in the presence of 3% Triton X-100 and only high-concentration $MgCl_2$ can dissociate PP-1δ from myosin [31]. Calponin appears to bind to actin and tropomyosin, and microfilament bundles (stress fibers) in cultured smooth muscle cells can be immunostained with anti-calponin antibodies [32, 33]. The co-localization of PP-1δ and calponin suggests that PP-1δ has also great physiological significance in the dephosphorylation of calponin. However, additional studies are needed to determine which phosphatase is actually responsible for regulation of the contraction of smooth muscle in vivo. We suggest that the difference in rates of dephosphorylation by PP-1δ or PP-2A between calponin and caldesmon may lead to

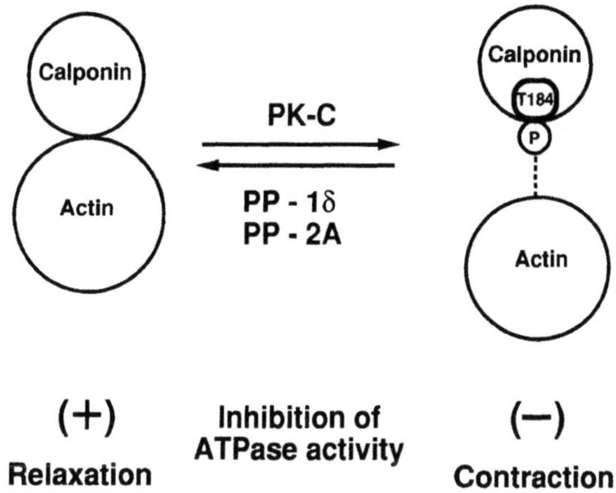


**Fig. 2.** Regulation of smooth muscle contraction by calponin phosphorylation by protein kinase C (*PK-C*) and dephosphorylation by protein phosphatase 1δ (*PP1δ*) and 2A (*PP-2A*)

a difference in phosphorylated states between calponin and caldesmon at any phase of the contraction–relaxation cycle of smooth muscle in vivo.

We have demonstrated that calponin can be rapidly dephosphorylated by PP-1δ, just as the 20-kDa myosin light chain can be dephosphorylated in vitro. These results suggest that type 1δ protein phosphatases cause the relaxation of smooth muscle by the dephosphorylation not only of myosin but also of calponin.

## Effects of Phosphorylated Calponin on $Ca^{2+}$-Induced Contraction in Skinned Artery

Calponin was found to inhibit calcium-induced contraction in skinned arterial smooth muscle. Moreover, calponin attenuated the $Ca^{2+}$-independent contraction in myosin light chain thiophosphorylated strips or on application of trypsin-treated myosin light chain kinase. These results suggest that calponin may play a physiological role in the inhibition of $Ca^{2+}$-regulated force of vascular smooth muscle. However, phosphorylated calponin by PKC had no effect on the $Ca^{2+}$-induced contraction in skinned smooth muscle [34].

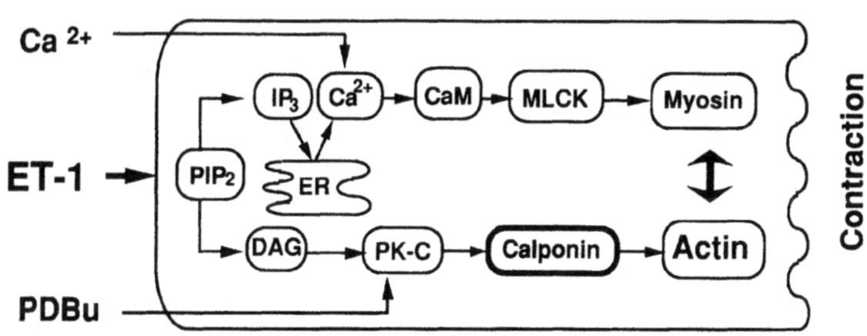


**Fig. 3.** Signal transduction in smooth muscle contraction. *PIP*$_2$, phosphatidylinositol bisphosphate; *IP*$_3$, inositol triphosphate; *CaM*, calmodulin; *MLCK*, myosin light chain kinase; *DAG*, diacylglycerol; *PK*-C, protein kinase C; *PDBu*, phorbol 12, 13-dibutyrate; *ET-1*, endothelin-1; *ER*, endoplasmic reticulum; *PDBu*, phorbol 12,13-dibutyrate; *ET-1*, endothelin-1

## Calponin Phosphorylation in Intact Porcine Coronary Artery

We investigated the calponin phosphorylation during endothelin-1 or PDBu stimulation of intact strips of porcine coronary artery. Strips were incubated in $^{32}P$ following exposure to the agonist, rapidly frozen, and homogenized. Calponin was immunoprecipitated from homogenates followed by SDS-PAGE. The radioactive bands corresponding to the autoradiogram were excised from gel, and $^{32}P$ incorporation of calponin was determined by liquid scintillation counting. Stimulation by endothelin-1 or PDBu resulted in a significant increase of $^{32}P$ incorporation into the calponin in association with development of force. These results suggest that calponin phosphorylation plays a potential role in the modulation of smooth muscle contraction by endothelin-1 or PDBu (Fig. 3).

*Acknowledgments.* This work was supported in part by Grants-in-Aid for Scientific Research from the Ministry of Education, Science, and Culture and from the Ministry of Health and Welfare, Japan. It was also supported in part by grants for the Cell Signaling Project of Mie University School of Medicine and from Bayer yakuhin, Ltd., Kuma Hospital, and Minami Hospital. We also thank Dr. T, Nakano, Dr. M. Ito, and their colleagues of the First Department of Internal Medicine, Mie University School of Medicine for cooperative collaboration; and Ms. H. Hashizume and Ms. I. Wada for skillful secretarial assistance.

## References

1. Berridge MJ, Irvine RF (1984) Inositol trisphosphate, a novel second messenger in cellular signal transduction. Nature 312:315–321
2. Nishizuka Y (1986) Studies and perspectives of protein kinase C. Science 233:305–312
3. Somlyo AP, Walker JW, Goldman YE, Trentham DR, Kobayashi S, Kitazawa T, Somlyo AV (1988) Inositol trisphosphate, calcium and muscle contraction. Philos Trans R Soc Lond [Biol] 320:399–414
4. Rasmussen H, Barrett PQ (1984) Calcium messenger system: an integrated view. Physiol Rev 64:938–984
5. Ebashi S, Endo M (1968) Calcium ion and muscle contraction. Prog Biophys Mol Biol 18:123–183
6. Hartshorne DJ (1987) Biochemistry of the contractile process in smooth muscle. In: Johnson LR (ed) Physiology of the gastrointestinal tract, 2nd edn. Raven, New York, pp 423–482
7. Adeltein RS, Eisenberg E (1980) Regulation and kinetics of the actin-myosin-ATP interaction. Annu Rev Biochem 49:921–956
8. Kamm KE, Stull JT (1985) The function of myosin and myosin light chain kinase phosphorylation in smooth muscle. Annu Rev Pharmacol Toxicol 25:593–620
9. Rembold CM, Murphy RA (1988) Myoplasmic [$Ca^{2+}$] determines myosin phosphorylation in agonist-stimulated swine arterial smooth muscle. Circ Res 63:593–603
10. Morgan JP, Morgan KG (1984) Stimulus-specific patterns of intracellular calcium levels in smooth muscle of ferret portal vein. J Physiol (Lond) 351:155–167
11. Takahashi K, Hiwada K, Kokubu T (1986) Isolation and characterization of a 34 000-dalton calmodulin- and F-actin-binding protein from chicken gizzard smooth muscle. Biochem Biophys Res Commun 141:20–26
12. Naka M, Kureishi Y, Muroga Y, Takahashi K, Ito M, Tanaka T (1990) Modulation of smooth muscle calponin by protein kinase C and calmodulin. Biochem Biophys Res Commun 171:933–937
13. Winder SJ, Walsh MP (1990) Smooth muscle calponin: inhibition of actomyosin MgATPase and regulation by phosphorylation. J Biol Chem 265:10148–10155

14. Okubo S, Erdödi F, Ito M, Ichikawa K, Konishi T, Nakano T, Kawamura T, Brautigan DL, Hartshone DJ (1993) Characterization of a myosin-bound phosphatase from smooth muscle. Adv Protein Phosphatases 7:293–312
15. Pato MD, Adelstein RS (1983) Purification and characterization of a multisubunit phosphatase from turkey gizzard smooth muscle: the effect of calmodulin binding to myosin light chain kinase on dephosphorylation. J Biol Chem 258:7047–7054
16. Bretscher A (1984) Smooth muscle caldesmon: rapid purification and F-actin cross-linking properties. J Biol Chem 259:12873–12880
17. Driska S, Hartshorne DJ (1975) The contractile proteins of smooth muscle: properties and components of a $Ca^{2+}$-sensitive actomyosin from chicken gizzard. Arch Biochem Biophys 167:203–212
18. Yazawa M, Sakuma M, Yagi K (1980) Calmodulins from muscles of marine inverstebrates, scallop and sea anemone. J Biochem 87:1313–1320
19. Persechini A, Hartshorne DJ (1981) Phosphorylation of smooth muscle myosin: evidence for cooperativity between the myosin heads. Science 213:1383–1385
20. Walsh MP, Hinkins S, Dabrowska R, Hartshorne DJ (1983) Smooth muscle myosin light chain kinase. Methods Enzymol 99:279–288
21. Haycock JW (1990) Phosphorylation of tyrosine hydroxylase in situ at serine 8, 19, 31, and 40. J Biol Chem 265:11682–11691
22. Kincaid RL, Vaughan M, Osbrone JC Jr, Tkachuk VA (1982) $Ca^{2+}$-dependent interaction of 5-dimethylaminonaphthalene-1-sulfonyl-calmodulin with cyclic nucleotide phosphodiesterase, calcineurin, and troponin I. J Biol Chem 257:10638–10643
23. Burtnick LD, Bhangu K (1986) Interaction of actin with dansyl-tropomyosin. FEBS Lett 198:307–310
24. Buss JE, Stull JT (1983) Measurement of chemical phosphate in proteins. Methods Enzymol 99:7–14
25. Laemmli UK (1970) Cleavage of structural proteins during the assenbly of the head of bacteriophage T4. Nature 227:680–685
26. Bradford MM (1976) A rapid and sensitive method for the quantitation of microgram quantities of protein utilizing the principle of protein-dye binding. Anal Biochem 72:248–254
27. Takahashi K, Nadal-Ginard B (1991) Molecular cloning and sequence analysis of smooth muscle calponin. J Biol Chem 266:13284–13288
28. Nakamura F, Mino T, Yamamoto J, Naka M, Tanaka T (1993) Identification of the regulatory site in smooth muscle calponin that is phosphorylated by protein kinase C. J Biol Chem 268:6194–6201
29. Winder SJ, Pato MD, Walsh MP (1992) Purification and characterization of calponin phosphatase from smooth muscle: effect of dephosphorylation on calponin function. Biochem J 286:197–203
30. Shimizu H, Ito M, Miyahara M, Ichikawa K, Okubo S, Konishi T, Naka M, Tanaka T, Hirano K, Hartshorne DJ, Nakano T (1994) Characterization of the myosin-binding subunit of smooth muscle myosin phosphatase. J Biol Chem 269:30407–30411
31. Mitsui T, Inagaki M, Ikebe M (1992) Purification and characterization of smooth muscle myosin-associated phosphatase from chicken gizzards. J Biol Chem 267:16727–16735
32. Gimona M, Herzog M, Vandekerckhove J, Small JV (1990) Smooth muscle specific expression of calponin. FEBS Lett 274:159–162
33. Takeuchi K, Takahashi K, Abe M, Nishida W, Hiwada K, Nabeya T, Maruyama K (1991) Co-localization of immunoreactive forms of calponin with actin cytoskeleton in platelets, fibroblasts, and vascular smooth muscle. J Biochem (Tokyo) 109:311–316
34. Itoh T, Suzuki S, Suzuki A, Nakamura F, Naka M, Tanaka T (1994) Effects of exogenously applied claponin on $Ca^{2+}$-regulated force in skinned smooth muscle of the rabbit mesenteric artery. Pfluegers Arch 427:301–308

# Smooth Muscle Cell Differentiation and Its Abnormality in Vascular Diseases

Ryozo Nagai, Masanori Aikawa, Makoto Kuro-o, Ichiro Manabe, Yasunari Sakomura, Masafumi Watanabe, and Yoshio Yazaki

*Summary.* The many isoforms of smooth muscle myosin heavy chains (MHC) are differentially expressed corresponding to smooth muscle cell phenotypes. Rabbit smooth muscles contain at least three MHC isoforms: SM1 (204 kDa), SM2 (200 kDa), and SMemb (200 kDa). SM1 and SM2 are specific to smooth muscles. SM1 is expressed from the early stage of fetal life into adulthood, but SM2 appears after birth. SMemb is a nonmuscle-type MHC abundantly expressed in the embryonic aorta. Dedifferentiation and redifferentiation of smooth muscles are involved in the cellular mechanisms of the vascular remodeling that occurs after mechanical injury and with atherosclerosis. Neointimal smooth muscles in rabbit resemble embryonic smooth muscles at the early stages after injury but regain the adult phenotype 4–6 weeks later in terms of MHC expression. In human arteries, MHC expression follows the pattern similar to that of the rabbit. One of the most characteristic features in human coronary artery is the formation of diffuse intimal thickening beginning at early adolescence. The thickened coronary intima is composed of smooth muscle cells, reacting equally to three antibodies for MHC isoforms during adolescence; with progression of atherosclerosis, however, there is diminished expression of SM1 and SM2, whereas smooth muscle α-actin is well preserved. The coronary restenotic lesion following percutaneous transluminal angioplasty was found to be composed of smooth muscles with the embryonic phenotype in which SMemb expression was pronounced. To understand the molecular mechanisms underlying smooth muscle cell proliferation, it is important to analyze the regulation of MHC genes. We have isolated and characterized the promoter regions of SMemb gene. SMemb gene contains a novel *cis*-element located −100 bp upstream of the transcription initiation site. More detailed analysis of these two MHC genes, especially characterization of their transcription factors, will lead to the understanding of smooth muscle cell differentiation and the pathogenesis of vascular diseases.

*Key Words.* Smooth muscle—Differentiation—SM1—SM2—SMemb

---

Third Department of Internal Medicine, University of Tokyo, 7-3-1 Hongo, Bunkyo-ku, Tokyo, 113 Japan

Present address: Second Department of Internal Medicine, University of Gunma, 3-39-22 Showa-machi, Maebashi, Gunma, 371 Japan

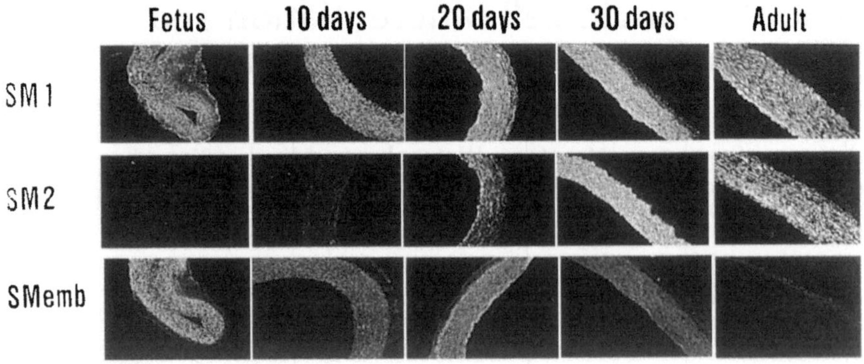


**Fig. 1.** Developmentally regulated expression of rabbit SM1, SM2, and SMemb in rabbit aorta on indirect immunofluorescence histochemistry. SM1 is consistently positive from the fetal stage to adulthood, but SM2 is positive after birth. SMemb, on the other hand, disappears in the adult

## Introduction

The diversity of myosin heavy chain (MHC) molecules, particularly in cardiac and skeletal muscles, has been investigated extensively. MHC isoforms not only are determinants of contractile properties of muscles, they also are molecular markers for muscle development and muscle diseases [1–6]. Compared to cardiac and skeletal muscles, little was known about smooth muscle and nonmuscle MHC isoforms until their cDNA clones were isolated and specific antibodies were developed [7–11]. We previously demonstrated that rabbit and rat smooth muscles contain at least three types of MHC: SM1 (204 kDa), SM2 (200 kDa), and SMemb (200 kDa). SM1 and SM2 are MHC isoforms arising from a single gene by alternative RNA splicing [8, 9]. Using cDNA probes and isoform-specific antibodies, we have shown that the expression of SM1 and SM2 in rabbit aorta is developmentally regulated; SM1 is constitutively expressed at all stages, whereas SM2 appears only after birth [10]. We have further revealed the presence of a third type of MHC isoform in rabbit embryonic aorta [11]. We referred to this novel MHC isoform as SMemb because it is present in greater abundance in embryonic aortas (Fig. 1). This MHC is identical with the MHC-B characterized in nonmuscle cells [12–14].

The presence of developmentally regulated MHC isoforms in vascular smooth muscles provides important molecular tools to investigate phenotypic modulation that occurs in arteriosclerosis or atherosclerosis. Vascular smooth muscles are known to change the phenotype during proliferation from the contractile to the synthetic state. In this study, we examined the immunohistologic expression of the MHC isoforms in arterial smooth muscles involved in vascular remodeling in both humans and animals. We report the presence of many varieties of phenotypically modulated smooth muscle cells in arterial lesions on the basis of MHC isoform expression.

## Materials and Methods

### Animal Models of Arteriosclerosis and Atherosclerosis

Intimal thickening was produced in 4-month-old white rabbits by endothelial cell denudation using a balloon catheter. Atherosclerosis was also induced in 4-month-old rabbits by feeding them a 1% cholesterol diet for 4 months. Balloon-injured rabbits were killed at 4, 7, 14, and 28 days after operation. Atherosclerotic rabbits were killed after 4 months of a high-cholesterol diet.

### Human Arteries

Human coronary arteries, including those of fetuses, were obtained from autopsied patients within 3 hours after death. These patients had had no clinical history of ischemic heart disease or major atherogenic factors (e.g., hyperlipidemia, diabetes, or smoking). An atherectomized tissue sample from a restenosis lesion was also examined for MHC expression. For immunohistologic studies, tissues were frozen in OCT compound or fixed in a solution of 95% ethanol and 1% acetic acid.

### Immunohistochemistry

The development of three monoclonal antibodies against SM1, SM2, and SMemb has been described elsewhere [15]. Arterial tissues were immunologically stained using frozen or ethanol-fixed specimens. Ethanol-fixed tissues were embedded with paraffin and sectioned in 3 μm slices. After deparaffinization and rehydration, immunoenzymatic staining was performed as previously described [15].

### Isolation and Characterization of SMemb Gene

The 5′-end cDNA sequence of rabbit SMemb was obtained by the RACE-PCR method. A rabbit genomic library constructed in EMBL3 was screened with this 5′-end cDNA probe. Rabbit-cultured vascular smooth muscle cells were plated in 60-mm dishes at a density of $2.5 \times 10^5$ cells/dish and transfected with 10 μg of SMemb/luciferase fusion plasmid. Luciferase activity in 10μl of the lysate was measured in a Lumat LB401 luminometer (Labo Prof Dr GmbH & Co., Parsfach, Germany).

## Results

### Phenotypes of Neointimal Cells in Injured Rabbit Arteries

Neointimal cells, recognized with anti-SMemb antibody, appeared as early as 4 days after vascular injury. The thickness of the neointima was regionally variable at 14 days after injury, when the neointimal cells were positive against anti-SM1 and anti-SMemb but almost negative against anti-SM2 antibody (Fig. 2). The proliferation of embryonic smooth muscle cells in the intima could be seen even at 8 weeks after the balloon-induced injury. At 4 weeks, however, in several rabbits we found that neointimal cells were positive against anti-SM1 and anti-SM2 but almost negative or only weakly positive against anti-SMemb antibody (Fig. 3). In rabbits fed the high-cholesterol diet, reexpression of SM2 did not occur.

### Non-Smooth-Muscle Cell Proliferation in the Medial Wall

At 4 weeks in the rabbits with balloon-induced injury, we found that medial smooth muscles were disappearing and were being replaced by a cluster of highly proliferative cells that were negative against both anti-SM1 and anti-SM2 but positive against anti-SMemb antibodies (Fig. 4). This finding indicates that a heterogeneous population of

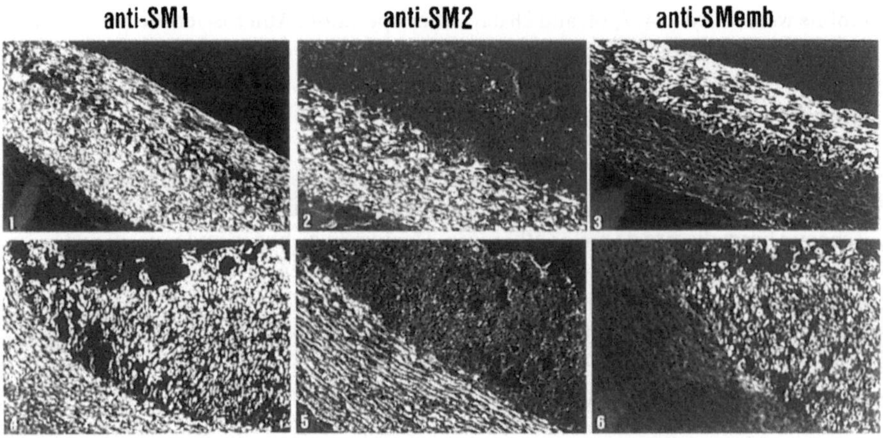


**Fig. 2.** Indirect immunofluorescence of injured rabbit aortas. Neointimal cells in the balloon-injured (1–3) or high-cholesterol-diet (4–6) aortas were positive for both SM1 and SMemb but negative for SM2, indicating that these cells are similar in phenotype to embryonic smooth muscle cells. Medial smooth muscles, on the other hand, maintain the adult phenotype because they are positive for SM1 and SM2

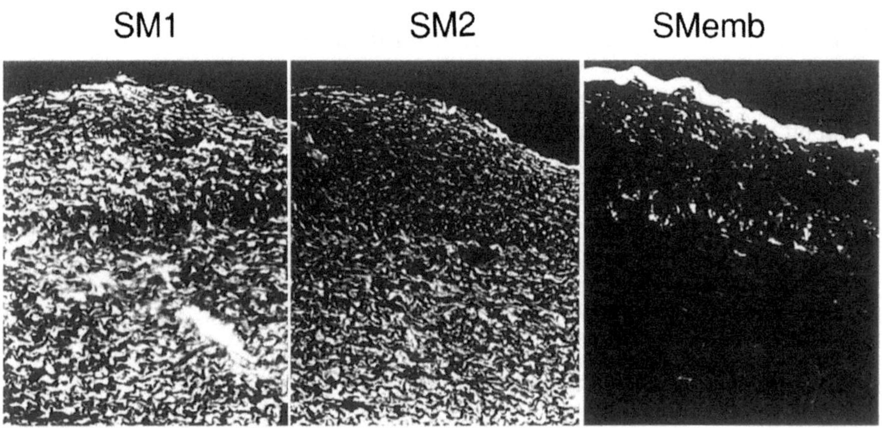


**Fig. 3.** Indirect immunofluorescence of the balloon-injured rabbit aorta at 8 weeks. Intimal smooth muscles are reexpressing SM2 but show reduced SMemb expression. This finding indicates that neointimal smooth muscle cells undergo redifferentiation

smooth muscle and nonmuscle cells are involved in the remodeling of the injured arterial wall.

## Immunohistochemistry of Smooth Muscles in Human Coronary Arteries Without Atherosclerotic Changes

Coronary arteries of autopsied patients were investigated by immunohistochemistry for MHC expression. At least five cases were examined from each of the following age groups: fetuses (gestational age 17–27 weeks), young adults (18–33 years old), and middle-aged to elderly adults (52–76 years old).

In coronary arteries, medial smooth muscles exhibited developmental transition in MHC expression in which SM2 was negative in the fetus but positive after birth (Fig. 5). Both SM1 and SMemb were positive through all ages, although the reactivities with anti-SMemb antibody decreased with aging.

It is known that human coronary arteries develop the intimal layer beginning during early adolescence, becoming several times thicker than the medial layer in the elderly. Our immunohistologic analyses indicated that the coronary intimal cells were for the most part smooth muscles because they reacted with both anti-SM1 and anti-$\alpha$-actin antibodies. In Fig. 6, we present a 37-year-old woman in whom smooth muscles in the highly thickened intima showed reduced reactivities to anti-SM2 antibody. This finding indicates that intimal smooth muscles undergo phenotypic modulation even before atheromatous changes develop.

## Immunohistochemistry of Smooth Muscles in Atherosclerotic Human Coronary Arteries

Intimal smooth muscles in atheromatous areas were usually positive for smooth muscle $\alpha$-actin and SM1 but almost negative for SM2. SMemb expression was often reduced in atherosclerotic plaques. With coronary atherosclerosis, an outstanding

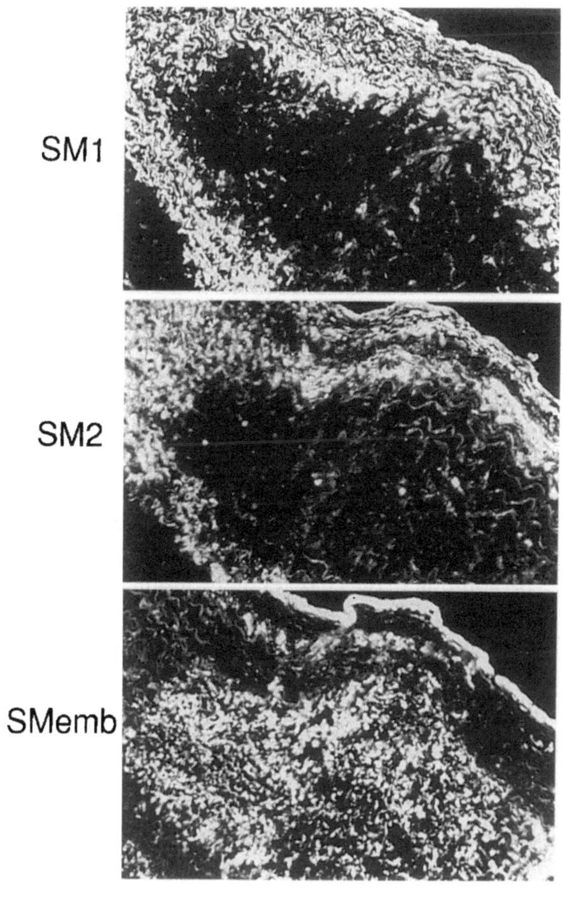


**Fig. 4.** Remodeling of the aortic meida at 4 weeks after balloon-induced injury. The proliferating cells are negative for both SM1 and SM2 and positive only for SMemb

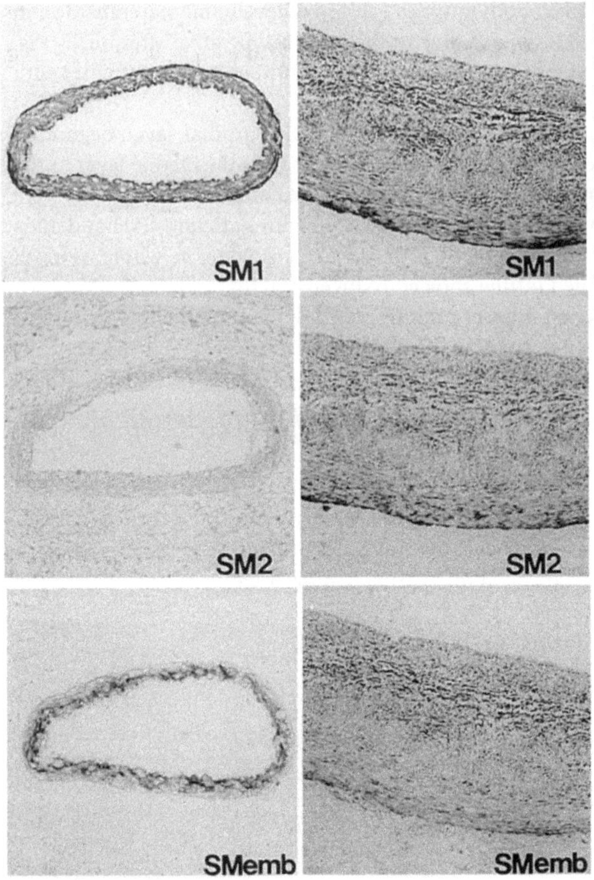


**Fig. 5.** Immunohistochemistry for SM1, SM2, and SMemb in the left anterior descending coronary arteries. No intima could be recognized in the fetal coronary artery at 21 weeks' gestation (**left panels**). SM2 expression in the fetal coronary artery was negative. Intimal cells were positive for SM1, SM2, and SMemb in the coronary artery of a 17-year-old boy (**right panels**)

pathologic feature is the presence of fibrohyaline or fibronecrotic plaques in the intima, smooth muscle-poor areas, and ground substance-rich regions. Figure 7 shows the left anterior descending coronary artery with a fibrohyaline region obtained from a 65-year-old patient. Although macrophages were not prominent in this lesion, both intimal and medial smooth muscles exhibited many varieties of modulated phenotypes in light of the expression of MHC isoforms. Particularly, smooth muscle cells in the fibrohyalinized areas, although small in number, were reduced in the expression of not only SM2 but also SM1, whereas most of cells were α-actin-positive.

In addition to the changes in the intimal layer, we noted phenotypic modulation of medial smooth muscles underlying the atherosclerotic plaques because SM1 and SM2 expression was reduced, whereas the expression of smooth muscle α-actin was preserved.

## Phenotype of Smooth Muscle Cells in Restenotic Lesions

Smooth muscle cells were found to proliferate in restenotic lesions following percutaneous transcatheter coronary angioplasty (PTCA). These smooth muscle cells obtained by atherectomy at 4 months after PTCA were postive for SM1 and SMemb

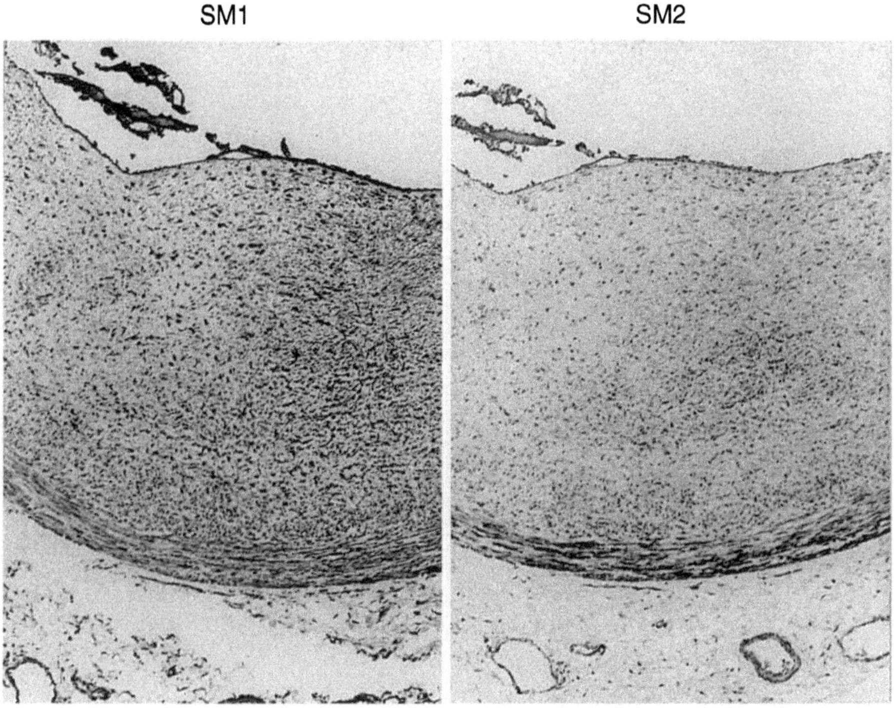


**Fig. 6.** Reduced expression of SM2 in the coronary intima of a 37-year-old woman. SM2 expression was reduced in the highly thickened intima without atheromatous changes, whereas SM1 expression was preserved

but negative for SMemb, indicating that dedifferentiation of smooth muscle cells also occurs in human coronary restenosis.

## Structure of the SMemb Gene

The 5′-flanking region of the rabbit SMemb gene lacked a canonical TATA box and A/T-rich sequences. A series of reporter plasmids with various 5′ deletions were constructed. We found that the region in or adjacent to the sequence around −100 bp was important for full promoter activity and may contain a positive *cis*-acting element. The detailed analysis of the SMemb gene will be reported elsewhere.

# Discussion

Proliferation and phenotypic modulation of vascular smooth muscle cells are outstanding phenomena that occur during the development of atherosclerosis [16–18]. In this study we have demonstrated that smooth muscle proliferation in experimental arteriosclerosis and atherosclerosis was associated with dedifferentiation of the smooth muscle cell toward the embryonic phenotype on the basis of the MHC expression. Human coronary arteries, however, present unique profiles of MHC expression at various stages of growth and age.

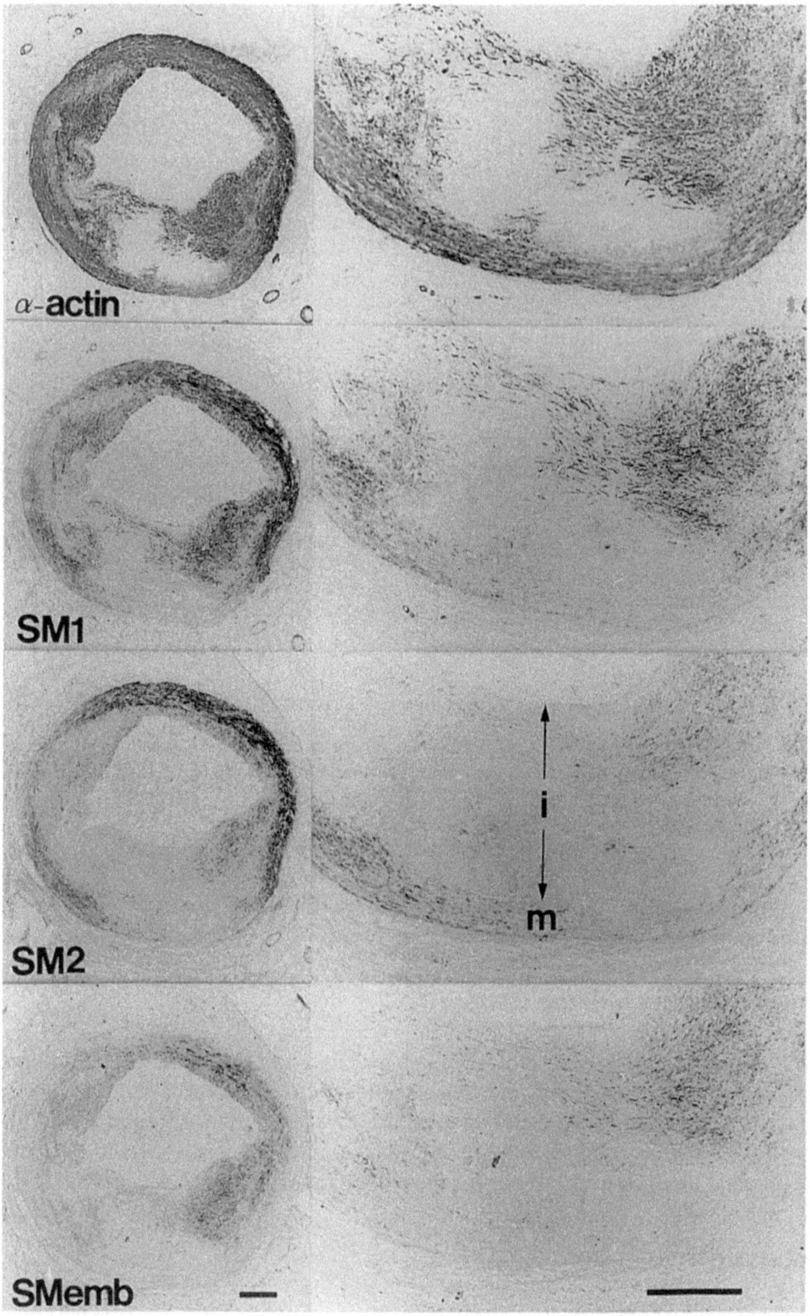


**Fig. 7.** Immunohistochemistry of the human left anterior descending coronary artery with an atherosclerotic eccentric intimal lesion from a 65-year-old man. Most of the cells surrounding the atheromatous plaque expressed smooth muscle α-actin. The number of SM1-positive cells in the atherosclerotic lesion was reduced compared to α-actin-positive cells. SM2 expression was almost negative in the fibrohyalinized area. Both SM1 and SM2 expression were reduced in the media underlying the atherosclerotic lesion, whereas the expression of smooth muscle α-actin was preserved. *i*, intima; *m*, media

1. *Both embryonic smooth muscles and non-smooth-muscle cells proliferate in the presence of experimental arterial injury.* The present study has revealed that arterial remodeling after vascular injury starts with proliferation of embryonic smooth muscle cells. The embryonic smooth muscle proliferation is mainly involved in the formation of neointima, continuing as long as 8 weeks. However, it does not seem to continue indefinitely because we observed redifferentiation to the adult phenotype in the neointima, which possibly indicates the cessation of proliferation. At the same time, the medial smooth muscle cells were replaced by SM1-negative cell proliferation. It is still not known whether these cells are highly dedifferentiated smooth muscle cells or fibroblasts. They seem to have originated from the medial wall because their proliferation occurred without a link to the adventitia or neointima.

2. *There is differential expression of human MHC isoforms during vascular development.* Vascular smooth muscle MHC isoforms are differentially expressed in humans depending on the developmental stage, age, and type of artery. On immunohistology, the medial wall of coronary arteries contained SM1 in large amounts beginning early in development, whereas SM2 was not detectable in fetal coronary arteries at 17–21 weeks' gestation.

3. *Intimal and medial smooth muscles in human atherosclerotic arteries undergo phenotypic modulation in light of MHC expression.* Immunohistochemical analyses of the present study have revealed several characteristic features in human coronary arteries. First, the major coronary arteries develop a thickened intimal layer, starting during early adolescence and becoming several times thicker than the media in mature adults. Most of the intimal cells are unequivocally smooth muscles because they reacted with anti-SM1 and smooth muscle anti-α-actin antibodies. Intimal thickening with smooth muscle cell accumulation may be considered to be physiologic in nature, but smooth muscles in the highly thickened intima modulated their phenotype so they were distinct from those in normal media or thin intima.

It has been assumed that coronary intimal thickening beginning during childhood is an integral part of atherosclerosis. We noted in this study that with the development of atherosclerosis the intimal smooth muscle was decreased, replaced by fibrous tissues. Smooth muscles in such areas displayed reduced expression of all MHC isoforms, first SM2 and then SM1, although α-actin expression was well maintained. It is noteworthy that medial smooth muscles underlying the atherosclerotic lesions often had decreased expression of SM1 and SM2. Therefore immunohistochemistry for MHC isoforms will be useful for elucidating the role of modulated smooth muscle cells during the progression of atherosclerosis.

The human intimal smooth muscles with reduced expression of SM2 in atherosclerotic lesions were found to be different from human embryonic smooth muscles in that the expression of SMemb was not prominent in the former. However, dedifferentiation of smooth muscle cells, as observed in the balloon-injured rabbit model, occurs in humans when smooth muscles rapidly proliferate in restenotic lesions after PTCA.

## Conclusion

We demonstrated that phenotypic modulation of arterial smooth muscles is an important feature characterizing the development of atherosclerosis and restenosis. We also revealed the importance of MHC isoforms as new molecular markers to study the process of human as well as experimental atherosclerosis. However, regulation of these MHC gene expression remains to be elucidated. We believe that future studies should be directed to identifying the transcription factors of the SMemb as well as the SM1 and SM2 genes, which may reveal molecular mechanisms underlying smooth muscle cell differentiation and proliferation.

## References


1. Huszar G (1972) Developmental changes of the primary structure and histidine methylation in rabbit skeletal muscle myosin. Nature 240:260–264
2. Pagani ED, Julian FJ (1984) Rabbit papillary muscle myosin isozymes and the velocity of muscle shortening. Circ Res 54:586–594
3. Alpert NR, Mulieri LA, Litten RZ (1983) Isoenzyme contribution to economy of contraction and relaxation in normal and hypertrophied hearts. In: Jacob R, Gulch RW, Kissling G (eds) Cardiac adaptation to hemodynamic overloaded, training, stress. Dr. Dietrich Steinkopff Verlag, Darmstadt, Germany, pp 147–157
4. Mahdavi V, Periasamy M, Nadal-Ginard B (1982) Molecular characterization of two myosin heavy chain genes expressed in the adult heart. Nature 297:659–665
5. Periasamy M, Wieczorec DF, Nadal-Ginard B (1984) Characterization of a developmentally regulated myosin heavy chain gene expressed in skeletal muscle. J Biol Chem 259:13573–13578
6. Nagai R, Pritzl N, Low RB, Stirewalt WS, Zak R, Alpert NR, Litten RZ (1987) Myosin isozyme synthesis and mRNA levels in pressure-overloaded rabbit hearts. Circ Res 60:692–699
7. Nagai R, Larson DT, Periasamy M (1988) Characterization of a mammalian smooth muscle myosin heavy chain cDNA clone and its expression in various smooth muscle types. Proc Natl Acad Sci USA 85:1047–1051
8. Nagai R, Kuro-o M, Babij P, Periasamy M (1989) Identification of two types of smooth muscle myosin heavy chain isoforms by cDNA cloning and immunoblot analysis. J Biol Chem 264:9734–9737
9. Babij P, Periasamy M (1989) Myosin heavy chain isoform diversity in smooth muscle is produced by differential RNA processing. J Mol Biol 210:673–679
10. Kuro-o M, Nagai R, Tsuchimochi H, Katoch H, Yazaki Y, Ohkubo A, Takaku F (1989) Developmentally regulated expression of vascular smooth muscle myosin heavy chain isoforms. J Biol Chem 264:18272–18275
11. Kuro-o M, Nagai R, Nakahara K, Katoh H, Tsai RC, Tsuchimochi H, Yazaki Y, Ohkubo A, Takaku F (1991) cDNA cloning of a myosin heavy chain isoform in embryonic smooth muscle and its expression during vascular development and in arteriosclerosis. J Biol Chem 266:3768–3773
12. Katsuragawa Y, Yanagisawa M, Inoue A, Masaki T (1989) Two distinct nonmuscle myosin-heavy-chain mRNAs are differentially expressed in the various chicken tissues. Eur J Biochem 184:611–616
13. Kawamoto S, Adelstein RS (1991) Chicken nonmuscle myosin heavy chains: differential expression of two mRNAs and evidence for two different polypeptides. J Cell Biol 112:915–924
14. Simons M, Wang M, MacBride OW, Kawamoto S, Yamanaka K, Gdula D, Adelstein RS, Weir L (1991) Human nonmuscle myosin heavy chains are encoded by two genes located on different chromosomes. Circ Res 69:530–539
15. Kim H-S, Aikawa M, Kimura K, Kuro-o M, Nakahara K, Suzuki T, Katoh H, Okamoto E, Yazaki Y, Nagai R (1993) Ductus arteriosus: advanced differentiation of smooth muscle cells demonstrated by myosin heavy chain isoform expression in rabbits. Circulation 88:1804–1810
16. Ross R (1986) The pathogenesis of atherosclerosis: an update. N Engl J Med 314:488–500
17. Schwartz SM, Campbell GR, Campbell JH (1986) Replication of smooth muscle cells in vascular disease. Circ Res 58:427–444
18. Campbell GR, Campbell JH (1985) Smooth muscle phenotypic change in arterial wall homeostasis: implications for the pathogenesis of atherosclerosis. Exp Mol Pathol 42:139–162

# Index

MIX
Papier aus verantwortungsvollen Quellen
Paper from responsible sources
FSC® C105338

If you have any concerns about our products,
you can contact us on
**ProductSafety@springernature.com**

In case Publisher is established outside the EU,
the EU authorized representative is:
**Springer Nature Customer Service Center GmbH**
**Europaplatz 3, 69115 Heidelberg, Germany**

Printed by Libri Plureos GmbH
in Hamburg, Germany